Ion Exchange and Solvent Extraction

A Series of Advances

Volume 20

ION EXCHANGE AND SOLVENT EXTRACTION SERIES

Series editors
Arup K. Sengupta
Bruce A. Moyer

Founding Editors
Jacob A Marinsky
Yizhak Marcus

Contents of Other Volumes

Ion Exchange and Solvent Extraction

A Series of Advances

Volume 20

Edited by

ARUP K. SENGUPTA

CRC Press
Taylor & Francis Group
Boca Raton   London   New York

CRC Press is an imprint of the
Taylor & Francis Group, an **informa** business

CRC Press
Taylor & Francis Group
6000 Broken Sound Parkway NW, Suite 300
Boca Raton, FL 33487-2742

First issued in paperback 2017

© 2011 by Taylor and Francis Group, LLC
CRC Press is an imprint of Taylor & Francis Group, an Informa business

No claim to original U.S. Government works

ISBN 13: 978-1-138-11167-7 (pbk)
ISBN 13: 978-1-4398-5539-3 (hbk)

Visit the Taylor & Francis Web site at
http://www.taylorandfrancis.com

and the CRC Press Web site at
http://www.crcpress.com

Contents

Preface

A wise man once said, "The further back I look, the further forward I can see." On the eve of printing the 20th volume of *Ion Exchange and Solvent Extraction: A Series of Advances*, it is only appropriate that we look closely at how the series began. Volume 1 was published in 1966. Titled *Ion Exchange* and edited by Jacob Marinsky, it contained nine chapters:

Chapter 1, Transport Processes in Membranes, S. Roy Caplan and Donald C. Mikulecky

Chapter 2, Ion Exchange Kinetics, F. Helfferich

Chapter 3, Ion Exchange Studies of Complex Formation, Y. Marcus

Chapter 4, Liquid Ion Exchangers, Erik Hogfeldt

Chapter 5, Precise Studies of Ion Exchange Systems Using Microscopy, David Freeman

Chapter 6, Heterogeneity and the Properties of Ion Exchange Resins, Lionel S. Goldring

Chapter 7, Ion Exchange Selectivity, D. Reichenberg

Chapter 8, Resin Selectivity in Dilute to Concentrated Solutions, R.M. Diamond and D.C. Witney

Chapter 9, Interpretation of Ion Exchange Phenomena, Jacob A. Marinsky

All these chapters essentially dealt with thermodynamics, kinetics, synthesis, and potential ways to apply ion exchange processes with greater efficiency. A wealth of knowledge subsequently emerged that, in turn, fueled new pursuits. The sheer fact that the 20th volume is now ready for printing testifies to the growth in the field of ion exchange and the synergies it has created with myriad scientific endeavors—from decontamination to deionization, mining to microelectronics, and drug delivery to desalination, to name just a few. Indeed, the adage, "True science does not fade away, it moves into new equilibrium," exemplifies the pace of change in ion exchange science.

Our scientific vocabulary has similarly kept pace with change. When the first volume was published, words such as "nanotechnology" and "drug delivery" were literally unborn, while analytical instruments for surface characterization—namely scanning electron microphotography (SEM), tunneling electron microphotography (TEM), and x-ray photoelectron spectroscopy (XPS)—did not exist.

No ingenuity grows in isolation. Over the past four decades, many scientists have made notable advancements in both the theory and application of ion exchange. Pioneers continue to break new ground by synthesizing novel materials and creating synergies with interdisciplinary fields of science and engineering. The seven chapters I have assembled in this latest volume reflect contributions from researchers across the globe who are making great strides in the field of ion exchange.

During the past decade, unique properties of various metals at nanoscale sizes have been unraveled and their potential applications highlighted. However, in solvents, including water, the metal nanoparticles tend to coalesce and the resulting aggregates fail to exhibit nanoscale properties, thus preventing potential application opportunities. Chapter 1 presents research studies showing how functionalized ion exchange polymers serve as supports to stabilize metal nanoparticles (MNPs) without forming aggregates. The resulting nanocomposites offer sensing and biosensing properties for a host of new applications.

Barring a few exceptions, both the application and practice of ion exchange have been limited to the aqueous phase of the process. Also, spherical beads and granular particles are the most common physical configurations of ion exchangers used today. Chapter 2 describes sorption of different gases from the air by ion exchange resins and fibrous ion exchangers. Ion exchange fibers have shorter diffusion path lengths and exhibit faster kinetics. Consequently, for high-volume air flow rates, ion exchange fibers have the potential to sorb water vapor, ammonia, carbon and sulfur dioxides, hydrogen chloride, and other contaminating gases present in minor concentrations.

Perchlorate contamination in groundwater and surface water is an emerging environmental problem in the United States and abroad. Apart from anthropogenic sources such as rocket fuels, explosives, and fireworks, natural perchlorate has also been found in the hyper-arid areas of northern Chile and in some areas of the United States. Chapter 3 discusses a selective ion exchange technology capable of removing and recovering perchlorate quantitatively through stable isotope ratio analysis of chlorine and oxygen atoms; the method allows forensic analysis of the origin of perchlorate present in contaminated water.

The effects of particle size and size distribution greatly influence the efficiency of chromatographic separation using ion exchange resins. This phenomenon is particularly applicable when the sorption affinities of solutes to be separated are nearly equal. Chapter 4 demonstrates how numerical simulations coupled with small-scale bench-top experiments can help tailor particle size distribution, enhancing the efficiency of each application.

Ion exchange processes invariably include a regeneration step that requires the use of highly concentrated chemicals. The spent regenerant also contains high concentrations of solutes desorbed from the ion exchanger—an ecological problem that is an inherent limitation of the ion exchange process. All ion exchange processes are either exothermic or endothermic, i.e., the equilibrium can shift due to a change in temperature. Chapter 5 reviews dual-temperature ion exchange processes in which sorption and desorption are carried out by varying the temperature without adding external chemicals.

Natural biopolymers from renewable sources, although desirable, often lack the durability and other physical attributes for sustained operation in ion exchange applications. Among the natural products, chitosan has long received significant attention because it can be modified chemically for improved efficiency and durability. Chapter 6 presents preparation and characterization of a new composite material in which microparticles of clinoptilolite, a naturally occurring zeolite, are embedded in a matrix of cross-linked chitosan. The capacity of the composite to remove toxic metal presents new opportunities for the natural biopolymer.

The polymeric ion exchange resins in wide use essentially have four composition variables: matrix, cross-linking, functional groups, and porosity. Recent studies demonstrate that the appropriate dispersion of metal oxide nanoparticles within the polymer phase (i.e., a fifth composition variable) greatly enhances the application opportunities of ion exchange resins. Chapter 7 presents preparation, characterization, and field-level experience of this emerging class of "hybrid ion exchangers."

As editor of the series, I am indebted to the researchers of the past, and immensely thankful to all the chapter contributors of this 20th volume. Finally, I acknowledge with thanks a great deal of assistance and unselfish effort of Prasun Chatterjee, a doctoral student at Lehigh University, for attending to the many details that brought this volume to its successful closure.

Arup K. SenGupta
Lehigh University, Bethlehem, Pennsylvania, USA

About the Editor

Dr. Arup K. SenGupta is currently the P. C. Rossin Professor of the Departments of Civil, Environmental, and Chemical Engineering at Lehigh University. He received his bachelor's degree in chemical engineering from Jadavpur University, India in 1973, served as a process development engineer from 1973–1980 for the Kuljian Corporation in Philadelphia, and in 1984 obtained a PhD in environmental engineering from the University of Houston. In 1985, he joined Lehigh University. He served as the editor of the *Reactive and Functional Polymers Journal* from 1996 to 2006.

Dr. SenGupta's research encompasses multiple areas of separation and environmental processes including ion exchange. For his research, Dr. SenGupta has received a variety of awards including, in part, the 2009 Lawrence K. Cecil Award from the American Institute of Chemical Engineers (AIChE) for outstanding contribution in the field of environmental engineering; the 2009 Astellas Foundation Award from the American Chemical Society (ACS) for scientific research that improves public health; the 2008 Dhirubhai Ambani Award from the Institution of Chemical Engineers (IChemE) in the United Kingdom for engineering innovation to provide potable water for resource-poor arsenic-affected people and, in 2007, the Grainger Silver Prize Award from the National Academy of Engineering (NAE) for providing a sustainable engineering solution for arsenic-contaminated drinking water in the Indian subcontinent. SenGupta is the inventor of the first regenerable arsenic-selective ion exchanger that is currently in use in multiple countries.

Dr. SenGupta is a frequent contributor to *Environmental Science and Technology, Reactive and Functional Polymers, the AIChE Journal, Water Research* and other peer-reviewed journals and has authored/co-authored nearly 70 articles as well as holding patents on seven inventions.

Contributors

Amanda Alonso
Department of Chemistry
Autonomous University of Barcelona
Barcelona, Spain

Abelardo D. Beloso, Jr.
Department of Earth and
*   Environmental Sciences*
University of Illinois at Chicago
Chicago, Illinois, USA

Yongrong Bian
Oak Ridge National Laboratory
Oak Ridge, Tennessee, USA
and
Institute of Soil Science
Chinese Academy of Sciences
Nanjing, China

John Karl Böhlke
United States Geological Survey
Reston, Virginia, USA

Gilbert M. Brown
Oak Ridge National Laboratory
Oak Ridge, Tennessee, USA

Maria Valentina Dinu
Institute of Macromolecular Chemistry
Bucharest, Romania

Ecaterina Stela Dragan
Institute of Macromolecular Chemistry
Bucharest, Romania

Baohua Gu
Oak Ridge National Laboratory
Oak Ridge, Tennessee, USA

Paul B. Hatzinger
Shaw Environmental, Inc.
Lawrenceville, New Jersey, USA

Linnea J. Heraty
Department of Earth and
*   Environmental Sciences*
University of Illinois at Chicago
Chicago, Illinois, USA

Vladimir A. Ivanov
M. V. Lomonosov
Moscow State University
Moscow, Russia

Andrew Jackson
Department of Civil and Environmental
*   Engineering*
Texas Tech University
Lubbock, Texas, USA

Xin Jiang
Institute of Soil Science
Chinese Academy of Sciences
Nanjing, China

Ari Kärki
Finex Oil
Kotka, Finland

Ruslan Kh. Khamisov
V. I. Vernadsky Institute of
*   Geochemistry and Analytical*
*   Chemistry*
Russian Academy of Sciences
Moscow, Russia

E. H. Kosandrovich
Institute of Physical Organic Chemistry
Belarus National Academy of Sciences
Minsk, Belarus

Jarmo Kuisma
Finex Oil
Kotka, Finland

Jorge Macanás
Department of Chemistry
Autonomous University of Barcelona
Barcelona, Spain

Marcela Mihai
Institute of Macromolecular Chemistry
Bucharest, Romania

Heikki Mononen
Finex Oil
Kotka, Finland

Maria Muñoz
Department of Chemistry
Autonomous University of Barcelona
Barcelona, Spain

Dmitri N. Muraviev
Department of Chemistry
Autonomous University of Barcelona
Barcelona, Spain

Prakhar Prakash
Chevron Energy Technology Company
Richmond, California, USA

Patricia Ruiz
Department of Chemistry
Autonomous University of Barcelona
Barcelona, Spain

Tuomo Sainio
Lappeenranta University of Technology
Lappeenranta, Finland

Sudipta Sarkar
Environmental Engineering Program
Lehigh University
Bethlehem, Pennsylvania, USA

Arup K. SenGupta
Environmental Engineering Program
Lehigh University
Bethlehem, Pennsylvania, USA

V. S. Soldatov
Institute of Physical Organic Chemistry
Belarus National Academy of Sciences
Minsk, Belarus
and
Lublin University of Technology
Lublin, Poland

Neil C. Sturchio
Department of Earth and
*   Environmental Sciences*
University of Illinois at Chicago
Chicago, Illinois, USA

Nikolay A. Tikhonov
M. V. Lomonosov
Moscow State University
Moscow, Russia

1 Ion Exchange-Assisted Synthesis of Polymer Stabilized Metal Nanoparticles

Jorge Macanás, Patricia Ruiz, Amanda Alonso, María Muñoz, and Dmitri N. Muraviev

CONTENTS

ABSTRACT

Stabilization of metal nanoparticles (MNPs) within various polymeric materials can overcome one of their main drawbacks: a trend toward aggregation. Stabilization also allows the solution of other important problems such as preventing uncontrollable growth of MNPs, controlling their size and growth rates, and providing MNP solubility in some organic solvents. Clearly, stabilization of MNPs in polymeric matrices is considered a very promising approach for solving MNP stability problems. The resulting polymer–metal nanocomposite materials find numerous applications in modern science and technology. This chapter summarizes recent results obtained by the authors in developing novel approaches for the inter-matrix synthesis (IMS) of polymer-stabilized MNPs. These approaches allowed us to synthesize MNPs in functionalized polymer matrices whose functional groups act somewhat as nanoreactors during synthesis of MNPs of various compositions and structures. The chapter discusses (1) the general principles of the IMS technique, (2) the mechanisms for stabilizing MNPs in polymers, (3) the synthesis of monometallic and polymetallic MNPs with core shell and core sandwich structures, and (4) the applications of polymer–metal nanocomposites in electroanalytical devices such as sensors, biosensors, and others.

1.1 INTRODUCTION

The ion exchange synthesis of metal nanoparticles (MNPs) involves a group of methods and is generally known as inter-matrix synthesis (IMS).[1] Its main feature is the dual function of the matrix that serves as both the medium for the synthesis of MNPs and as a stabilizer that prevents their uncontrollable growth and aggregation.[2]

IMS was essentially the first method used in the production of nanocomposite inorganic materials such as MNP-containing glasses and ceramics in which MNPs played the roles of very stable decorative pigments and dyes.[3] The oldest known such object is thought to be the Lycurgus Cup that dates back to Rome in the late fourth century BCE[4,5] and can still be seen at the British Museum. It is made of a type of glass that appears green in reflected light and appears translucent red when light is shone directly on it. Analysis of the glass reveals that it contains a small number of Ag–Au MNPs with diameters of approximately 70 nm and its approximate Ag:Au ratio is 14:1. Indeed, these MNPs give the Lycurgus Cup its special color characteristic.[6] It is remarkable that Greco–Roman techniques have been used up to modern times; related formulas were described by Arabian authors during the medieval period[7] and the Renaissance as practical applications of alchemical knowledge,[8] and by modern chemists, from the *Encyclopedie* of Diderot and d'Alembert[9] to the present day.[10]

Glass colored by MNPs was widely used to decorate the stained glass windows of the European cathedrals in medieval times.[11–13] Bright examples of the early applications of MNP-based nanotechnology can be still seen in many European countries (for example, in the Cathedral of Santiago de Compostela in Spain and the Convento di Santa Croce in Florence, Italy among many others). In both Asia and Europe, the

IMS technique was used simultaneously in the production of luster pottery decorated with various MNPs such as Cu and Ag. Samples of these ceramics that exhibit a typical metallic luster due to the presence of MNPs have been intensively studied in recent years.[14–18] The IMS technique used to produce this pottery required two main steps: (1) the immobilization of metal cations (MNP precursors) inside the ceramic matrix, and (2) the reduction of metal ions to the zero-valent state with carbon monoxide, leading to the formation of MNPs.[18] Although these examples of the ancient nanotechnology look really impressive, they were created based only on pure intuition.

The fundamentals of modern nanoscience and nanotechnology appeared around the mid-nineteenth century. Scientists began to understand that objects in the nanometer size range exhibited special properties and behavior that gave rise to the beginning of colloid chemistry, the precursor of modern nanoscience and nanotechnology. Scottish scientist Thomas Graham devised the *colloid* name in 1861 to describe materials in the nanometric size range (1 to 1000 nm in any dimension) that due to the dominance of surface forces tended to be *sticky* or *adhesive*. In fact, a colloid or colloidal dispersion is a form of intermediate (suspension) between a true solution and a mixture.

The colloid sciences saw intensive progress in the first decades of the twentieth century. The achievements garnered four Nobel Prizes between 1924 and 1932. The recipients were Richard A. Zsigmondy, chemistry, 1924–1925; Theodor Svedberg, chemistry, 1925; Jean Baptiste Perrin, physics, 1926; and Irving Langmuir, chemistry, 1932.[19]

The first communication about the IMS of MNPs in ion exchange resins is from 1949.[20] Mills and Dickinson described the preparation of an anionic resin containing Cu MNPs ("colloid copper") and the use of this nanocomposite material to remove oxygen from water based on its interaction with Cu MNPs. Since then, many studies of the modification of ion exchange resins with MNPs (mainly Cu MNPs) resulted in the development of a new class of ion exchange materials combining ion exchange and redox properties; they are known also as redoxites and electron ion exchangers.[21–24]

The IMS preparation of such materials involves two consecutive stages: (1) the immobilization of metal ions in the polymer matrix and (2) their reduction inside the matrix. The immobilization of metal cations inside the ion exchange resins was carried out by using either conventional ion exchange reactions (in the case of cation exchange resins) or the formation of metal complexes (in weak base anion exchangers). The redoxites have found wide application in the complex water treatment processes at power stations for the removal of hardness ions (by ion exchange) and oxygen (by redox). However, essentially no information about the sizes and structures of MNPs synthesized in redoxite matrices can be found in the literature.

In the past two decades, practitioners intended to elucidate and control nanometer-size objects in new interdisciplinary fields known as nanoscience and nanotechnology. The result is a new wave of intensive and more detailed studies of MNPs and various nanocomposites. Figure 1.1 shows the exponential increase in publications about nanotechnology due to its wide application in medicine, chemistry, physics, and other fields. Moreover, a huge number of patents have been issued in the past decade—more than 5,000 in 2008 alone.

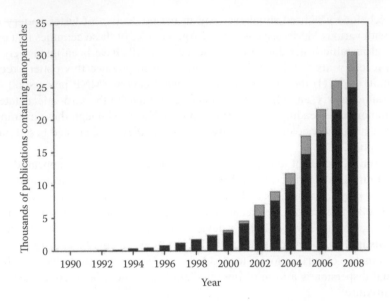

FIGURE 1.1   Bibliographic analysis based on search of nanoparticles in *Scifinder Scholar*. Black represents articles; gray represents patents.

The main goals of nanoscience and nanotechnology are (1) the creation of functional materials, devices, and systems through control of matter on the nanometer length scale and (2) exploitation of novel phenomena and properties (physical, chemical, and biological) at that scale. To achieve these goals, it is necessary to use multidisciplinary approaches; inputs from physicists, biologists, chemists, and engineers are required to advance our understanding of the preparations, applications, and impacts of new nanotechnologies.[25] This chapter summarizes the results of the authors' studies over the past five years on the further development of IMS and its application for the synthesis of MNPs of various structure and composition.

1.2   METAL NANOPARTICLES (MNPS)

1.2.1   GENERAL PROPERTIES

The most commonly accepted definition for a nanomaterial is "a material that has a structure in which at least one of its phases has a nanometer size in at least one dimension (D)."[1,25] Nano objects can be subdivided into three groups: (1) 1D nanometer-size objects (e.g., thin films); (2) 2D nanometer-size objects (e.g., nanowires, nanorods, and nanotubes), and (3) 3D nanometer-size objects (e.g., nanoparticles and/or nanoclusters). This chapter focuses on the synthesis and characterization of the third group: nanoparticles.

Nanometer-sized metal particles are objects of great interest in modern research[26,27] due to their unique physical and chemical properties (electrical, magnetic, optical, ionization potentials, etc.), which substantially differ from those of both bulk material and single atoms.[26–31] This can be illustrated, for example, by the dependence of gold melting temperature on the sizes of gold nanoparticles[32] (Figure 1.2). Similarly, suspensions of Ag nanoparticles with sizes ranging from 40 to 100 nm show different colors.[33] In addition, certain physical phenomena do not exist in materials with larger grain sizes. The general quantum-size effect for optical transitions in semiconductor nanocrystals is a case in point.[34] Very small nanoparticles (<10 nm) are especially interesting for the quantum confinement effects inherent in particles of that size.[35]

Typically, MNPs have a large surface-area-to-volume ratio as compared to the bulk equivalents, making them particularly attractive candidates for catalytic applications[25,36–38] (see Table 1.1). Nanoparticles can also be used for many other practical applications such as electrocatalysis-based processes, photochemistry, optics, medicine, and others.[39–48] Indeed, research centered on nanoscopic materials extends from the semiconductor industry, where the ability to produce nanometer-scale features leads to faster and less expensive transistors[49] to biotechnology, where luminescent nanoparticles are extremely interesting as bioprobes.[50] Other examples include catalysts for fuel cells[51] and electrocatalysts used in sensing devices with enhanced properties. In this sense, nanoparticles have already exerted major impacts on electrochemical biosensors ranging from glucose enzyme electrodes to genoelectronic sensors.[52]

1.2.2 MNP Preparation: Stability Challenges

At present, many different methods are used to produce inorganic nanoparticles[34,38,75–85] (Table 1.2 and Table 1.3). Both chemical and physical techniques can be used, but chemical processes are usually less expensive because they normally do not require complex and costly equipment.[27] Frequently, the reduction of metal salts dissolved in appropriate solvents produces small metal particles of varying size distributions.[30,105] A variety of reducing agents used include alcohols, glycols, metal borohydrides, and certain specialized reagents such as tetrakis (hydroxymethyl) phosphonium chloride.

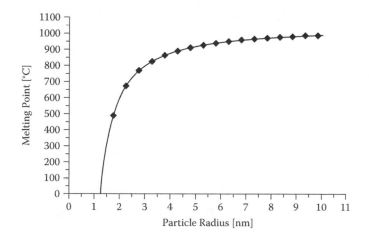

FIGURE 1.2 Dependence of melting point temperature on size for gold nanoparticles.

TABLE 1.1
Some Catalytic Applications of MNPs

Metal Type	Metal	Applications	Refs.
Noble	Pd	C–C bond formation, hydrogenation, oxidation, hydrodechlorination, C–H activation	53
	Pt	Hydrogenation, electrocatalytic oxidation (fuel cells)	54–58
	Ag	Biological and medical antimicrobial applications, selective oxidation, dehydrogenation, hydrogenation	59–62
Transition	Ru	Highly active and selective in various catalytic processes	63–66
	Ni	Hydrogenation of aromatic compounds, fatty acid ethyl esters, etc.; hydrogen storage	67–71
	Cu	Azidiration, cyclopropanation, reduction of NO, dehydrogenation of methanol	72–74

TABLE 1.2
Physical Routes to Nanoparticle Preparation

Method	Basis	Advantages	Disadvantages	Refs.
Sonochemistry	Use of sonic waves on chemical systems	Enhances mass transport, no heating necessary; environmentally friendly, controllable MNP size distribution	Requires additional reducing agents	86,87
Microwave irradiation	Microwave-assisted deposition of metal salts as precursors	Short reaction times, small particle sizes, narrow size distributions, high purity	MNP size and distribution difficult to control	88,89
Pulsed laser ablation	Vaporization of metals and deposition on support	Simple and environmentally friendly, strong adsorption of MNPs onto supports, no by-products, can be scaled up for industry, MNP composition can be adjusted with most metals or mixtures	Specialized equipment required	88,90
Plasma	Plasma reduction	Environmentally friendly, fast, simple alternative to H_2 reduction at high temperatures	Specialized equipment required	91

TABLE 1.3
Chemical Routes to Nanoparticle Preparation

Method	Basis	Advantages	Disadvantages	Refs.
Traditional	Impregnation: wetting of porous solid support with MNP precursor dissolved in minimum solvent Co-precipitation: Simultaneous precipitation of metal and support Precipitation–deposition: Dissolution of metal precursor followed by pH adjustment to achieve complete precipitation of metal hydroxide; hydroxide formed is calcined and reduced to elemental metal	Ease, no specialized equipment	Broad NP size distribution, poor control of NP size, particle agglomeration	92–96
Microemulsions	Solid support impregnated with microemulsion containing dissolved metal salt precursor	Environmentally friendly, narrow crystallite distribution		97,98
Photochemistry	Photo-assisted deposition	Environmentally friendly	Control of size and distribution of MNPs remains unclear	99,100
Chemical vapor deposition (CVD)	Vaporization of metal and growth of MNPs under high vacuum in excess stabilizing organic solvents	Controllable, reproducible, relatively narrow size distribution, MNP preparation on wide range of organic and inorganic supports under mild conditions	Often limited by vapor pressure of precursor and mass transfer-limited kinetics	101,102
Other chemical	Ion exchange/reduction	Better control of growth and distribution of NPs	Excess additional reductant needed to ensure complete reduction	103,104

However, the main drawback that still limits the wide application of MNPs is their insufficient stability arising from their tendency to self-aggregate.[106] Indeed, MNPs can be so fragile and unstable that if their surfaces touch, the particles will fuse together and thus lose their nanometric size and special properties[107] (the properties of nanoparticles are size- and shape-dependent). These features of nanoparticles, in part determined by the conditions of synthesis, create enormous difficulties in their fabrication and application.[107]

In many instances, nanoparticles are dispersed after synthesis in a liquid or solid medium by using different mechanochemical approaches, including sonication, but the scope of such approaches for dispersing the nanoparticles is limited by their re-aggregation. Moreover, nanoparticles can aggregate as a result of a further manipulation and also during their growth. A typical mechanism of aggregation is Ostwald ripening—a growth mechanism by which small particles dissolve and are consumed by larger particles.[108,109] As a result, the average nanoparticle size increases over time while particle concentration decreases. Simultaneously, as particles augment in size, their solubility diminishes. Figure 1.3 illustrates the main stages of transforming an individual atom into a bulk metal through cluster, nanosize, and colloidal particles according to the literature.[80,110]

Therefore, the stabilization of MNPs is required to: (1) prevent the uncontrollable growth of particles; (2) prevent particle aggregation; (3) control particle growth rate and final particle size; (4) allow particle solubility in various solvents; and (5) terminate the particle growth reaction.

1.2.3 STABILIZATION MECHANISMS OF MNPs IN POLYMERS

The successful synthesis of nanocrystals has three steps: nucleation, growth, and stabilization by a ligand (or stabilizing agent) via colloidal forces.[82] The three main types of colloidal forces are van der Waals interactions, electrical double-layer interactions, and steric interactions. In addition, hydrophobic and solvation forces may be important.[111] Electrostatic repulsion stabilizes colloids through the adsorption of ions to MNP surfaces and creation of an electrical double layer[112] that results in a Coulombic repulsion force between individual particles (Figure 1.4a). The magnitude of the repulsion depends on the surface potential and the electrolyte concentration

Particle size	0.1 nm	1 nm	~10 nm	~10^2 nm	>>10^3 nm
Name	Metal atom	Cluster	Nanoparticle	Colloid	Bulk metal
Scheme					

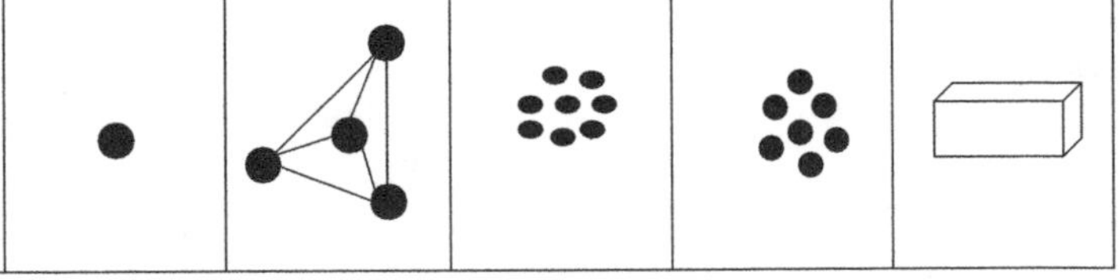

FIGURE 1.3 Main stages of transformation of metal atoms into bulk metal.

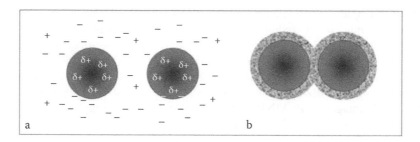

FIGURE 1.4   Stabilization mechanisms of MNP: (a) electrostatic stabilization; (b) steric stabilization.

and valence.[113] This happens, for instance, in stabilizing gold nanoparticles with citrate anions.[114,115] When low molecular weight stabilizing agents or polymers are used, the stabilization mechanism can be qualified as a steric stabilization because nanoparticles appear to be surrounded by layers of material that are sterically bulky (Figure 1.4b).

Rozenberg et al.[107] and many other authors[1,42,116–122] have stated that polymer-assisted fabrication of inorganic nanoparticles is the most efficient and universal way to overcome the stability problem of MNPs and to retain their properties. Metal nanoparticles synthesized by this approach exhibit long-time stability against aggregation and oxidation while nanoparticles prepared in the absence of polymers are prone to quick aggregation and oxidation.[107,123]

Stabilizing polymer shells can be obtained by synthesizing the inorganic nanoparticles and then dispersing them in a polymer solution. This can be called an ex situ approach and it is the most general one because it sets no limitations (at least theoretically) on the kinds of nanoparticles and polymers used.[78,81,107] The presence of such a shell increases the compatibility of particles and polymer and simplifies their homogeneous dispersion inside the matrix. However, significant challenges are associated with blending polymers and nanoparticles to afford homogeneous and well dispersed inorganic material in the polymer. Indeed, the integration of nanoparticles into polymers has always been of significant theoretical and experimental interest. Inorganic fillers have been used in conjunction with organic polymer materials, largely in an effort to enhance the physical and mechanical properties over those of the polymers alone. For instance, in the middle nineteenth century, research efforts by Charles and Nelson Goodyear showed that vulcanized rubber can be toughened significantly by the addition of zinc oxide and magnesium sulfate.[124,125]

To cope with these difficulties, a different strategy can be considered: protective polymer coating formation and MNP preparation can be combined into a single process or performed as a series of consecutive processes in one reactor (the in situ approach).[1,2,118,120,121,126–128] MNPs are generated inside a polymer matrix by precursors that are transformed into the desirable MNPs by appropriate reactions.[129–132] In situ approaches are currently getting much attention because of their obvious technological advantages over ex situ methods, permitting exquisite control of particle size and morphology. Indeed, the in situ method can be extended to the preparation of many metal–polymer nanocomposites.

As an example, we can cite the investigation of Ag MNP formations of various molecular weights by the reduction of AgNO$_3$ with formaldehyde in the presence of polyvinylpyrollidone. The study revealed that polymer stabilization efficiency depends on both the molecular weight of the polymer stabilizer and the rate of MNP formation.[133] Another example illustrates the results of MNP synthesis with and without polymer stabilization. Figures 1.5a and b are transmission electron microscopy (TEM) images of Pt MNPs synthesized inside a polyvinyl chloride (PVC) matrix via the solid phase incorporated reagents technique[1,134,135] after reduction with aqueous formaldehyde solution. Note that the size of platinum MNPs does not exceed 5 to 7 nm.

The particles are well separated because of stabilization with PVC chains inside the polymeric matrix that prevents aggregation. Figures 1.5c and d illustrate platinum MNP formation outside the PVC matrix, i.e., in the reducing agent (formaldehyde) solution in the course of the same experiment due to the partial

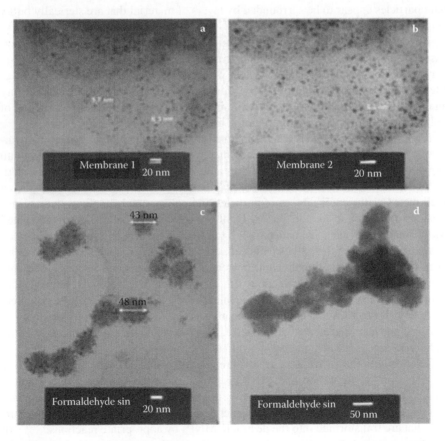

FIGURE 1.5 Formation of Pt MNP with (a, b) and without (c, d) stabilization: (a, b) synthesis of Pt MNP by Sphiner technique inside PVC matrix; (c) formation of large Pt MNP in aqueous solution of reducing agent (formaldehyde); (d) further aggregation of Pt MNP in formaldehyde solution.

extraction of platinum ions from the polymer surface by transfer to the form-aldehyde. As shown in Figure 1.5c, in the absence of a stabilizing agent, the size of the Pt MNPs exceeds by almost one order of magnitude the size of the Pt PSMNP (40 to 50 nm versus 5 to 7 nm). Moreover, non-stabilized Pt MNPs tend to aggregate in the solution phase, then finally form bigger particle clusters (Figure 1.5d).

Despite the methodological differences of ex situ and in situ preparation methods, understanding the processes occurring in polymer interactions with nanoparticles is crucial. The mechanisms of MNP stabilization with polymers can be explained by two approaches that run simultaneously in the system and influence each other.

The first approach is the *substantial increase of viscosity of the immobilizing media (polymer matrix).* As follows from the Smoluchowsky equation,[80] the rate constant of particle coagulation k_c is inversely proportional to the viscosity of the media η (k = Boltzmann constant and T = temperature):

$$k_c = \frac{8kT}{\eta} \tag{1.1}$$

A similar conclusion follows from the Stokes–Einstein equation that allows deter-mination of the diffusion coefficient D of a spherical particle of radius r in a viscous medium.[80]

$$D = \frac{kT}{6\pi\eta r} \tag{1.2}$$

Nevertheless, as shown by Cole et al.,[135,136] the mobility of Au nanoparticles in a poly(t-butyl acrylate) and gold composite decreases by 2 to 3 orders of magnitude compared with the mobility predicted by the Stokes–Einstein equation. The authors ascribe this discrepancy to a strong bridging interaction between Au MNP and the segments of chains of the stabilizing polymeric matrix. As the result, the mobility of nanoparticles inside the polymer substantially decreases and the matrix, in turn, appears somewhat cross-linked so that effective viscosity of the polymer increases by a factor of ~4.[137]

The second approach is the *substantial decrease of the energy of particle–particle interactions in PSMNP systems versus non-stabilized MNP dispersions.* The potential energy of attraction U_r between two spherical particles of radius r can be approxi-mately described by the following simplified expression:[138]

$$U_r \approx \frac{Ar}{12 l_o} \ at \ r \gg l_o \tag{1.3}$$

where A is the effective Hamaker constant having the dimensions of energy and l_o is the minimum distance between particle surfaces. The value of A is known to be close to kT for polymer particles (6.3×10^{-20} J for polystyrene), while for the metal dispersions it is far higher (40×10^{-20} J for silver).[139]

1.3 GENERAL PRINCIPLES OF INTER-MATRIX SYNTHESIS IN ION EXCHANGE MATRICES

This section focuses on only one version of the IMS technique, namely IMS in ion exchange matrices. Considering the in situ approach above, the combination of the wide number of matrices available and the different types of MNPs that can be prepared gives rise to a huge number of possible metal–polymer nanocomposites that can be produced by this technique.

Even if the number of polymers is reduced to those that have ion exchange properties, the number of variables remains and a great number of different materials can be used and produced. For example, the scheme shown in Figure 1.6 considers four main parameters for the IMS technique: (1) the nature of the ion exchange matrix, (2) the type of MNP precursor (metal ions or metal complexes), (3) the reaction that forms the MNP (reduction, oxidation, precipitation), and (4) the MNP composition (metal compounds or zero-valent metals). These possible bases roughly illustrate the huge number of combinations of polymer stabilized MNPs (PSMNPs) and nanocomposites that can be obtained.

The general principles of IMS apply to all types of polymer matrices and nanoparticles. In summary: (1) polymer molecules serve as nanoreactors and provide a confined medium for the synthesis (thus controlling particle size and distribution); and, (2) polymer molecules stabilize and isolate the generated nanoparticles, thus preventing their aggregation. Moreover, this method frequently dictates a well ordered spatial arrangement of the generated nanoparticles.

In the case of ion exchange matrices, the functional groups that can immobilize metal ions and metal ion complexes are the key points for IMS because they are homogeneously distributed in the ion exchange matrix and behave as combinations of single isolated nanoreactors generating homogeneous nanocomposites. This is the case for ion exchange membranes (IEMs) that are characterized by their ionizable functional groups and permeability to electrolytes in aqueous solutions. Depending on the ions with which they interact, IEMs can be classified as (1) cation exchange membranes (CEMs) containing negatively charged groups (e.g., SO_3^-), or (2) anion exchange membranes (AEMs) containing positively charged groups (e.g., NR_3^+).

1.3.1 REQUIREMENTS FOR PARENT POLYMERS

When using IMS, it is important to consider both the polymer properties and the final application of the nanocomposite, since these points dictate certain requirements for the parent matrix (inside which the MNPs must be synthesized). For example, when using a nanocomposite for sensor or biosensor constructions to be applied in analysis of aqueous solutions, the polymer must be insoluble in water. Conversely, the polymer must provide sufficient permeability of the PSMNP-containing membrane to the ions or molecules of the analyte under study. The polymer must be hydrophilic or swell slightly in water to enhance the sensor and response rates. The solubility of nanocomposites in some organic solvents allows the preparation of homogeneous PSMNP solutions (PSMNPs "inks") that may be deposited onto the desired surfaces (e.g., of electrodes) to modify their properties. This solubility also allows characterization of MNPs via microscopic analysis, electrochemical study, and other techniques.[126]

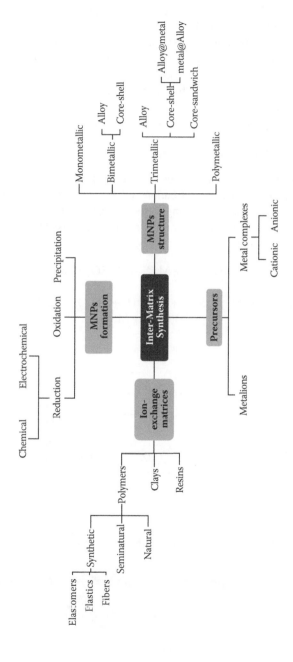

FIGURE 1.6   IMS possibilities for different ion exchange matrices.

One polymer that fulfills these requirements is sulfonated poly(ether ether ketone) or SPEEK.[128,140] It attracts great interest for fabricating membranes for fuel cells due to its thermoplastic properties, chemical strength, stability under oxidation, and low cost.[141–144] Figure 1.7 shows the structures of the initial poly(ether ether ketone) (PEEK) and sulfonated PEEK (SPEEK) polymers. For noncross-linked functionalized polymers, the sulfonation degree (SD) is the most important factor because it determines the solubility of SPEEK in water and organic solvents.[145,146] Complete sulfonation of PEEK (one sulfonic group per benzene ring = 100% sulfonation) results in good solubility of SPEEK in hot water[145] but is unsuitable for sensor applications in aqueous solutions. The insolubility in water and solubility in some organic solvents (e.g., dimethylformamide or DMF) can be achieved at an SD of ~60%.[126]

1.3.2 Synthesis of Monometallic MNPs

PSMNP formation involves two simple consecutives stages (Figure 1.8): the loading of ion exchange groups with metal cations (Cu^{2+}, Co^{2+}, Ni^{2+}, etc.), followed by metal reduction inside the matrix resulting in the formation of monometallic MNPs. Reduction can be done by using $NaBH_4$ or other reducing agents. These two steps may be described by the following equations:

$$\text{Metal loading: } 2R\text{-}SO_3^-Na^+ + M_1^{2+} \rightarrow (R\text{-}SO_3^-)_2\, M_1^{2+} + 2\,Na^+ \tag{1.4}$$

$$\text{Metal reduction: } (R\text{-}SO_3^-)_2\, M_1^{2+} + 2NaBH_4 + 6H_2O \rightarrow 2R\text{-}SO_3^-Na^+ +$$
$$7H_2 + 2B(OH)_3 + M_1^0 \tag{1.5}$$

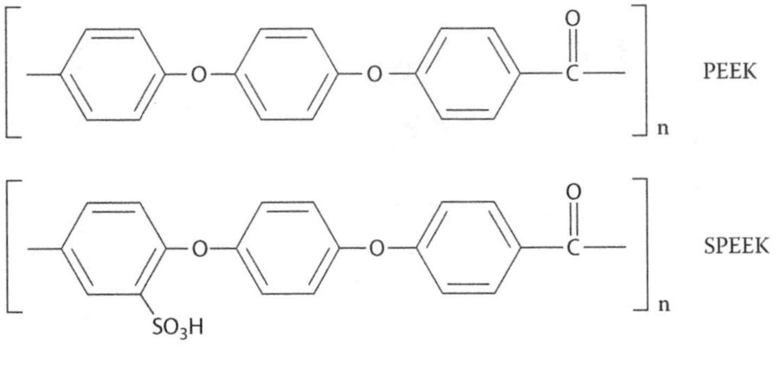

FIGURE 1.7   Structures of repetitive units of PEEK and SPEEK.

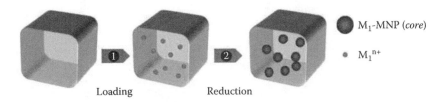

Loading   Reduction

M_1-MNP (*core*)

M_1^{n+}

FIGURE 1.8   Monometallic MNP preparation.

Since reduction of metal ions to zero-valent MNPs takes place in the solution boundary close to the ion exchange groups, the metal loading reaction (1.5) can be better depicted as the combination of an ion exchange reaction (1.6) and a reduction reaction (1.7):

$$(\text{R-SO}_3^-)_2 \, M_1^{2+} + 2Na^+ \rightarrow 2R\text{-SO}_3^-Na^+ + M_1^{2+} \tag{1.6}$$

$$M_1^{2+} + 2BH_4^- + 6H_2O \rightarrow 2B(OH)_3 + M_1^0 + 7H_2 \tag{1.7}$$

The preparation of monometallic MNPs via this simple technique with different metals (Cu, Pt, Pd, Co, etc.) has been recently reported.[128] The characterization of nanoparticles by TEM confirms that they are well separated and do not form visible aggregates. Most MNPs (except Pt MNPs) obtained by IMS in a SPEEK matrix had spherical shapes (Figures 1.9 and 1.10).

A nanocomposite can also be characterized without dissolution by cutting a thin slice of the sample with an ultramicrotome and submitting it to TEM analysis. This procedure is useful because it demonstrates that nanoparticle formation does not occur throughout a sample as metal reduction (by borohydride anions), but occurs only near the surface of the material (Figure 1.11).

This type of nanoparticle distribution occurs because the polymer matrix bears negative charges due to the presence of well dissociated functional groups.[147,148] Therefore, borohydride anions (BH_4^-) bearing the same charge cannot deeply penetrate inside the matrix due to electrostatic repulsion, also known as the Donnan exclusion effect.[149,150] Thus, anion penetration inside the matrix is balanced by the sum of two oppositely directed driving forces: (1) the gradient of borohydride concentration and (2) the Donnan exclusion effect. The action of these two driving forces results in the formation of MNPs mainly in the surface part of the polymer (Figure 1.12). The results of the microscopic studies of the polymer–metal nanocomposites shown in Figure 1.12 confirm the validity of this conclusion. Such distribution of MNPs inside a supporting polymer is of particular

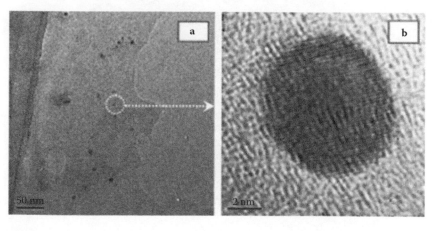

FIGURE 1.9 Transmission electron microscopy (a) and high resolution transmission electron microscopy images (b) of Cu PSMNP synthesized in SPEEK matrix.

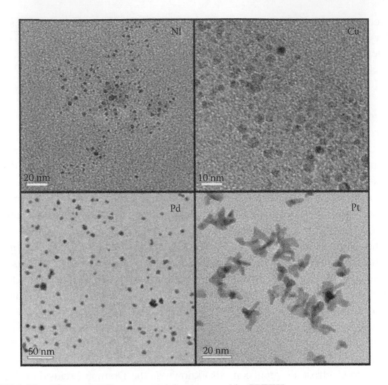

FIGURE 1.10   Monometallic nanoparticles prepared in SPEEK.

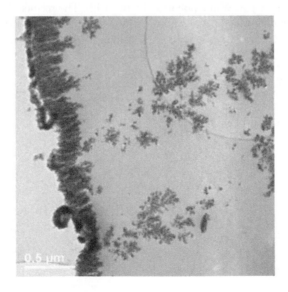

FIGURE 1.11   Transmission electron microscopy image corresponding to cross-section of SPEEK–Cu PSMNPs.

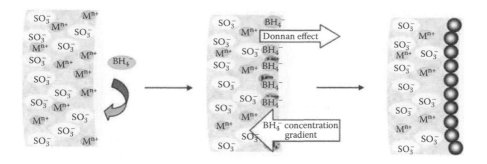

FIGURE 1.12 Influence of Donnan exclusion effect on nanoparticle distribution in nanocomposite.

importance in, for example, catalytic applications of polymer–metal nanocomposites, as the catalytically active MNPs are more accessible to reagents.[151]

Based on the (1.5) and (1.6) reactions, after the metal loading-reduction cycle that results in the formation of MNPs, the functional groups of the polymer are converted back into the Na form, and as a result become regenerated. This allows repeat of the metal loading and reduction cycles (IMS), with the aim to increase PSMNP content in the polymer,[2] that appears to be not limited by the intrinsic ion exchange capacity of the matrix.

In certain cases this procedure appears more complex. For example, when carrying out the second (repetitive) Cu loading cycles (by agitation for several hours of a Cu MNP-containing SPEEK membrane with an aliquot of $CuSO_4$ solution) a gradual change of the membrane color during the second copper loading from a bright copper metal color to green is observed.[2] Reduction of the reloaded membrane again yields the typical nanocomposite membrane as shown in Figure 1.13. In this case, the change of membrane color is most probably associated with the oxidation of Cu MNPs with Cu^{2+} ions due to the decrease (negative shift) of Cu MNP redox potential in comparison with that of the bulk copper.[28] The second metal loading reduction cycle must proceed by the following reaction schemes:

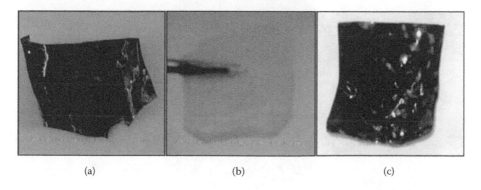

(a)   (b)   (c)

FIGURE 1.13 Images corresponding to Cu PSMNP–SPEEK membrane after (a) first metal loading-reduction cycle, (b) after second metal loading, and (c) after second metal reduction.

$$\text{Oxidation of Cu MNPs with } Cu^{2+} \text{ ions: } Cu^0 + Cu^{2+} \rightarrow 2Cu^+ \qquad (1.8)$$

$$\text{Loading of SPEEK membrane with } Cu^+ \text{ ions: } R\text{-}SO_3^-Na^+ + Cu^+ \rightarrow$$
$$R\text{-}SO_3\text{-}Cu^+ + Na^+ \qquad (1.9)$$

$$\text{Reduction of } Cu^+ \text{ to } Cu^0 \text{ with } NaBH_4 \text{ solution: } 2R\text{-}SO_3^-Cu^+ +$$
$$2NaBH_4 + 6H_2O \rightarrow 2R\text{-}SO_3^-Na^+ + 7H_2 + 2B(OH)_3 + 2Cu^0 \qquad (1.10)$$

The second copper loading reduction cycle makes it possible to double the quantity of Cu inside the polymer matrix due to the coproportionation reaction (1.8) that appears to be a stage of the IMS process.[152] Moreover, the results indicate that parameters such as metal concentration, reducing agent concentration, and time of the second loading can be used as "tuning parameters" for the synthesis of polymer–metal nanocomposites with specific desired levels of Cu MNPs.

1.3.3   SYNTHESIS OF POLYMETALLIC CORE SHELL MNPS

Sequential metal loading reduction cycles make possible the synthesis of MNPs of more complex compositions including bimetallic MNPs with core shell structures and polymetallic MNPs.[153–156] Thus, use of a different metal within the repetitive (sequential) metal loading-reduction cycle allows the coating of preformed MNPs with the second metal shell.[127,128] In other words, two sequential metal loading–reduction cycles with two different metals (M_1 and M_2) permits the formation of core shell (M_2–M_1) MNPs with the desired composition (Figure 1.14).

The formation of such a shell allows modification of charge and functionality; improves MNP stability (i.e., against oxidation); or combines the properties of core and shell to make their future applications more efficient. For instance, the synthesis of bimetallic MNPs with magnetic cores (e.g., Co MNPs) coated with shells having catalytic properties (e.g., Pt) allows production of easily recoverable and/or

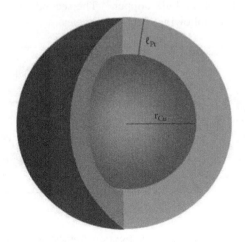

FIGURE 1.14   Core shell MNPs.

recyclable nanocatalysts. Conversely, by using a less expensive core metal such as Cu, one can substantially reduce nanocatalyst cost. Indeed, the main advantage of core shell MNPs with platinum group metal (PGM) shells and cheap metal cores (Pt–Cu or Pd–Cu) is the far lower PGM loading (at the same catalytic activity) in comparison with monometallic PGM MNPs. The main requirement for MNPs of this type is a good compatibility of core and shell that mainly depends on the crystalline forms of the metals used.

Figure 1.15 shows the scheme of IMS of Pt–Cu MNPs by a simple procedure that produces bimetallic core shell nanoparticles and the corresponding polymer–metal nanocomposites on their bases. This straightforward method can also produce core shell MNPs of other types by using other metal pairs. To better understand the procedure shown in the figure, let us consider the reactions corresponding to the IMS of PSMNPs inside a parent polymeric matrix after the first loading and reduction cycle (M_1-SPEEK in this case):

$$\text{Second metal loading: } 2R\text{-}SO_3^-Na^+ + M_1^0 + M_2^{2+} \rightarrow (R\text{-}SO_3^-)_2\, M_2^{2+} + M_1^0 + 2\,Na^+ \tag{1.11}$$

$$\text{Second metal reduction: } (R\text{-}SO_3^-)_2\, M_2^{2+} + M_1^0 + 2NaBH_4 + 6H_2O \rightarrow 2R\text{-}SO_3^-Na^+ + 7H_2 + 2B(OH)_3 + 4NH_3 + M_2\text{-}M_1 \tag{1.12}$$

According to some authors,[157] the second metal ions act as an oxidizing agent for the core metal (M_1^0), resulting in oxidation of the first metal by the following transmetallation reaction (17):

$$M_2^{2+} + 2M_1^0 \rightarrow M_1^{2+} + M_2^0\text{-}M_1^0 \tag{1.13}$$

The number of such cycles is not limited to two and can be, for example extended to three or more to make it possible to synthesize MNPs of more complex architectures (e.g., core sandwich and others; Figure 1.16). Until now, the IMS of MNPs in ion exchange materials has been successfully applied to synthesize bimetallic alloy MNPs (Co–Ni MNPs), bimetallic core shell MNPs (Co, Ni, or Cu core MNPs coated with Pd or Pt shells), and trimetallic core shell MNPs (bimetallic Co and Ni core coated with a Pt or Ru shell, Ru–Pt–Cu and Pt–Ru–Cu). As shown in Figures 1.17 and 1.18, the spherical MNPs synthesized are well separated from each other.[127]

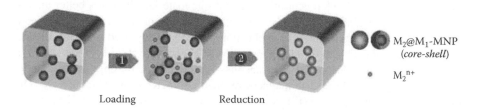

Loading   Reduction

$M_2@M_1$-MNP
(*core-shell*)

M_2^{n+}

FIGURE 1.15   IMS for preparation of bimetallic core shell MNPs.

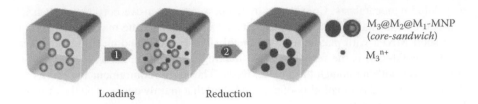

FIGURE 1.16   IMS for preparation of polymetallic core sandwich MNPs.

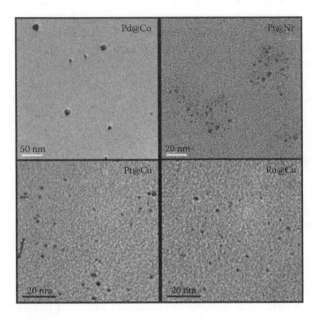

FIGURE 1.17   Transmission electron microscopy images of bimetallic nanoparticles synthesized in SPEEK.

1.4   CHARACTERIZATION OF POLYMER-STABILIZED NANOPARTICLES (PSMNPs)

The detailed characterization of PSMNPs and nanocomposites on their bases plays an important role in the development of nanocomposite materials. It allows better under-standing of the main features of their synthesis, adequately explains their properties, and determines areas for their practical application. Several widely used material science techniques are applicable to the characterization of nanocomposites.[158,159] They include scanning electron microscopy (SEM), transmission electron microscopy (TEM), atomic force microscopy (AFM), nuclear magnetic resonance (NMR), x-ray diffraction (XRD), impedance spectroscopy[160,161] (IS), x-ray photoelectron spectroscopy[162] (XPS), x-ray energy dispersion spectroscopy (EDS), infrared attenuated total refraction (IR-ATR), and many others.

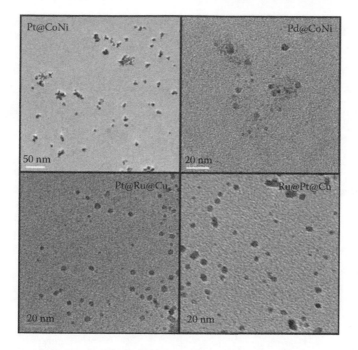

FIGURE 1.18 Transmission electron microscopy images of trimetallic nanoparticles prepared in SPEEK.

The main parameters that usually characterize nanocomposite materials include composition, size and size distribution of MNPs, nanocomposite morphology, and special (magnetic, catalytic, electrocatalytic, etc.) properties. These parameters can be studied by using the techniques listed above or by combinations of techniques that can yield better and more quantitative results of sample characterization. For example, the composition of an MNP (even that of a single nanoparticle) can be determined by high resolution TEM analysis coupled with EDS (see Figure 1.19). The microscopic technique allows the selection of the nano objects and the EDS provides the composition analysis. This combination usually gives quantitative or semi-quantitative results that can be useful for estimating the composition differences in a single sample.

Conversely, the quantitative compositions of PSMNPs can be determined by using inductively coupled plasma optic emission spectroscopy (ICP-OES) or inductively coupled plasma mass spectrometry (ICP-MS)[1,127] for the analysis of a solution obtained after the treatment of a known quantity of nanocomposite with aqua regia to completely dissolve the MNPs. Electrochemical characterizations of nanocomposites can be achieved by modifying the surfaces of electrodes by deposition of PSMNP solutions, also known as PSMNP inks, followed by solvent evaporation. The modified electrodes can be used to characterize the electrocatalytic properties of metal–polymer nanocomposites via cyclic voltametry[134] or chronoamperometry[126] measurements.

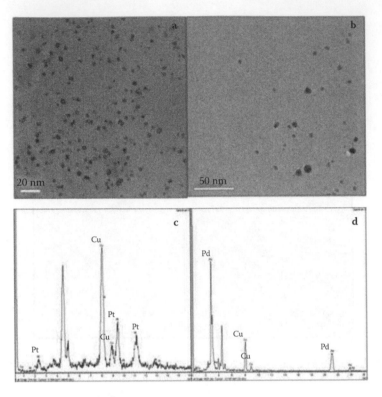

FIGURE 1.19 Transmission electron microscopy images (a, b) and EDS spectra (c, d) of Pt–Cu-(a, c) and Pd–Cu PSMNPs (b, d).

1.4.1 CHARACTERIZATION OF CORE SHELL NANOPARTICLES

One of the important points for characterizing core shell MNPs is to achieve a good coating of the core by the functional shell, as in many instances this coating prevents oxidation of the core metal. The simplest way to determine the quality of the coating is with the use of a simple procedure based on the treatment of a piece of the nanocomposite with an acid, followed by the analysis of the solution to determine the concentrations of core metal ions leached from the sample.

Figure 1.20 shows the dissolution kinetics of Cu and Pt–Cu-MNP SPEEK nano-composites in 0.01 M HNO$_3$ and HCl solutions. Copper nanoparticles are rapidly dissolved in HNO$_3$ and even partially in HCl. Most probably, the dissolution of Cu MNPs in HCl (which is quite surprising) proceeds through the partial oxidation of copper by dissolved O$_2$, followed by reaction of the formed copper oxide layer with the acid. It is also possible that Cu MNPs are much more reactive than bulk copper due to the changes in redox potential as stated by Plieth.[27] This point requires further investigation. Conversely, Pt-coated Cu MNPs appear completely stable against HCl treatment—indicating the complete platinum coating of the surfaces of Cu nanoparticles.

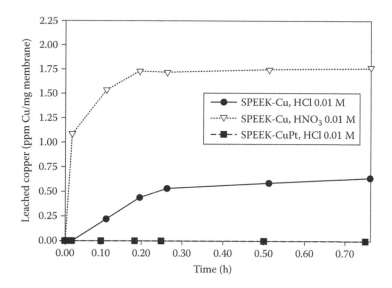

FIGURE 1.20   Dissolution kinetics of Cu and Pt–Cu PSMNPs in acid solutions.

1.4.2 Nanoparticle Size and Size Distribution: PSMNP Stability

The sizes and shapes of MNPs, their size distribution, and possible changes in time are important parameters for evaluating their stabilities, lifetimes and other nano-composite characteristics. Further development of nanotechnology requires complete understanding of nanomaterial properties—achieved by better characterizations of the above parameters that are usually evaluated via TEM.[163] Thanks to the solubility properties of SPEEK polymeric matrices, it is possible to prepare MNP suspensions or "inks" (5 mass % solution) in polar solvents such as dimethylformamide (DMF) for deposit onto a TEM grid for microscopic characterization.

Unfortunately, in practice, the quality of TEM images may be poor because of high noise and low contrast levels, making such processing a challenging task. Thus, the quantitative treatment of TEM images is often carried out by manual measure-ments of high numbers of nanoparticles—a highly subjective and expensive task. In recent years, several computer imaging particle analysis software tools have allowed researchers to automatically (or semi-automatically) extract useful data from images that may allow more accurate assessments of the sizes and frequencies (size distribu-tions) of nanoparticles.[164–166]

Figure 1.21 shows typical TEM images of Cu PSMNPs evaluated by Eykes automatic software, which rapidly characterizes MNPs. Through the automatic or manual image analysis of TEM micrographs, it is possible to build histograms showing the size distributions of MNPs, such as the one in Figure 1.22. The obtained data can be used fitted to a three-parameter Gaussian curve [Equation (1.14)] where a is the height of Gaussian peak, d_m is the position of the center of the peak (corresponding to the most frequent diameter), and σ is the standard deviation.

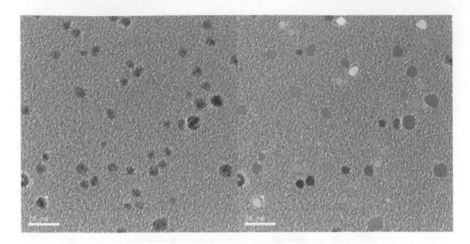

FIGURE 1.21 Typical transmission electron microscopy images of polymer-stabilized Cu nanoparticles. Original image (left) and software analyzed image (right).

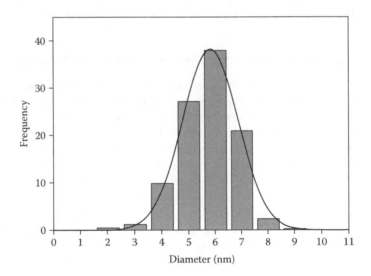

FIGURE 1.22 Typical histogram of Cu PSMNP. Solid line corresponds to fitting with three-parameter Gaussian curve.

$$y = a \cdot \exp\left[-0.5\left(\frac{d - d_m}{\sigma}\right)^2\right] \tag{1.14}$$

By performing periodical analysis of such inks stored under laboratory conditions, it is possible to evaluate MNP suspension stability. As shown in Figure 1.23, the shape of a histogram remains essentially unchanged even after several months of storage of a PSMNP solution. This indicates a very high stabilizing efficiency of

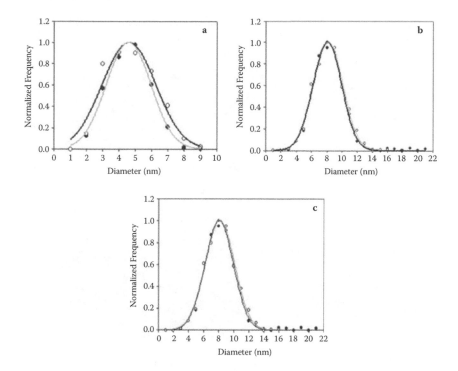

FIGURE 1.23 Size distribution histograms for Pd–Cu- (a), Pd (b) and Pt–Cu PSMNP (c) samples, (a, b): ● = freshly prepared ink. ○ = after 7 months' storage, (c). ● = freshly prepared ink. ○ = after 3.5 months' storage. ▼ = after 10 months' storage.

the SPEEK matrix for MNPs.[127] Another conclusion from these results concerns the recommendations for PSMNPs storage conditions. Based on the results shown in Figure 1.23, the PSMNP inks appear to be stable at least for half a year and the stabilities (lifetimes) of the nanoparticles inside solid SPEEK membranes must be far higher. For this reason, PSMNPs are best stored in the form of solid membranes and dissolved before use.

1.4.3 METAL–POLYMER NANOCOMPOSITE MORPHOLOGY

Repetitive metal loading–reduction cycles lead to accumulations of additional MNPs in membranes (in the case of monometallic MNPs) and also cause dramatic changes of the membrane morphology. Indeed, SEM analysis of a sample surface (Figure 1.24) clearly shows that the initially smooth gel structure of a SPEEK membrane surface transforms to a "worm-like" structure after the first and second Cu MNP-loading cycles.

These morphological changes are most likely associated with the inter-membrane mechanical stress resulting from strong interactions between Cu nanoparticles and the SPEEK chains similar to the reactions reported by Cole et al.[136,137] and Kim et al.[167] These changes may substantially increase the number of practical applications of these materials (e.g., for catalysis) due to increased mass transfer properties of SPEEK–Cu

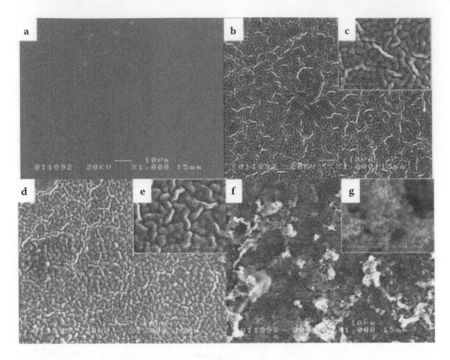

FIGURE 1.24   Scanning electron microscopy images of initial SPEEK membrane (a), same membrane after first (b, c) and second (d, e) metal (Cu) loading–reduction cycle (f, g). Note internal macroporous structure of Cu PSMNP-loaded membrane.

PSMNP composite membranes in comparison with the initial (Cu-free) polymer. Figure 1.25 shows membrane nanoporosity inside a matrix.

The results of BET measurements of nanoparticle-induced porosity in a SPEEK membrane (Figure 1.26) confirm this conclusion. The major portion of the pores that appear in the SPEEK membranes as the result of accumulation of Cu PSMNPs have diameters of nanometric size (>30 nm), and the more the Cu PSMNP content, the higher the number of pores of small diameter. It seems important to emphasize that the nanoporosity simultaneously appears in SPEEK membranes.

1.4.4   ELECTRICAL PROPERTIES

The electrical conductivity of a polymer–metal nanocomposite is another important parameter that determines the utility of such materials in various fields. Conductivity appears to depend on various parameters of a system, but mainly on those of the nano-component.[2] The overall electrical conductivity of metal–SPEEK nanocomposite membranes consists of (1) the ionic conductivity associated with the sodium ions fixed on the functional groups of the polymer (highly dissociated), and (2) the electronic conductivity due to the presence of MNPs.

The ionic conductivity can be considered to be constant for all membrane samples as the number of functional groups (ion exchange capacity of SPEEK)

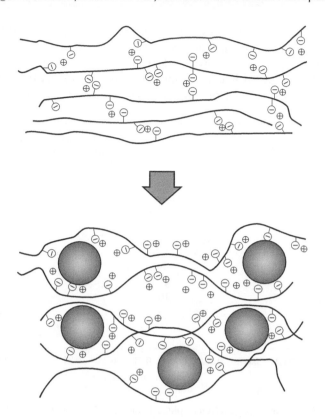

FIGURE 1.25   Nanoporosity formation in SPEEK matrix after IMS of MNPs.

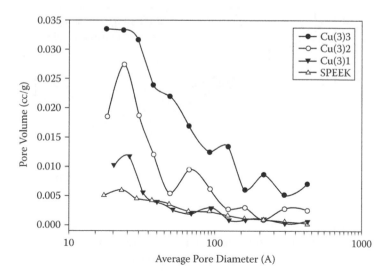

FIGURE 1.26   Results of BET measurements of nanoporosity in SPEEK membranes loaded with different amounts of Cu PSMNP.

is the same in all cases. The presence of Cu PSMNPs or Pt–Cu PSMNPs in the membrane gives rise to the appearance of the electronic aspect of conductivity. Figure 1.27 and Figure 1.28 show the dependence of the electrical conductivity of a Pt–Cu PSMNP–SPEEK nanocomposite membrane versus Pt content and Pt shell thickness, respectively. Note that the conductivity linearly decreases with Pt

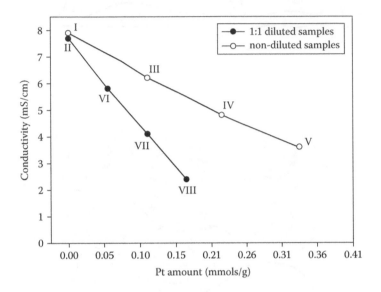

FIGURE 1.27 Electric conductivity plotted against Pt amount in Pt–Cu–SPEEK nanocomposites.

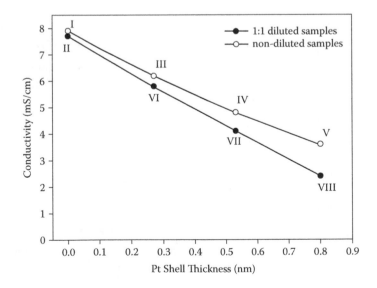

FIGURE 1.28 Electric conductivity plotted against Pt shell thickness of Pt–Cu–SPEEK nanocomposites.

loading for both diluted (with MNP-free SPEEK) and undiluted membrane samples. Moreover, the conductivity of a Cu PSMNP-containing membrane does not essentially change after dilution with metal-free SPEEK (see the first points on the curves corresponding to the zero thickness of the Pt shell). Another conclusion based on Figure 1.28 is a decrease of membrane conductivity with the increase of Pt layer thickness that may be attributed to the difference between the electrical conductivities of the metals: 590 mS/cm for Cu and 90 mS/cm for Pt. Conversely, the dependencies shown in these figures can serve somewhat as calibration curves to estimate the shell thicknesses of the core shell MNPs. It seems important to emphasize that relatively simple conductivity measurements appear sensitive to this parameter even at its low variations (>1 nm).

To conclude this section, it seems important to note that the mechanical properties of undiluted membrane samples appear poor due to the internal stress resulting from strong bridging interactions of MNPs and the polymer chains of the stabilizing matrix.[1,126] Dilution of the membranes with a metal-free polymer substantially enhances mechanical stability without remarkably changing other properties—a particularly important factor in electrochemical applications of metal–polymer nanocomposites.

1.4.5 MAGNETIC PROPERTIES

The synthesis of a core shell MNP consisting of a ferromagnetic (or superparamagnetic) core coated with a functional shell (having, for example, catalytic or bactericidal properties) is of particular interest for two reasons. First, due to the magnetic properties of MNPs, the nanocomposite can be easily recovered and reused in sequential catalytic cycles. Second, the ferromagnetic nature allows easy recovery and recycling of MNPs—particularly important for nanoparticles containing PGMs.

IMS makes it possible to prepare both monometallic and bimetallic Co compounds such as Pd–Co within different cation exchange matrices (SPEEK membranes and FIBAN fibers). In all cases, the polymer–metal nanocomposites appeared to be ferromagnetic and easily attached to permanent magnets. Some parameters, such as nanoparticle size and spatial distribution, have been studied to determine the sample cross sections and magnetic properties of nanocomposite materials (Figure 1.29). Similar results were also obtained for Ag–Co PSMNPs synthesized in a FIBAN–K-1 polymer matrix (see Section 1.5.2). In all cases, nanoparticles were distributed heterogeneously inside polymer matrices at far higher concentrations on the surfaces of the nanocomposites that appear to be favorable for catalytic applications.

1.5 APPLICATIONS OF POLYMER–METAL NANOCOMPOSITES

The practical applications of metal-polymer composites cover wide fields because the composites combine the properties of the MNPs and polymers. Such combinations give rise to many possibilities for preparing nanomaterials that can handle industrial challenges (Figure 1.30). Metal–polymer composites find applications in electroconductive pastes and glues, special coatings, paints and varnishes, rheomagnetic fluids, antifrictional polymeric coatings, aviation and

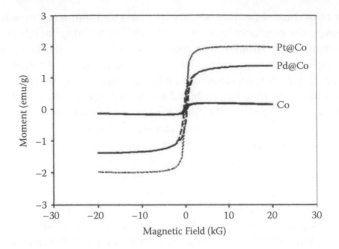

FIGURE 1.29   Magnetic measurements of Pt–Co, Pd–Co, and Co PSMNPs via SQUID technique.

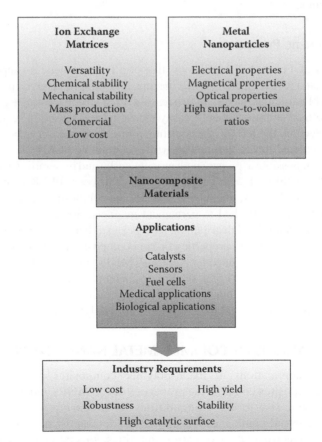

FIGURE 1.30   MNP-based nanocomposite materials and their potential practical applications.

space technology, catalysis, and many other uses. PSMNPs containing polymeric membranes are used as electroconductive and optical materials, supported catalysts, and active elements in sensors and biosensors.[25,34,80]

Ion exchange nanocomposite materials containing PSMNPs find industrial applications; they are known as resin-stabilized, catalytically active MNPs[41–43] obtained via IMS (by anchoring the suitable MNP precursors to a functional resin followed by their reduction inside a polymeric network). One of these nanocomposites represents a bifunctional catalyst composed of an acid as well as hydrogenation-active centers arising from the presence of protonated sulfonic groups and Pd MNPs. Such catalysts are currently used in at least four industrial processes. Nevertheless, the potential applications of PSMNPs can be widened substantially due to unique combinations and a great deal of scientific experience in this particular area.[116,118,119,168,169]

In general, the catalytic (or electrocatalytic) efficiency of supported catalysts (polymeric and carbon or inorganic supports) is primarily determined by the accessibility of the catalytically active MNP to reactants. The physicochemical features of a polymeric support must be designed and controlled to enhance the mass transfer inside the polymer gel microreactor. This principle must also be considered when designing sensor or biosensor systems based on PSMNPs.

1.5.1 Amperometric Sensors and Biosensors

A chemical sensor is an analytical device that can provide chemical information about a sample. Such sensors are widely used to detect and quantify both organic and inorganic substances in clinical,[170] biomedical,[171] environmental,[172] and other fields. The main requirements for a sensor include good sensitivity, selectivity, high response speed, long life, low consumption of analytes and power, and low cost when used for mass production in industrial applications.

This section focuses on the application of polymer–metal nanocomposites to amperometric sensor and biosensor designs. Amperometric devices are based on the application of an external potential, leading to an electronic transfer between a working electrode and species in solution. The current passing through an electrochemical cell containing electroactive species is proportional to the analyte concentration. The measurement of this current allows the quantitative determination of many analytes (organic and inorganic) at trace levels.

For this quantification, the main requirement is the ability of the analyte to be oxidized or reduced electrochemically onto the electrode surface. However, the disadvantage of amperometric sensors is limited selectivity and relatively long response times. The interdependence of the current intensity and analyte concentration is described by the following:

$$i = \frac{nFAD_aC_a}{\delta} = k_A \cdot C_a \tag{1.15}$$

where i corresponds to the current intensity, A is the surface area, n is the number of mols of electrons implied in the electrochemical reaction per mol of analyte, F is the

Faraday constant, D_a is the diffusion coefficient, C_a is the concentration of electroactive species, δ is the thickness of the diffusion layer, and $k_A = nFAD_a/\delta$.

The use of PSMNPs for modification of amperometric sensors yields a number of advantages based on the enhancement of both the rate of mass transfer inside the nanocomposite membrane (sensing element), and the electrocatalytic activity arising from the great specific surface area of the nanocatalyst in comparison with that of the bulk material.[173] The conductive polymer–metal nanocomposites containing PSMNPs can be used to modify electrode surfaces via a simple deposition technique that mainly determines their attractiveness for development of new sensors and biosensors. PGM PSMNPs and nanocomposites represent some of the most promising materials for this kind of electrode modification. For example, Pt and Pd MNPs are widely used (despite their relatively high cost) because of their good stability both in acidic and alkaline media, resistance to oxidation,[174] and high catalytic activity in various types of chemical reactions.[175]

Our recent publications[126-128,134] describe the modification of graphite–epoxy composite electrodes (GECEs)[176] with Pt and Pd PSMNPs of various types (monometallic and polymetallic). For such modification, the PSMNPs inks (solutions of nanocomposite membranes in appropriate solvents) are deposited drop by drop on the surface of a GECE, followed by air drying at room temperature (see Figure 1. 31).

The calibration curves[128] illustrating the performance of a GECE modified with Pt–Cu, Pd–Cu, Pt–Co, Pd–Co, Pt–Ni, Pd–Ni, and Pd–Co–Ni PSMNP composite membranes in amperometric detection of H_2O_2 are shown in Figure 1.32. The sensitivity of the sensors toward the analyte under study increased in the following order: Pd–Cu > Pd–Co > Pd–Ni > Pt–Ni > Pt–Co > Pt–Co–Ni > Pt–Cu. These results indicate that Pd PSMNPs manifest higher catalytic activities in H_2O_2 decomposition than Pt PSMNPs under the same experimental conditions.

A similar conclusion follows from the comparative study of sensitivity of sensors modified with Pt–Cu and Pd–Cu MNPs. Figure 1.33 demonstrates the calibration curves from a comparative study of sensors modified with monometallic Pd, Pt, Cu, and bimetallic Pt–Cu and Pd–Cu PSMNPs. The sensors modified with bimetallic core shell MNPs showed greater sensitivities than those modified with monometallic MNPs. These results correlate well with the sizes of corresponding MNPs. The bimetallic MNPs (Pt–Cu and Pd–Cu) are characterized by diameters of 3 to 4 nm; their monometallic analogs have diameters around 7 nm, and therefore the bimetallic

FIGURE 1.31   Supramolecular construction of glucose biosensor based on a SPEEK–Pt–Cu PSMNP nanocomposite.

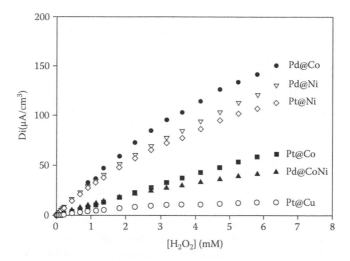

FIGURE 1.32 Calibration curves of electrochemical detection of H_2O_2 concentration with Pd–Co, Pd–Co, Pt–Ni, Pd–Ni, Pd–Co–Ni, and Pt–Cu PSMNP composite membranes at –250 mV using a 0.1 M acetate buffer at pH 5.0.

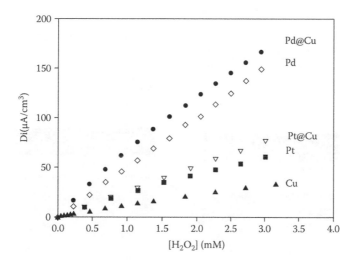

FIGURE 1.33 Calibration curves of electrochemical detection of H_2O_2 concentration with Pt–Cu, Pd–Cu, Pt, Pd, and Cu PSMNP composite membranes at –250 mV using a 0.1 M acetate buffer at pH 5.0.

MNPs have greater specific surface areas. The breakdown of the second metal loading–reduction stage (leading to shell formation) into several steps results in the formation of a shell with higher purity of the respective metal (e.g., Pt). This technique allows further increase of the sensitivities of sensors modified with Pt–Cu PSMNPs.[177]

PSMNP-based nanocomposites can be also used to enhance biosensor performance. This clearly follows from the possibility of substantially improving the electron transfers in amperometric biosensors. The electron transfer mechanisms in such biosensors were reviewed by Habermüller et al.[177] The group detected several possible electron transfer mechanisms, such as direct electron transfer (tunneling; see Figure 1.34) and mediated electron transfer with the use of redox mediators that serve as "electron transfer shuttles" or "electron trampolines" in cases of electron hopping mechanisms. Thus, the introduction of PSMNPs into an enzyme immobilizing matrix can substantially improve the electron conductivity of the matrix and, therefore, enhance electron transfer from the enzyme molecule to the surface of the electrode as shown in Figure 1.34. The same effect can be expected in the case of mediated electron transfer systems due to the shortening of the electron hopping distance.

We recently reported[128] the development of PSMNP-based glucose biosensors. The biosensors were prepared by using the layer-by-layer deposition technique. The GECE surface was first modified with a SPEEK–Pt–Cu PSMNP membrane, followed by sequential treatment with polyethyleneimine (PEI) and the glucose oxidase (GOX) enzyme that bear opposite charges and, as a consequence, undergo an electrostatic interaction. The electrode was immersed sequentially in solutions containing PEI or GOX to form intermittent monolayers of "sandwich-like" structures as shown in Figure 1.35. This permits the accumulation of the desired amount of enzyme inside the top of the biosensor membrane.

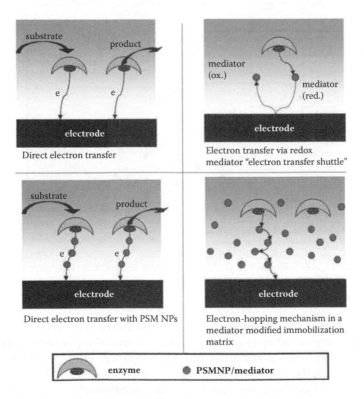

FIGURE 1.34   Electron transfer mechanisms in amperometric biosensors.

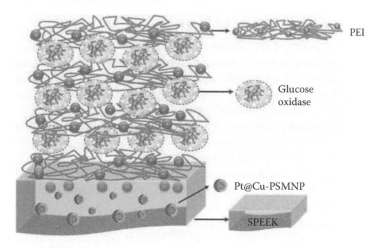

FIGURE 1.35 Supramolecular construction of glucose biosensor based on SPEEK-Pt–Cu PSMNP nanocomposite.

As seen in Figure 1.36, the sensitivity of two biosensors with the same enzyme content appears to strongly depend on the amount of Pt in the shells of Pt–Cu core shell MNPs. This conclusion is based on the high sensitivity of Pt–Cu toward hydrogen peroxide, which is one of the by-products of the enzymatic reaction in the course of glucose detection. Even better results (higher biosensor sensitivity) can be expected for biosensors prepared by using Pd–Cu PSMNPs (see Figure 1.34) and we will continue our study in this direction.

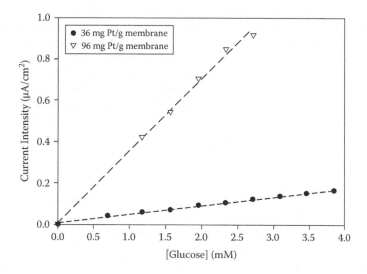

FIGURE 1.36 Calibration curves of amperometric detection of glucose with GECE biosensors modified with SPEEK–Pt–Cu PSMNP and six PEI GOX "sandwiches" at −250 mV using 0.1 M KCl in a 0.1 M acetate buffer at pH 7.0.

1.5.2   PSMNP-Based Bactericide Filters

Another group of practical applications of polymer–metal nanocomposite materials relates to water treatment. Ion exchange materials modified with MNPs that display antimicrobial activity, for example, silver nanoparticles,[25,179–181] are of particular interest. Indeed, ion exchange materials are widely used for various water treatment processes, mainly to eliminate undesired or toxic ionic impurities such as hardness ions, iron, heavy metals, and other materials. Their modification with bactericide MNPs enables water utilities to combine traditional water treatment with disinfection to eliminate microbiological contaminants.

However, in the case of water treatment, and drinking water treatment in particular, one must consider the environmental safety of the nanocomposite. This issue is one of the hottest nanotechnology topics of the past few years.[80] The main concerns dealing with the rapid development and commercialization of various nanomaterials (NMs) are (1) the approved higher toxicities of many NMs in comparison with their larger counterparts, (2) the absence of adequate analytical techniques for detection of NMs in the environment, and (3) the lack of legislation normative for permitted levels of NMs in water and air.[182] Obviously, the safety of NMs is of particular importance.

A possible solution appears in this section, which describes the results obtained by developing environmentally safe polymer–metal nanocomposite materials exhibiting bactericidal activities useful for reagent-less water disinfection and complex water treatment. One material is a fibrous functional polymer (e.g., FIBAN-K1 or FIBAN-K4) with immobilized core shell Ag–Co MNPs distributed mainly at the surfaces of the fibers so that they are maximally accessible for the elimination of bacteria. MNPs consist of superparamagnetic cores coated with silver metal shells of minimal thickness that provide maximal bactericidal activity. The MNPs are captured inside a polymer matrix that prevents their escape into the water under treatment.

The superparamagnetic nature of MNPs provides additional material safety, as MNPs leached from a polymer matrix can be easily captured by magnetic traps to completely prevent post-contamination of the treated water. The MNPs do not block the functional (e.g., sulfonic) groups of the polymer so that the polymer–metal nanocomposite can be also used to remove undesired ions (e. g., hardness ions, iron, etc.) from the water. The preliminary results of tests of the bactericidal activities of the nanocomposite material indicate its high antibacterial efficiency against the following bacteria that are considered the most dangerous to humans: *Staphylococcus aureus, Escherichia coli, Klebsiella* and *Enterobacter* spp, *Pseudomonas aeruginosa*, and *Salmonella* spp.

Figure 1.37 shows the results from disinfection of water containing *E. coli* at a concentration of 10^3 cfu/ml. The results were obtained by passing the water through a nanocomposite filter with thickness of 0.5 mm at a constant flow rate. The samples of treated water were collected periodically and subjected to Petri dish recounts. Note that the concentrations of bacteria colonies rapidly decreased with treatment time. More quantitative results are shown in Figure 1.38, which shows the kinetic curve of disinfection treatment of water contaminated with *E. coli*. The results in Figures 1.37 and 1.38 reveal the high bactericidal activities of nanocomposite materials against *E. coli*. Similar results were obtained with other types of bacteria listed above.

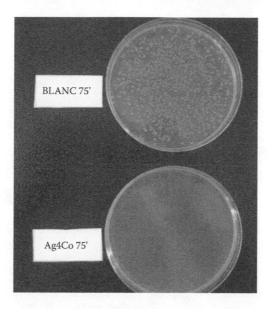

FIGURE 1.37 Antibacterial activity of FIBAN–Ag–Co PSMNP nanocomposite filter against *Escherichia coli*. Top: control. Bottom: 75 minutes after bactericidal treatment.

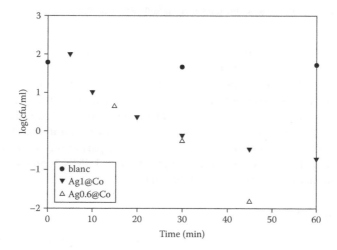

FIGURE 1.38 Kinetics of bactericidal treatment of *Escherichia coli*-contaminated water with FIBAN–Ag–Co PSMNP filter.

1.5.3 NANOCATALYSTS FOR ORGANIC SYNTHESIS

PSMNPs of certain types synthesized by IMS can be also applied to catalyze some organic reactions.[4,25,36,37,183] MNP–polymer nanocomposites with superparamagnetic properties present a great advantage: the magnetic properties permit reuse of the catalytic material. After the first reaction cycle, the nanocatalyst can be recovered

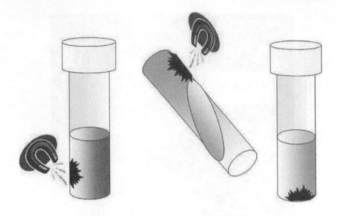

FIGURE 1.39 Recovery and reuse of nanocatalyst by permanent magnet.

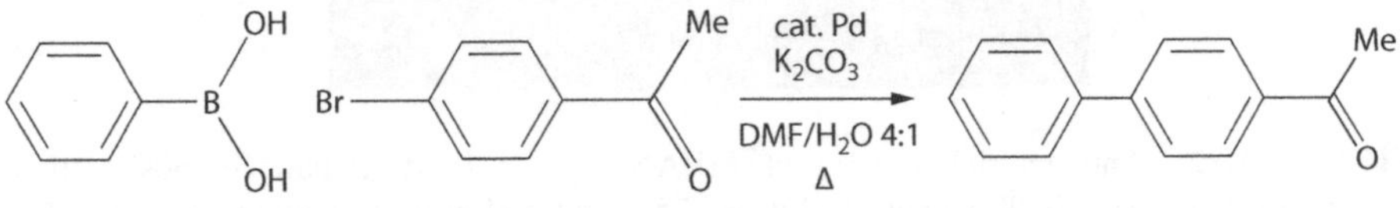

FIGURE 1.40 Suzuki cross-coupling reaction and experimental conditions.

and reused. Furthermore, these properties aid in preventing nanoparticle leakage through use of an electromagnetic trap (Figure 1.39).

We synthesized a catalytically active nanocomposite for a Suzuki reaction—a cross-coupling reaction between arylboronic acids and aryl bromides leading to the formation of biphenyls[148,183] (Figure 1.40). The nanocomposite material consisted of a functional polymer in the form of a membrane or fiber with immobilized Pd–Co core shell PSMNPs distributed mainly on the surface part of the polymeric matrix. This distribution in the case of catalytic applications of PSMNPs provides the maximum accessibility of reactants to catalytic centers.[183]

1.6   CONCLUSIONS

The ion-exchange assisted IMS of PSMNPs represents one of the most promising techniques, because it can produce a high variety of polymer-metal nanocomposites, which offer practical applications in different fields of the modern science and technology. The attractiveness of this technique is determined not only by its relative simplicity but also by its flexibility and the possibility to purposely tune the properties of the final nanocomposites to the requirements of its final application.

Moreover, the IMS technique is based on the metal-loading-reduction cycles, which are carried out inside the polymer matrices. Therefore, the repetitive cycles can be carried out either with the same or different metal (or metals), which results, in the first case, in the increase of the density of PSMNPs population inside the polymer matrix and in the variation of the MNP diameters and other parameters of the system. In the second case, one can easily obtain bimetallic or polymetallic MNPs with core-shell or core-sandwich structures.

The IMS of PSMNPs can be carried out inside both cross-linked and noncross-linked polymers. In the latter case, the final polymer-metal nanocomposite can be dissolved in an appropriate solvent to form a sort of PSMNP ink, which can be easily deposited on the surface of electrodes to produce MNP-based sensors or biosensors. The MNP-modified sensors can be used as a tool for characterization of electro-chemical parameters of nanocomposites.

It is important to note that the properties of the metal-polymer nanocomposite are not only determined by those of MNPs. The formation of MNPs inside polymer matrices results in quite strong modification of the polymer morphology, in the appearance of nanoporosity, and in other changes, which enhance the rate of mass-transfer and other parameters of nanocomposite in comparison with the initial polymer.

The nature of the reducing agent used for the reduction of metal ions to MNPs plays an important part in IMS because it determines not only the conditions of MNPs formation inside the polymer matrix (metal reduction stage) but also its distribution inside the matrix. Thus, the use of anionic reducing agents (*e.g.*, borohydride anion) for IMS of PSMNPs in cation-exchange matrices results in their formation mainly by the surface part of the polymer. This type of MNPs distribution appears to be effective in many practical applications of polymer-metal nanocomposites (water disinfection, catalysis, and others).

The IMS of core-shell PSMNPs consisting of a superparamagnetic core coated with a functional shell (having *e.g.*, catalytic or bactericide properties) permits the production of polymer-metal nanocomposites with additional advantages. In the case of catalytic application, the nanocomposite can be easily recovered from the reaction mixture and reused. In the case of water treatment applications, the magnetic nature of MNPs easily prevents their uncontrollable escape into the treated water by using simple magnetic traps.

ACKNOWLEDGMENTS

We are sincerely grateful to all our associates cited throughout the text for making this publication possible. Part of this work was supported by Research Grant MAT2006-03745, 2006–2009 from the Ministry of Science and Technology of Spain, which is also acknowledged for the financial support of Dmitri N. Muraviev and Ramon y Cajal. J. Macanás thanks the Departament d'Innovació, Universitats i Empresa de la Generalitat de Catalunya for predoctoral and postdoctoral grants.

REFERENCES

1. Muraviev, D.N. *Contrib. Sci.* 2005, 3, 19–32.
2. Muraviev, D.N., Macanás, J., Parrondo, J. et al. *React. Funct. Polym.* 2007, 67, 1612–1621.
3. Walter, P., Welcomme, E., Hallégot, P. et al. *Nano Lett.* 2006, 6, 2215–2219.
4. Astruc, M-C. *Chem. Rev.* 2004, 104, 293–346.
5. Min'ko, N.I. and Nartsev, V.M. *Glass Ceram.* 2008, 65, 148–153.
6. Liz-Marzán, L-M. *Matl Today* 2004, 7 26–31.

7. Bultan, I. Le Taqwim al-Sihha (Tacuini Sanitatis): *Un traité médical du XIe siècle*, p. 276 (translated and published by Académie Royale de Belgique, 1990).
8. Meudrac, M. *La Chymie charitable et facile en faveur des dames*, p. 1666, CNRS, Paris, 1999.
9. Diderot, D. and D'Alembert, M. *Encyclopédie ou Dictionnaire raisonné des sciences, des arts et des métiers*, Tome troisième, Paris, 1751, p. 319.
10. Raber, L. *Chem. Eng. News* 2000, 78, 52.
11. Vilarigues, M. and da Silva, R.C. *Appl. Phys. A*, 2004, 79, 373–378.
12. Fernandes, P., Vilarigues, M., Alves, L.C. et al. *J. Cultural Heritage* 2008, 9, e5–e9.
13. Vilarigues, M., Fernandes, P., Alves, L.C. et al. *Nucl. Instr. Meth. B,* 2009, 10.1016/j. nimb.2009.03.049.
14. Pradell, T., Molera, J., and Smith, A.D. *J. Archaeol. Sci.* 2008, 35, 1201–1215.
15. Sciau, Ph., Salles, Ph., Roucau, C. et al. *Micron.* http://dx.doi.org/10.1016/j.micron.2009.02.012
16. Padovani1, S., Borgia, I., Brunetti, B. et al. *Appl. Phys. A* 2004, 79, 229–233.
17. Roque, J., Pradell, T., Molera, J. et al. *J. Non-Cryst. Solids* 2005, 351, 568–575.
18. Mirguet, C., Fredrickx, P., Sciau P. et al. *Phase Trans.* 2008, 81, 253–266. http://nobelprize.org/
19. Mills, G.F. and Dickinson, B.N. *Ind. Eng. Chem.* 1949, 41, 2842–2844.
20. Ergozhin, E.E. and Shostak, F.T. *Russ. Chem. Rev.*, 1965, 34, 949–964.
21. Kozhevnikov, A.V. *Electron Ion Exchangers: A New Group of Redoxites*. Wiley, New York, 1975.
22. Kravchenko, T.A. and Nikolaev, N.I. *Kinetika I dinamoka protsessov v redoxitakh* (*Kinetics and Dynamics of Processes in Redoxites*). Khimia, Moscow, 1982.
23. Kravchenko, T.A. and Aristov, I.V. Kinetics and dynamics of redox sorption. In *Ion Exchange. Highlights of Russian Science*, Vol. 1, D. Muraviev et al., Eds., Marcel Dekker, New York, 2000, pp. 691–764.
24. Campelo, J.M., Luna, D., Luque, R. et al. *Chem. Sus. Chem*, 2009, 2(1), 18–45.
25. Schmid, G. *Clusters and Colloids: From Theory to Applications*, VCH, Weinheim, 1994.
26. Fendler, J.H. *Nanoparticles and Nanostructured Films*. Wiley-VCH, Weinheim, 1998.
27. Plieth, W.J. *J. Phys. Chem.* 1982, 86(16), 3166–3170.
28. Edwards, P.P., Johnston, R.L., and Rao, C.N.R. In *Metal Clusters in Chemistry*. P. Braunstain et al., Eds., Wiley-VCH, Weinheim, 1998.
29. de Jongh, L.J. *Physics and Chemistry of Metal Cluster Compounds*. Kluwer, Dordrecht, 1994.
30. Haberland, H. *Clusters of Atoms and Molecules*. Springer, Berlin, 1994.
31. Mulvaney, P. *Nanoscale Materials in Chemistry*, 2001, 121–167.
32. Mirkin, C.A. *Small* 2005, 1(1), 14–16.
33. Schmid, G., Ed. *Nanoparticles*. Wiley Interscience, New York, 2004.
34. Brus, L. *J. Phys. Chem. Solids* 1998, 59(4), 459–465.
35. Astruc, D. *Nanoparticles and Catalysis*. Wiley-VCH, Weinheim, 2008.
36. Astruc, D., Lu, F., and Aranzaes, J.R., *Angew. Chem. Intl.* 2005, 44(48), 7852–7872.
37. Rao, C.N.R., Müller, A., and Cheetham, A.K. *Chemistry of Nanomaterials: Synthesis, Properties and Applications*. Wiley-VCH, Weinheim, 2004.
38. Henglein, A. *Chem. Rev.* 1989, 89(8), 1861–1873.
39. Schmid, G. *Chem. Rev.* 1992, 92(8), 1709–1727.
40. Lewis, L.N. *Chem. Rev.* 1993, 93(8), 2693–2730.
41. Corain, B. and Kralik, M. *J. Mol. Catal. A.: Chem.* 2000, 159(2), 153–162.
42. Biffis, A., D'Archivio, A.A., Jerabek, K. et al. *Adv. Mater. Sci.* 2000, 12(24), 1909–1912.
43. Choi, W.C., Jeon, M.K, Kim, W.J. et al. *Catal. Today* 2004, 93–95, 517–522.
44. Tartaj, P., Morales, M. Del Puerto, et al. *J. Phys. D: Appl. Phys.* 2003, 36(13), R182–R197.

45. Teng, C-H., Ho, K-C., Lin, Y-S. et al. *Anal. Chem.* 2004, 76(15), 4337–4342.
46. Vazquez, M., Luna, C., Morales, M.P. et al. *Physica B* 2004, 354(1-4), 71–79.
47. Pivkina, A., Ulyanova, P., Frolov, Y. et al. J. *Propellants, Explosives, Pyrotechnics* 2004, 29(1), 39–48.
48. Wang, F.C., Zhang, W.E., Yang, C.H. et al. *Appl. Phys. Lett.* 1997, 70(22), 3005–3007.
49. Gerion, D., Parak, W.J., Williams, S.C. et al. *J. Am. Chem.* Soc. 2002, 124(24), 7070–7074.
50. Steele, B.C.H. and Heinzel, A.K., *Nature* 2001, 414(6861), 345–352.
51. Wang, J. *Analyst* 2005, 130, 421–426.
52. Tsuji, J. *Palladium Reagents and Catalysts*, Wiley-VCH, Weinheim, 2004.
53. Molnar, E., Tasi, G., Konya, Z. et al. *Catal. Lett.* 2005, 101(3–4), 159–167.
54. Savastenko, N., Volpp, H.R., Gerlach, O. et al. *Nanopart. Res.* 2008, 10(2), 277–287.
55. Rioux, R.M., Komor, R., Song, H. et al. *J. Catal.* 2008, 254 (1), 1–11.
56. Guo, G., Qin, F., Yang, D. et al. *Chem. Mater.* 2008, 20(6), 2291–2297.
57. Ding, J., Chan, K.Y., Ren, J. et al. *Electrochim. Acta* 2005, 50(15), 3131–3141.
58. Campelo, J.M., Conesa, T.D., Gracia, M.J. et al. *Green Chem.* 2008, 10(8), 853–858.
59. Sambhy, V., MacBride, M.M., Peterson, B.R. et al. *J. Am. Chem. Soc.* 2006, 128(30), 9798–9808,
60. Kenawy, E.R., Worley, S.D., and Broughton, R. *Biomacromolecules* 2007, 8(5), 1359–1384.
61. Grant, R.B. and Lambert, R.M. *J. Catal.* 1985, 92(2), 364–375.
62. Shelef, M. and Gandhi, H.S. *Ind. Eng. Chem. Prod. Res. Dev.* 1972, 11(4), 393–396.
63. Sun, C., Peltre, M.J., Briend, M. et al. *Appl. Catal. A: Mol.* 2003, 245(2), 245–256.
64. Marconi, G., Pertici, P., Evangelisti, C. et al. *J. Organomet. Chem.* 2004, 689(3), 639–646.
65. Nijs, H., Jacobs, P.A., and Uytterhoeven, J.B. *Chem. Soc. Chem. Commun.* 1979, (4), 180–181.
66. Heiz, U., Vanolli, F., Sanchez, A. et al. *J. Am. Chem. Soc.* 1998, 120(37), 9668–9671.
67. Wojcieszak, R., Zielinski, M., Monteverdi, S. et al. *J. Colloid Interface Sci.* 2006, 299(1), 238–248.
68. Boudjahem, A.G., Monteverdi, S., Mercy, M. et al. *J. Catal.* 2004, 221(2), 325–334.
69. Bautista, F.M., Campelo, J.M., Garcia, A. et al. *J. Mol. Catal.* 1991, 67(1), 91–104.
70. Kim, H.S., Lee, H., Han, K.S. et al. *Phys. Chem. B* 2005, 109(18), 8983–8986.
71. Haq, S., Carew, A., and Raval, R., *J. Catal.* 2004, 221(1), 204–212.
72. Tada, M., Bal, R., Namba, S. et al. *Appl. Catal. A: Mol.* 2006, 307(1), 78–84.
73. Kantam, M.L., Jaya, V.S., Lakshmi, M.J. et al. *Catal. Commun.* 2007, 8(12), 1963–1968.
74. Dahl, J.A., Maddux, B.L., and Hutchison, J.E., *Chem. Rev.* 2007, 107(6), 2228–2269.
75. Nalwa, H.S., Ed. *Handbook of Nanostructured Materials and Nanotechnology.* Academic Press, New York, 2000.
76. Cao, G., *Nanostructures and Nanomaterials.* Imperial College Press, London, 2004.
77. Ajayan, P.M., Schadler L.S, and Braun P.V. *Nanocomposite Science and Technology.* Wiley-VCH, Weinheim, 2003.
78. Bhushan, B., Ed. *Handbook of Nanotechnology.* Springer, Berlin, 2004.
79. Pomogailo, A.D. and Kestelman V. *Metallopolymer Nanocomposites.* Springer, Berlin, 2005.
80. Liu, T., Burger, C., and Chu, B. *Progr. Polymer Sci.* 2003, 28, 5–26.
81. Nicolais, L. and Carotenuto, G., Eds. *Metal–Polymer Nanocomposites.* Wiley Interscience, New York, 2005.
82. Adamson, A.W. and Gast, A.P. *Physical Chemistry of Surfaces.* Wiley, New York, 1997.

83. Castelvetro, V. and De Vita, C. *Adv. Colloid Interface Sci.* 2004, 108–109, 167–185.
84. Shi, X., Shen, M., and Monwald, H. *Progr. Polym Sci.* 2004, 29(10), 987–1019.
85. Li, H., Wang, R., Hong, Q. et al. *Langmuir* 2004, 20(19), 8352–8356.
86. Baranchikov, A.Y., Ivanov, V.K., and Tretyakov, Y.D., *Russ. Chem. Rev.* 2007, 76(2), 133–151.
87. Glaspell, G., Hassan, H.M., El Zatahry, A. et al. *Topics Catal.* 2008, 47(1-2), 22–31.
88. Glaspell, G., Fuoco, L., and El Shall, M.S. *J. Phys. Chem. B* 2005, 109, 17350–17355.
89. Senkan, S., Kahn, M., Duan, S. et al. *Catal. Today* 2006, 117(1-3), 291–296.
90. Legrand, J.C., Diamy, A.M., Riahi, G. et al. *Catal. Today* 2004, 89(1-2), 177–182.
91. Campelo, J.M., Lee, A.F., Luna, D. et al. *Chem. Eur. J.* 2008, 14(19), 5988–5995.
92. Chen, X., Zhu, H.Y., Zhao, J.C. et al. *Angew. Chem. Intl.* 2008, 47(29), 5353–5356.
93. Chiang, C.W., Wang, A., and Mou, C.Y. *Catal. Today* 2006, 117(1-3), 220–227.
94. Liu, H., Ma, D., Blackley, R.A. et al. *Chem. Commun.* 2008, (23), 2677–2679.
95. Date, M., Okumura, M., Tsubota, S. et al. *Angew. Chem. Intl.* 2004, 43(16), 2129–2132.
96. Cushing, B.L., Kolesnichenko, V.L., and O'Connor, C.J. *Chem. Rev.* 2004, 104(9), 3893–3946.
97. Hoar, T.P. and Shulman, J.H. *Nature* 1943, 152, 102–103.
98. Kohsuke, M., Miura, Y., Shironita, S. et al. *Stud. Surf. Sci. Catal.* 2007, 170B, 1319–1324.
99. He, P., Zhang, M., Yang, D. et al. *J. Surf. Rev. Lett.* 2006, 13(1), 51–55.
100. Serp, P., Kalck, P., and Feurer, R. *Chem. Rev.* 2002, 102(9), 3085–3128.
101. Liang, C., Xia, W., Soltani-Ahmadi, H. et al. *Chem. Commun.* 2005, (2), 282–284.
102. Sunagawa, Y., Yamamoto, K., Takahashi, H. et al. *Catal. Today* 2008, 132(1-4), 81–87.
103. Shin, J.Y., Lee, B.S., Jung, Y. et al. *Chem. Commun.* 2007, (48), 5238–5240.
104. Schmid, G. *Clusters and Colloids: From Theory to Applications.* VCH, Weinheim, 1994.
105. Mei, Y., Sharma, G., Lu, Y. et al. *Langmuir* 2005, 21(26), 12229–12234.
106. Rozenberg, B.A. and Tenne, R. *Progr. Polym. Sci.* 2008, 33(1), 40–112.
107. Imre, Á., Beke, D.L., Gontier-Moya, E. et al. *Appl. Phys. A* 2000, 71, 19–22.
108. Houk, L.R., Challa, S.R, Grayson, B., et al. *Langmuir* 2009, 25(19), 11225–11227.
109. Furstner, A. *Active Metals.* VCH, Weinheim, 1996.
110. Luckham, P.F. *Adv. Colloid Interface Sci.* 2004, 111(1-2), 29–47.
111. Kruyt, H.R. *Colloid Science.* Elsevier, Amsterdam, 1952.
112. Verwey, E.J. and Overbeek, J.T. *Theory of Stability of Lyophobic Colloids.* Elsevier, Amsterdam, 1948.
113. Grabar, K.C., Freeman, R.G., Hommer, M.B. et al. *J. Anal. Chem.* 1995, 67(4), 735–743.
114. Dotzauer D.M., Dai, J., Sun, L. et al. *Nano Lett.* 2006, 6(10), 2268–2272.
115. Pomogailo, A.D., Rozenberg, A.S., and Uflyand. I. E. *Metal Nanoparticles in Polymers,* Khimia, Moscow, 2000 (in Russian).
116. Pomogailo. A.D. *Russ. Chem. Rev.* 2000, 69(1), 53–80.
117. Pomogailo, A.D., Dzhardimalieva, G.I., Rozenberg, A.S. et al. *Nanoparticle Res.* 2003, 5(5-6), 497–519.
118. Pomogailo, A.D. *Uspekhi Khimii (Russ. Chem. Rev.).* 1997, 66, 679–716.
119. Muraviev, D.N., Macanas, J., Esplandiu, M.J. et al. *Phys. Status Solidi A* 2007, 204(6), 1686–1692.
120. Macanas, J., Parrondo, J., Muñoz, M. et al. *Phys. Status Solidi A* 2007, 204(6), 1699–1705.
121. Antonietti, M., Wenz, E., Bronstein, L. et al. *Adv. Mat. Sci.* 1995, 7(12), 1000–1005.
122. Litmanovich, O.E., Litmanovich, A.A., and Papisov, I.M. *Vysokomolek Soed A* 2000, 42, 670–675.
123. Goodyear, C. Improvement in India Rubber Fabrics. U.S. Patent 3633, June 15, 1844.

124. Sill, K., Yoo, S., and Emrick, T. Polymer–nanoparticle composites. *In Encyclopedia of Nanoscience and Nanotechnology*. 2009, Marcel Dekker, New York. DOI:10.1081/E-ENN 120013728.
125. Muraviev, D.N., Macanás, J., Farre, M. et al. *Sensor Actuat. B Chem.* 2006, B118(1–2), 408–417.
126. Muraviev, D.N., Macanás, J., Ruiz, P. et al. *Phys. Status Solidi A* 2008, 205(6), 1460–1464.
127. Muraviev, D.N., Ruiz, P., Muñoz, M. et al. *J. Pure Appl. Chem.* 2008, 80(11) 2425–2437.
128. Sun, H., Ning, Y., Zhang, H. et al. *J. Nanosci. Nanotechnol.* 2009, 9(12), 7374–7378.
129. Feng, C., Gu, L., Yang, D. et al. *Polymer* 2009, 50(16), 3990–3996.
130. Kumar, R., Pandey, A.K., Tyagi, A.K. et al. *J. Colloid Interface Sci.* 2009, 337(2), 523–530.
131. Zhao, X., Yu, J., Tang, H. et al. *J. Colloid Interface Sci.* 2007, 311(1), 89–93.
132. Chou, K-S. and Lai, Y-S. *Mater. Chem. Phys.* 2004, 83(1), 82–88.
133. Macanas, J., Farre, M., Muñoz, M. et al. *Phys. Status Solidi A* 2006, 203(6), 1194–1200.
134. Macanás, J., Muraviev, D.N., Oleinikova, M. et al. *Solvent Extract. Ion Exch.* 2006, 24, 565–587.
135. Cole, D.H., Shull, K.R., Rehl, L.E. et al. *Phys. Rev. Lett.* 1997, 78(26), 5006–5009.
136. Cole, D.H., Shull, K.R., Baldo, P. et al. *Macromolecules* 1999, 32(3), 771–779.
137. Toshima, N., Teranishi, T., Asanuma, H. et al. *Chem. Lett.* 1990, 5, 819–822.
138. Visser, J. *Adv. Colloid Interface Sci.* 1972, 3, 331.
139. Korchev, A.S., Bozack, M.J., Slaten, B.L. et al. *J. Am. Chem. Soc.* 2004, 126, 10–11.
140. Li, X., Liu, C., Lu, H. et al. *J. Membr. Sci.* 2005, 255, 149–155.
141. Jagur-Grodzinski, *J. Polym. Adv. Technol.* 2007, 18, 875–799.
142. Karthikeyan, C.S., Nunes, S.P., Prado, L.A. et al. *J. Membr. Sci.* 2005, 254, 139–146.
143. He, T., Frank, M., Mulder, M.H. et al. *J. Membr. Sci.* 2008, 307, 62–72.
144. Huang, R.Y.M., Shao, P., Burns, C.M. et al. *Appl. Polym. Sci.* 2001, 82(11), 2651–2660.
145. Mikhailenko, S.D., Zaidi, S.M., and Kaliaguine S. *Catal. Today* 2001, 67(1–3), 225–236.
146. Ruiz, P., Muñoz, M., Macanás, J. et al. *Dalton Trans*, 2010, 39(7), 1751–1757.
147. Alonso, A., Macanás, J., Shafir, A. et al. *Dalton Trans.* 2010, 39, 2579–2586.
148. Koros, W.J., Ma, Y.H., Shimidzu, T. *Pure Appl. Chem.* 1996, 68(7), 1479.
149. Donnan, F.G. *J. Membr. Sci.* 1995, 100, 45.
150. Wang, S., Liu, P., Wang, X. et al. *Langmuir* 2005, 21, 11969–11973.
151. Ruiz, P., Muñoz, M., Macanás, J. et al. Manuscript in preparation.
152. Zhou, S., Varughese, B., Eichhorn, B. et al. *Angew. Chem. Intl.* 2005, 44, 4539–4543.
153. Toshima, N. and Yonezawa, T. *New J. Chem.* 1998, 22(11), 1179–1201.
154. Son, U., Jang, Y., Park, J. et al. *J. Am. Chem Soc.* 2004, 126(16), 5026–5027.
155. Lee, W., Kim, M.G., Choi, J. et al. *J. Am. Chem Soc.* 2005, 127(46), 16090–16097.
156. Park, J. and Cheon, J. *J. Am. Chem Soc.* 2001, 123(24), 5743–5746.
157. Ruiz, P., Muñoz, M., Macanás, J.et al. *Book of Abstracts: Eurosensors XXII*. 319, 2008.
158. Smith-Sørensen, T. *Surface Chemistry and Electrochemistry of Membranes*. Surfactant Science Series 79, Marcel Dekker, New York, 1999.
159. McDonald, R., Ed. *Impedance Spectroscopy*, Wiley, New York, 1987.
160. Benavente, J., Oleinikova, M., Muñoz, M. et al. *Electroanal. Chem.* 1998, 451, 173–180.
161. Ariza, M.J., Rodríguez-Castellón, E., Rico, R. et al. *J. Colloid Interface Sci.* 2000, 226, 151–158.
162. Van Tendeloo, G., Geuens, P., Colomer, J.F. et al. Proceedings of Sixth Multinational Congress on Microscopy, 2003, Pula, Croatia.
163. Ruiz, P., Courteille, F., and Macanás, J. *Book of Abstracts: TNT 2009*, Barcelona.
164. Reetz, M.T., Maase, M., Schilling, T. et al. *J. Phys. Chem. B* 2000, 104, 8779–8781.
165. Pyrz, W.D. and Buttrey, D.J. *Langmuir* 2008, 24, 11350–11360.

166. Kim, Y.J., Choi, X.C., Woo, S.I. et al. *Electrochim. Acta* 2004, 49, 3227–3234.
167. Pomogailo, D. *Uspekhi Khimii (Russ. Chem. Rev.)*, 2000, 69, 53–80.
168. Kickelbrick, G. *Progr. Polym. Sci.* 2003, 28, 83–114.
169. Mottram, T., Rudnitskaya, A., Legin, A. et al. *Biosensors Bioelectron.* 2007, 22, 2689–2693.
170. Lin, M.S. and Leu, H.J. *Electroanalysis* 2005, 17(22), 2068–2073.
171. Endoa, T., Yanagida, Y., and Hatsuzawa, T. *Sensors Actuators B* 2007, 125, 589–595.
172. Shenhar, R., Norsten, T.B., and Rotello, V.M. *Adv. Mater.* 2005, 17, 657–669.
173. Welch, C.M. and Compton, R.G. *Anal. Bioanal. Chem.* 2006, 384, 601–619.
174. Hrapovic, S., Liu, Y., Male, K.B. et al. *Anal. Chem.* 2004, 76(4), 1083–1088.
175. Céspedes, F., Martinez-Fabregas, E., and Alegret, S. *Trends Anal. Chem.* 1996, 15, 296–304.
176. Ruiz, P., Macanás, J., Muñoz, M. et al. Manuscript in preparation.
177. Habermuller, K., Mosbach, M., and Schuhmann, W. Fresenius *J. Anal. Chem.* 2000, 366, 560–568.
178. Young, S. *J. Mater. Sci.* 2003, 38, 2143.
179. Novack, B. *Environ. Pollut.* 2007, 150, 5.
180. Kong, H. and Jang, J. *Langmuir* 2008, 24(5), 2051–2056.
181. Muraviev, D.N., Ruiz, P., Alonso, A. et al. Proceedings of Nanotech Europe 2009, Berlin.
182. Fernández, F., Cordero, B., Durand, J. Et al. *Dalton Trans.* 2007, 5572–5581.
183. Niembro, S., Shafir, A., Vallribera, A. et al. *Org. Lett.* 2008, 10, 3215.

2 Ion Exchangers for Air Purification

V. S. Soldatov and E. H. Kosandrovich

CONTENTS

ABSTRACT

A review of works on sorptions of different gases from the air by ion exchange resins and fibrous ion exchangers is presented. The sorptions of water vapor, ammonia, carbon and sulfur dioxides, hydrogen chloride, hydrogen fluoride, silicon tetrafluoride, nitrogen oxides, hydrogen sulfide, elemental halogens, and others are considered. We present data revealing the influences of different factors such as gas concentration, relative air humidity, acid–base strengths of sorbent and sorbate functional groups, air flow rate, and thickness of filtering layers on the dynamics of sorption of these substances. The filtering layers of fibrous ion exchange materials have significantly higher sorption characteristics than those of granular ion exchangers. Theoretical models allowing calculation of water sorption isotherms and sorption of gases according to acid–base mechanisms (exemplified by ammonia sorption on cation exchangers) are presented. The devices used in air purification by ion exchangers are described.

2.1   INTRODUCTION

The applicability of ion exchangers for extracting of acid and base compounds from gas media directly follows from their chemical nature. The first publication on this topic appeared in 1955[1]. Since then, numerous original papers, patents, and reviews[2–14] reporting successful attempts to apply ion exchangers to extract different compounds from gases were published. Despite those activities, ion exchangers did not until now find wide applications in practice. Traditional technologies for gas treatment continue to dominate in certain industries because conventional ion exchangers, such as resins and granular inorganics with excellent chemical properties also carry inherent features restricting their applicability to practical gas treatment. These features will be discussed in this chapter. Better prospects for practical gas treatment with fibrous ion exchangers have been known for about 50 years.

Ion exchangers can find applications in different aspects of gas treatment such as purification, separation and drying of synthetic and natural gases, treatment of furnace and vent gases, processing of exhaust air from cattle farms, gas analyses, and deep gas purification. These gaseous media differ widely in concentrations of their target components and other components based on relative humidity and temperature. The volumes of emissions can vary from hundreds to millions of cubic meters per hour. Because of that, the technologies of gas treatment also differ; they require different ion exchange materials, equipment, and process sequences.

The large scale processes are possible only with ion exchange resins produced in large amounts. In many cases, refluxing the sorbent with acid or alkali solutions is recommended. In fact, in this case the ion exchanger plays the role of an active packed bed. The process can be continuous or cyclic. The concentration of ionizable components in the initial gas is usually 10^0 to 10^1 g/m^3 and their concentration in the purified air is 10^{-1} to 10^{-2} g/m^3. We restrict our consideration to application of ion exchangers for purification of the air by removal of toxic, malodorous, and corrosive compounds; we will cover only diluted solutions of contaminants in the air.

Air purification in small and medium systems requiring filtering units with working capacities of 1,000 to 50,000 m^3/h can be handled by ion exchangers. For better reader orientation, note that the capacity of a laboratory fume hood is usually

1,000 to 3,000 m³/h. The capacity required for an average galvanic workshop can be 10,000 to 20,000 m³/h. The level of air contamination suitable for ion exchange technologies is usually below 300 mg/m³; the required concentration of contaminant at the outlet of a filtering apparatus is usually about one MPC level or lower. In this category are emissions from several industrial processes such as wood pulping, paper manufacture, petroleum and metal refining and smelting, galvanic production, chemical laboratories, food industries, cattle farms, butchering operations, and many other sources of atmospheric and indoor air contamination. The best solution involves air purification based on automatically regenerable filtering elements made of fibrous ion exchange materials installed in compact and silent filtering plants capable of continuous operation and low energy consumption. Another important task is purification of input ventilation in private and public housing; hospitals, especially in large cities; and ensuring adequate ventilation during natural disasters such as forest fires and chemical accidents. A special problem is pollution of indoor air by tobacco smoke.

To achieve air purification in the clean rooms of electronic, optical, and other industries, the inlet concentration corresponds to natural contents of the impurity in the air of the working zone (usually 10^1 to 10^2 ppb). The outlet concentration should usually be 1 to 10 ppb. Due to the low level of contamination of the inlet air, filter service periods generally range from 6 to 18 months; disposable cassette filters are used. Another important field suitable for ion exchange air purification is individual protection for humans—gas masks, respirators, escape masks, and protective clothing. The task of ion exchange materials in this case is short term temporary purification of air in contact with the human respiratory system and skin to acceptable levels.

It is necessary to discuss some features of air as a sorption medium. Ambient air is a multicomponent system. In addition to a target substance interacting with an ion exchanger, ambient air contains at least two other active components that directly or indirectly influence sorption of the target substance: water vapors and carbon dioxide. The third component, in many cases important for the interaction of target substance and ion exchanger, is oxygen that can oxidize the sorbate or sorbent.

This chapter focuses mainly on experimental and theoretical studies of the parameters of sorption of different gases and vapors in shallow beds of ion exchangers under external conditions similar to those in real applications and allowing their extrapolation to full scale air purification systems. We will also discuss the main applications of ion exchangers in air purification. Most of the relevant information was published in Russian, Polish, and Japanese languages. We hope that we have included the most important literature references in this chapter. We apologize to the authors whose papers we could not find and include in our review.

2.1.1 GENERAL CHARACTERISTICS OF ION EXCHANGE AIR PURIFICATION SYSTEMS

Filtering plants intended for continuous operation in large and medium air purification systems must be based on certain features:

1. Maximal working capacity of the purified air per volume unit of the filtering chamber
2. Fast regeneration procedure
3. Minimal pressure loss of filtering chamber
4. Maximum use periods between changes of filter materials

Ion exchange filtering plants with working capacities of 1,000 to 20,000 m^3/h on fibrous ion exchange materials used in industry require the following characteristics:

1. Working capacity per cubic meter of filtering chamber: 5,000 m^3/h
2. Continuous or periodic regeneration without termination of air purification process
3. Maximum pressure loss on filtering chamber: 700 Pa
4. Minimum of 3 years of continuous use between changes of filter materials

Details of construction of filtering devices will be covered in Section 2.4. The requirements listed are unachievable for the filtering plants using granular ion exchangers.

The characteristics of air purification units stated above are attainable because the fibrous sorbent used is a form of nonwoven fabric with a thickness of ~10 mm and density ~0.1 g/cm^3. About 30 m^2 of the filtering material can be installed in a 1 m^3 filtering chamber. This allows a working capacity of 5,000 m^3/h at a moderate linear flow rate of air across the filtering layer of ~0.1 m/s. The pressure drop under these conditions is ~500 Pa.

For comparison purposes, we present the parameters of a filtering unit of the same volume that utilizes granular ion exchangers. The minimal thickness of the filtering bed with 1 to 2 mm beads is 100 to 200 mm. At a linear air flow rate across the layer of 0.1 to 0.3 m/s, the pressure drop is 1,500 to 4,500 Pa. A 10-shelf filtering unit can reach a working capacity of 2,000 to 6,000 m^3/h. Ion exchangers with smaller beads or thicker filtering layers caused increased pressure drop—one of the most critical parameters of filtering plants. Plants with granular ion exchangers, such as systems with fluidized beds or continuous spraying of alkali or acid on an ion exchanger,[15] improve some parameters of a filtering system but require more complicated construction, lose more pressure, and consume more energy.

The problem of regeneration has been solved easily for plants using fibrous materials. The solution is arranging periodic cross flow of the regenerant through the layers of the filtering materials as described in Section 2.4. Air filtering does not cease during regeneration.

2.1.2 FEATURES OF FIBROUS ION EXCHANGERS

Most types of ion exchangers are in the form of fibrous materials. The methods of their syntheses and basic properties were described in recent reviews[16, 17]. Most often the materials are formed from staple fibers with filaments of uniform 5 to 50 μ effective diameter and length of 30 to 80 mm. Different devices can be made from ion exchange staples such as nonwoven canvases, threads, and cloth. The largest

FIGURE 2.1 Ion exchangers in different physical forms. 1. Granular and fibrous forms (diameter of granular ion exchanger = 0.49 mm). 2. Nonwoven needle punctured Fiban material.

application is a nonwoven needle-punctured material with a surface density of 150 to 1,000 g/m² and thickness of 1.5 to 10 mm. Only a few tons of these materials are produced annually by several companies in Belarus (Fiban*), Russia (Vion), Japan (Ionex), and the Peoples' Republic of China. Figure 2.1 shows different ion exchange materials.

The main advantage of filtering layers of fibrous ion exchangers is a unique combination of properties: extremely high sorption and desorption processes; possible control of aerodynamic and/or hydrodynamic resistances by predetermined density of the fibers; integrity of the filtering layer under vibration and changing orientation; and extremely high osmotic stability allowing multicycling processes without destruction of the filaments.

The high rate of sorption is caused exclusively by the small effective diameters of the filaments (5 to 50 μ) compared to the diameters of beads (300 to 2,000 μ) of conventional ion exchange systems. Sorption is controlled by internal diffusion of the sorbate molecules within the filaments of the ion exchanger in direct proportion to the diffusion coefficient and inversely proportional to the squared diameter of the particle[18]. The filament diameter can be 10 to 100 times smaller than that of industrial resins. The increase in the overall sorption is usually only 10 to 100 times faster because diffusion in the fibers is usually slower than in the resin beads (diffusion coefficients are smaller by two powers of ten[19]). The common statement that the rate of sorption on fibrous ion exchangers is faster in resins because they have larger specific surfaces is incorrect because the ion exchange fibers are not porous and their specific surfaces are equal to geometrical surfaces of the filaments (about 1 m²/g). Macroporous ion exchange resins, having specific surfaces hundreds of square meters per gram do not exhibit such fast sorption because the diffusion path through their beads is much greater than the path through ion exchange fibers.

---

* Institute of Physical Organic Chemistry of the National Academy of Sciences of Belarus, Republic of Belarus.

2.2 THEORY

2.2.1 Water Sorption Isotherms

The concentration of water vapor in ambient air is usually 1 to 25 g/m^3, i.e., about three powers of magnitude greater than the concentration of target impurities to be removed via air purification. Water plays manifold roles in sorption. Water sorption accompanies sorptions of different substances from the air and determines kinetics and equilibria of the processes. It causes dissociation of the sorbate and the functional group of the ion exchanger; it can be a direct participant of reaction between the functional group and the sorbate molecule or be a product of such reaction. The "free" water is the diffusion medium for the sorbate molecules in the phases of ion exchange and determines accessibility of the sorption centers for the sorbate molecules. Water sorption from the gas phase is also important because ion exchangers can be used as gas drying agents[20–24]. At high relative humidity ($\alpha = P/P_0 > 0.2$), ion exchangers exhibit greater sorption capacity for water than zeolites or silica gels.

The literature contains little information about the kinetics and dynamics of water vapor sorption and desorption. The kinetics of water vapor sorption by the AV-17 strong base anion exchange resin containing benzyltrimethylammonium groups and its Fiban A-1 fibrous analogue were studied using McBain balances[19]. From linear dependence

$$F = F\left(\sqrt{t}\right)$$

(relative sorption value of time), the average effective diffusion (D) coefficients of water molecules in the ion exchanger were calculated. They appeared dependent on the relative humidity (water content in ion exchanger), decreasing with increasing α for the fibrous ion exchanger from $1.7 \cdot 10^{-13}$ m^2/s for $\alpha = 0.07$ to $0.3 \cdot 10^{-13}$ m^2/s at $\alpha = 0.92$. Diffusion in the gel-type resin was significantly faster ($D = 8.8 \cdot 10^{-12}$ to $1.4 \cdot 10^{-12}$ m^2/s). This proves that the faster water sorption by ion exchange fibers than by ion exchange resins is caused exclusively by the shorter diffusion path in the fiber.

The equilibrium of water vapor with ion exchangers was extensively studied by many authors and described in monographs, reviews, and original papers[12,19,25–34]. Most works investigated strong acid and strong base ion exchangers; data about other ion exchangers are scarce and not systematic. The sorption isotherms (amounts of absorbed water versus relative gas phase humidity, α) were used to calculate thermodynamic functions of this process. As we will show in this chapter, they provide data about interionic and ion–molecular interactions during ion exchange; such data are needed for quantitative interpretation of the interactions between the sorbate molecules and the ion exchanger.

The presence of air usually does not affect water sorption isotherms for gel-type ion exchangers; they do not exhibit hysteresis evidencing the absence of porous structures in the sorbent. The amount of adsorbed water depends on the nature of the ion exchanger (type of functional group), counterion, type of polymer matrix, and partial pressure of water vapor. The sorption isotherm $W = f(\alpha)$ usually has sigmoid form similar to the dependences of many other substances in different sorption systems, for example:

Physical adsorbents	Gases (physical multilayer adsorption
Hydrophilic polymers (proteins, cotton, gelatin), sulfuric and phosphoric acids, inorganic salts	Water vapors

Much attention has been paid to deriving mathematical forms for dependence $W = f(\alpha)$.

The most used equation was derived by Brunauer, Emmett, and Teller (and known as the BET equation) for physical polymolecular adsorption[35]. This equation was successfully applied to systems whose physical and chemical natures were completely different from those stated in the original BET model. Later several other models suggesting different mathematical forms for description of the same dependence with the same or better accuracy were developed. Five such models were compared using as an example the water sorption levels of 18 anionic forms of the fibrous strong base anion exchanger Fiban A-1[19]. The models compared were as follows (here we use the symbols used in the original papers).

BET model:

$$n_w = \frac{\hat{n}_w \cdot c \cdot a\left[1 - (n+1) \cdot a_w^n + n \cdot a_w^{n+1}\right]}{(1 - a_w) \cdot \left[1 + (c-1) \cdot a_w - c \cdot a_w^{n+1}\right]} \tag{2.1}$$

where $\hat{n}_w$ is the amount of sorbate needed to cover the sorbent surface with a monomolecular layer; a_w is water vapor activity; n is the maximum number of adsorption layers; c is a constant characterizing the difference between heats of adsorption and condensation (net heat of sorption). $\hat{n}_w$, n, and c are constants that have clear physical meanings but their estimations in independent experiments or correct theoretical calculations are practically impossible. In fact, they are fitting parameters, allowing us to fit Equation (2.1) to experimental data independently of the physical meaning of the constants.

Iovanovich model[36]:

$$n_w = \hat{n}_w \cdot \left[1 - \exp(-b \cdot a)\right] \cdot \exp(c \cdot a_w) \tag{2.2}$$

where a and b are constants dependent on the molecular area of the sorbate, its residence time in the first, second, and subsequent adsorption layers, mass of the sorbate molecule, temperature and saturation pressure. In practice $\hat{n}_w$, a, and b are fitting parameters.

White and Eyring model[37]:

$$(\alpha - x) \cdot (B - x) = \beta \cdot x^2 \tag{2.3}$$

$$\ln \frac{\alpha - x}{\alpha} + \gamma \cdot \upsilon_2^{\frac{1}{3}} = \ln \frac{p}{p_0} \tag{2.4}$$

$$\upsilon_2 = \frac{Z \cdot \nu}{\alpha + Z \cdot \nu} \tag{2.5}$$

where α is a total amount of water absorbed at a given P/P_0; x is water absorbed on the localized sites at a given P/P_0; B is a number of localized sorption sites per gram of sorbent (not a constant). Fitting the equations to experimental data gives an average value of $B = B(\alpha)$; β and γ are empirical constants; Z is a number of polymer segments equal in volume to a sorbate molecule per chain; and υ is a number of chains per unit volume. The fitting parameters are x, B, β, γ, Z, and υ. The model accounts for swelling (deformation of matrix) of the ion exchanger during sorption. It has been successfully used to describe the sorption isotherms of different ion exchangers[26] for a range water relative humidity levels.

Model of Sosinovich, Novitskaya, Hogfeld, Soldatov[38]:

$$\omega = \left. \sum_{i=1}^{n} i . K_i . a_{H_2O}^i \middle/ \left(1 + \sum_{i=1}^{n} K_i . a_{H_2O}^i \right) \right. \tag{2.6}$$

All water absorbed by the ion exchanger is considered present in hydrates with numbers of water molecules i and stability constant K_i. The number of fitting parameters is not limited and depends only on the accuracy of the experimental data and the theoretically calculated $W = f(\alpha)$.

Gantman and Veshev model[32]:

$$In\left[a_w . \left(1 + \frac{a}{z . f} \right) \right] = \gamma . f \tag{2.7}$$

$$\gamma \equiv \frac{k . V_w^2}{R . T} \tag{2.8}$$

$$\alpha = 0,009.K_d .(h + f).\left\{ \left(1 + \frac{222}{\left[K_d .(h + f).z \right]} \right)^{0,5} - 1 \right\} . z \tag{2.9}$$

$$h = h_o . a_w^n \tag{2.10}$$

where h is the number of "bound" water moles per equivalent of the functional groups; h_0 is the degree of hydration of the functional groups in a completely swollen ion exchanger; α is the degree of dissociation of the functional groups; z equals the charge of the counterion; f is the content of the "free" water in moles per equivalent of the functional groups; V_w is the partial mole volume of water in the ion exchanger; and k represents the matrix elasticity coefficient. The fitting parameters are h, h_0, α, f, and k.

The application of all these models to the same ion exchanger in different ionic forms indicates that they all produce experimental data with satisfactory accuracy. The mean average deviation of experimental points from the calculated sorption curves was the largest for the BET model (4 to 8.5%); in the other cases it was usually less than 5%.

The study of different models based on different statements shows that the description is formal and the only criterion in favor of some models is the number of fitting parameters used for calculation of the theoretical sorption curve. The smallest number of fitting parameters (three) applies to the BET and Jovanovich models. Nevertheless, the BET model is most often cited in the literature, possibly because of tradition rather than its better descriptive power or clarity of its fitting parameters. The simple conclusion is that the fewer uncertain parameters a model has, the better it is in practical use.

Note that among the many models of water sorption isotherms, 11 are mentioned in a recent publication[39]. The article states that by simple mathematical combinations of the main equations of different pairs of models, the equation of sorption isotherm of a new model can be obtained ("If we have a set of water sorption isotherms models, then any product of them represents a water sorption isotherm model."). Clearly, the main equations of these models are only empirical and suitable to describe the families of sigmoid curves with appropriate choices of constants. The models do not have a predictive force and the meaning imposed on the fitting parameters in application to ion exchangers may be completely different from that postulated by the authors of the models when they were derived. The main information obtained from comparisons of the models from the experimental data on water sorption is contained in the values of the fitting parameters. For example, applying the BET model to activated carbon, we learn the value of specific surface of the sorbent, the effective number of sorption layers, and the energy of the sorption process. Applying the same model to an ion exchanger or sulfuric acid (also in a good agreement with experimental sorption data), the authors interpret the original parameters of the model as the other quantities, like hydration numbers, secondary hydration, surface of polymer matrix, etc.

Our goal was to develop a model of sorption of ionizable impurities (acids or bases) from the air by ion exchangers in alkaline or acidic forms, depending on relative humidity (see Section 2.3). The structure of this model requires knowledge of concentration of the free water in the ion exchange (water in which the target component of the gas phase dissolves).

None of the models considered gives an unambiguous answer to this question. Therefore we also suggest a new model for water sorption, developed particularly for ion exchangers and having fitting parameters easily understood in terms of physical chemistry of ion exchange. With the help of these two models it was possible to suggest a scheme for the description of acid–base equilibria in the system (ion exchanger–water vapor–sorbate) and establish mathematical relations of the value of sorption, concentration of the sorbate in the gas, relative humidity, solubility of the sorbate in water, acid–base strengths of the sorbate and ion exchanger, its capacity and water uptake. This scheme will be exemplified by ammonia sorption on strong and weak acid ion exchangers.

2.3 ACID–BASE EQUILIBRIA IN ION EXCHANGER–WATER VAPOR–SORBATE SYSTEM

2.3.1 MODEL OF WATER SORPTION BY ION EXCHANGERS

This model was briefly described in publications[40,41]. Its principal concepts are:

1. The gel type ion exchanger in equilibrium with water vapors is a homogeneous phase corresponding to the classic phase definition in Gibbs' thermodynamics. We do not account for uncontrolled macroscopic imperfections of the particles in the ion exchanger.
2. The water in the ion exchanger can be virtually separated into hydrate water and free water. The hydrate water cannot dissolve foreign substances. The free water forms an ideal or nonideal mixture with the hydrate particles and can dissolve foreign substances.

If the ion exchanger contains i types of hydrates containing q molecules of water per functional group, the total number of moles of water in the ion exchanger is defined as:

$$W = \sum_i q_i \cdot n_{h,i} + n_w \qquad (2.11)$$

where $n_{h,i}$ and n_w represent numbers of moles in the hydrate i containing q_i water molecules and free water, respectively. Hydrate formation is described by the constant of equilibrium:

$$RH + q_i H_2O \leftrightarrow RH \times q_i H_2O \qquad (2.12)$$

$$K_i = \frac{n_{h,j}}{n_{R,j} \cdot \alpha^{qi}} \qquad (2.13)$$

Consider a system containing one mole of functional groups; $n_{R,i} + n_{h,i} = 1$ and it follows from Equation (2.13) that the total amount of water in the hydrates is:

$$n_h = \sum_i q_i \cdot n_{h,i} = \sum_i \frac{q_i \cdot K_i \cdot \alpha^{qi}}{1 + K_i \cdot \alpha^{qi}} \qquad (2.14)$$

We assume in the first approximation that the ion exchanger is an ideal mixture of hydrates and free water molecules and the system should obey Rault's law. This law in its classic form is not applicable to ion exchangers (or other systems with restricted swelling) because $\alpha_0 = 1$ does not correspond to infinitely diluted solutions of polyelectrolytes, but corresponds to complete swelling of an ion exchanger in water. In this state it contains n_{0w} moles of free water per mole of functional group. The total water uptake is W_0 mol/eq. To bring into agreement this property of the ion exchanger and Rault's law, we assume that the relation between the activity of water in the vapor phase and the mole fraction of the free water in the ion exchanger may be expressed by:

$$\alpha = X_W \times f_W/(X_{0W} \times f_{0W}) \qquad (2.15)$$

where X_W, X_{0W}, and f_W, f_{0W} are, respectively, mole fractions and rational activity coefficients of free water in a partially and completely swollen ion exchanger ($\alpha = 1$).

$$X_W = n_W/(n_W + 1) \qquad (2.16)$$

$$X_{0W} = n_{0W}/(n_{0W} + 1) \qquad (2.17)$$

If the reference state is chosen so that $f_{0W} = 1$ at $X_W = X_{0W}$, the amount of free water in the ion exchanger is:

$$n_w = \frac{\alpha \cdot \left(W_o - \sum_i q_i \cdot n_{h,i} \right) \cdot f_w}{1 + W_o - \sum_i q_i \cdot n_{h,i} - \alpha \left(W_o - \sum_i q_i \cdot n_{h,i} \right)} \qquad (2.18)$$

The new equation of the water sorption isotherm can be obtained by combining Equations (2.11), (2.14), and (2.18).

$$W = \sum_i \frac{q_i \cdot k_i \cdot \alpha^{qi}}{1 + K_i \cdot \alpha^{qi}} + \frac{\alpha \cdot \left(W_o - \sum_i q_i \cdot n_{h,i} \right) \cdot f_w}{1 + W_o - \sum_i q_i \cdot n_{h,i} - \alpha \left(W_o - \sum_i q_i \cdot n_{h,i} \right)} \qquad (2.19)$$

It appears that this equation accurately describes experimental data on water sorption from the gas phase by different ion exchangers at a proper choice of fitting parameters. We verify this with published data for the Dowex ion exchanger[27] and for KU-2[42]. In the most cases, a reasonable agreement between the experimental and calculated data was obtained if only one hydrate was considered and the system was described as pseudo-ideal ($f_W = 1$) as illustrated by the data in Reference[41] and Figure 2.2. In this case Equation (2.19) is reduced to:

$$W = \frac{q \cdot k \cdot \alpha^q}{1 + K \cdot \alpha^q} + \frac{\alpha \cdot \left(W_o - q \cdot n_h \right)}{1 + W_o - q \cdot n_h - \alpha \cdot \left(W_o - q \cdot n_h \right)} \qquad (2.20)$$

The two fitting parameters q and K in Equation (2.19) have clear physical meanings. The number of water molecules in the hydrate q can take on several discrete meanings (e.g., 1/2, 1, 2, 3, 4). For ion exchangers in the forms of strongly hydrated counter ions, two hydrates must be understood to accurately describe the sorption data: (1) a sulfonic cation exchanger in H^+ or Li^+ form and (2) a strong base anion exchanger in OH^- and F^- forms (Figure 2.3). The figure and Table 2.2 indicate that accounting for the second hydrate markedly improves the accuracy of these theoretical calculations.

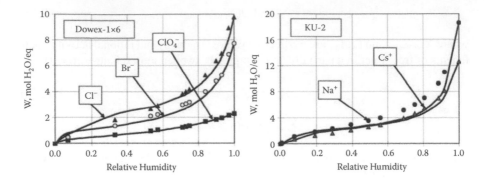

FIGURE 2.2   Water sorption isotherms for ion exchangers Dowex-1 × 6 and KU-2 in different ionic forms. The points are experimental data (see *Sources*). The lines are calculated from the model and assume one stable hydrate with parameters from Table 2.1. (*Sources:* Boyd, G.E. and Soldano, B.A. *Z. Electrochem.* 57, 162, 1953; Kirgintsev, A.N. et al. In *Ion Exchangers and Ion Exchange (Ionity i ionnyj obmen)*, Voronezh University, 1966, 34. With permission.)

Accounting for formation of more hydrates can give a perfect formal agreement between the experimental and calculated sorption isotherms[38] but only cases with one or two hydrates may yield a practical meaning because increasing the number of hydrates by one leads to appearance of two additional fitting parameters. We neglected the activity coefficient of water in Equation (2.20) and treated the ion exchanger as an ideal binary mixture of hydrates and free water.

The nonideality of water not bound into these hydrates is formally expressed by its activity coefficient in Equation (2.19). That can be of practical use only if function $f_W = f(\alpha)$ is known. Analyzing a large number of sorption isotherms of different ion exchangers convinced us that the following empirical equation well describes experimental data.

$$f_W = \alpha^\lambda \qquad (2.21)$$

where λ is a constant and λ is a measure of nonideal behavior of the free water; $\lambda = 0$ for ideal behavior and $\lambda > 0$ for a negative deviation from the ideal; $\lambda < 0$ for a positive one. The absolute value of λ raises with increasing deviation of the system from ideal behavior.

TABLE 2.1
Parameters of Equation (2.20) for Dowex-1 × 6 and KU-2 Ion Exchangers

Ion Exchanger	Ionic Form	K	q	W_0 (mol/eq)
Dowex 1 × 6	Cl^-	$1.0 \cdot 10^2$	2	9.75
Dowex 1 × 6	Br^-	$5.0 \cdot 10^1$	1	7.69
Dowex 1 × 6	ClO_4^-	$5.0 \cdot 10^0$	0.5	2.31
KU-2	Cs^+	$7.0 \cdot 10^1$	2	12.62
KU-2	Na^+	$2.0 \cdot 10^2$	2	18.63

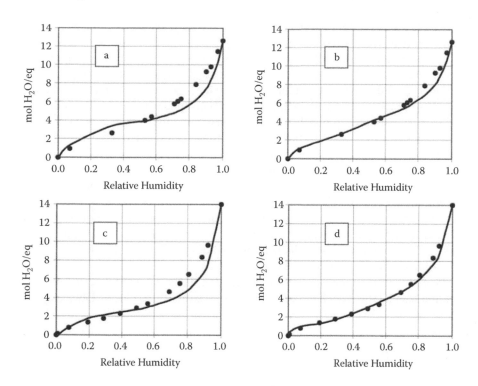

FIGURE 2.3 Water sorption isotherms for ion exchanger Dowex-1 × 6 in F⁻ ionic form (curves a and b) and KU-2 in Li⁺ ionic form (curves c and d). Curves a and c: one hydrate. Curves b and d: two hydrates. The points are experimental data (see *Sources*). The lines are calculated from the model with parameters from Table 2.2. (*Sources:* Boyd, G.E. and Soldano, B.A. *Z. Electrochem.* 57, 162, 1953; Kirgintsev, A.N. et al. In *Ion Exchangers and Ion Exchange (Ionity i ionnyj obmen).* Voronezh University, 1966, 34. With permission.)

TABLE 2.2

Parameters of Equation (2.19) for Dowex 1 × 6 (Fluoride) and KU-2 (Lithium)

Curve in Figure 2.3	K_1	q_1	K_2	q_2	R^2
a	$2.0 \cdot 10^3$	3	–	–	0.9241
b	$1.0 \cdot 10^3$	1	$2.0 \cdot 10^1$	3	0.9748
c	$1.0 \cdot 10^2$	2	–	–	0.9280
d	$1.0 \cdot 10^2$	1	$6.0 \cdot 10^0$	3	0.9909

Lines a and b: Dowex 1x6 in F⁻ ionic form; $W_0 = 12.6$ mol H_2O/g.
Lines c and d: KU-2 in Li⁺ ionic form; $W_0 = 14.0$ mol H_2O/g).
R^2 = standard deviation.

Quantity λ is a fitting parameter whose absolute value usually ranges between 0 and 10. It strongly affects part of the sorption isotherm at $\alpha > 0.5$. Accounting for f_W in some cases greatly improves the accuracy of theoretical descriptions of experimental data. We met one important case in which the sorption isotherm could not be described in the assumption of formation of two stable hydrates: a carboxylic acid ion exchanger in hydrogen form. Its water uptake at $\alpha = 1$ is 3.5 mol H_2O/mol COOH and this is probably too small to consider the system an ideal mixture of hydrates and free water. In this case, assuming the presence of only one or two hydrates is not sufficient for quantitative description dependence $W = f(\alpha)$. A good agreement of experimental and theoretical isopiestic curves was obtained by assuming that the activity coefficient in Equations (2.18) and (2.19) was approximated by the empirical Equation (2.21) with $\lambda = 3.5$ (Figure 2.5). In practical calculations, like those presented in the next section, it is more convenient to use Equation (2.21) in order to calculate nonideality caused by formation of the second hydrate, rather than in the explicit form with the help of Equation (2.13), $q_2 > q_1$. Also, it reduces the number of fitting parameters by one.

We can see from Figure 2.4 and Table 2.3 that the accuracy of description of the sorption data by the model in these cases is approximately the same. Obviously, the

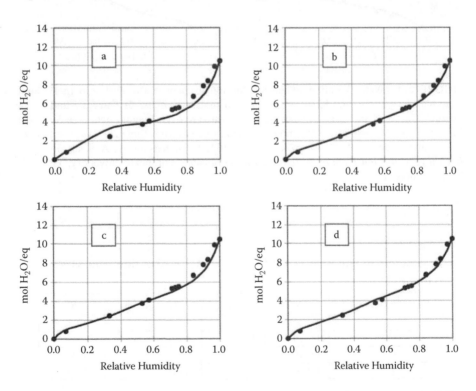

FIGURE 2.4 Isopiestic curves of ion exchanger Dowex-1 × 6 (OH⁻ ionic form) in different approximations. The points are experimental data (see *Source*). The lines are calculated from the model with parameters from Table 2.3. (*Source:* Boyd, G.E. and Soldano, B.A. Z. *Electrochem.* 57, 162, 1953. With permission.)

TABLE 2.3

Parameters of Equation (2.19) for Dowex-1 × 6 (Hydroxide)

Curve in Figure 2.4	K_1	q_1	K_2	q_2	λ	R^2
a	$1.0 \cdot 10^3$	3	–	–	0	0.9616
b	$1.0 \cdot 10^2$	1	$1.5 \cdot 10^1$	3	0	0.9933
c	$1.7 \cdot 10^1$	3	–	–	–1.0	0.9853
d	$1.0 \cdot 10^2$	1	$1.5 \cdot 10^1$	3	–0.2	0.9942

OH^- ionic form; $W_0 = 10.5$ mol H_2O/g.

R^2 = standard deviation.

best precision is obtained when two hydrates and a water activity coefficient are used at the same time.

Figure 2.5 presents the experimental and theoretical isopiestic curves for two fibrous cation exchangers known as Fiban K-1 and Fiban K-4 (polysulfostyrene and polyacrylic acid grafted into polypropylene fibers) and cited in the next section in relation to ammonia sorption. The syntheses and properties of these ion exchangers are described elsewhere[17]. We will show in the following section that the sorption of ammonia occurs when the fraction of the free water exceeds ~7% of the total content independently of the latter (in our case 20 and 7 g H_2O/eq for Fiban K-1 and Fiban K-4, respectively).

2.3.2 MODEL OF ACID–BASE EQUILIBRIA IN ION EXCHANGER VIA AMMONIA SORPTION

The sorption of ammonia is the simplest case of interaction of a single sorbate in gas phase with an ion exchanger because it is not complicated by side processes such as oxidation, catalytic conversion, or association. It is the most convenient case for theoretical consideration of acid–base equilibrium in a gas ion exchanger system. Therefore, we will use this system to illustrate the applicability of a theoretical model establishing mathematical relations among the most important properties of the system such as acid–base strength of the functional groups of the ion exchanger and sorbate molecules, concentration of the sorbate, and relative air humidity. The model suggests a mathematical form for the dependence of sorption of gaseous sorbate by the ion exchanger (g) as a function of the following properties:

1. Concentration of sorbate in gas phase: $[NH_3]_G$
2. Concentration of water vapor in gas phase (relative humidity): α
3. Solubility of gas in water (via Henry's constant): K_H
4. Dissociation constant of sorbate in aqueous phase: K_D
5. Acidity parameters of ion exchanger as described in References[43] and[44]
6. Exchange capacity of ion exchanger: E
7. Uptake of liquid water by ion exchanger: W_0

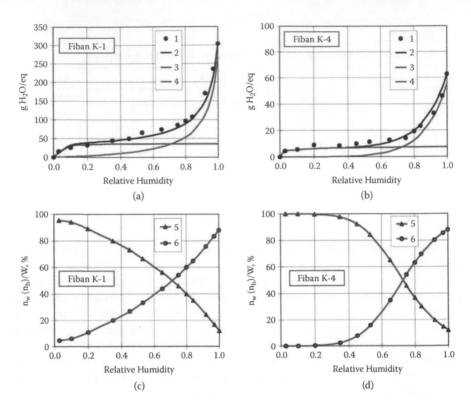

FIGURE 2.5 Isopiestic curves of ion exchangers Fiban K-1 and Fiban K-4. The symbols in a and b represent experimental data. 1 = Experimental points. 2 = Total water. 3 = Amount of hydrate water. 4 = Amount of free water. 5 = Percentage of hydrate water. 6 = Percentage of free water. Note that 2, 3, and 4 are theoretical curves computed from Equations (2.14), (2.18), and (2.19) with the following parameters: (a) q = 2, K = 500, λ = 0; (b) q = ½, K = 5, λ = 3.5).

The values of all these quantities can be determined in independent experiments. To confirm these concepts, we consider a practical case: equilibrium in a system cation exchanger in H^+ form involving ammonia gas and water vapor. The interaction of the ion exchanger with the components of the gas phase can be represented for this example as a totality of the following elementary equilibrium processes. The subscripts G and R define the gas phase and ion exchanger, respectively.

1. Sorption of water vapor by ion exchanger:

$$(H_2O)_G \leftrightarrow (H_2O)_R \tag{2.22}$$

2. Dissolution of molecular ammonia in free water absorbed by ion exchanger:

$$(NH_3)_G \leftrightarrow (NH_3)_R \tag{2.23}$$

3. Ionization of ammonia in ion exchange phase:

$$(NH_3)_R + (H_2O)_R \leftrightarrow (NH_4^+)_R + (OH^-)_R \tag{2.24}$$

4. Hydrolysis of ammonia form of ion exchanger:

$$RNH_4 + (H_2O)_R \leftrightarrow RH + (NH_4^+)_R + (OH^-)_R \tag{2.25}$$

5. Neutralization:

$$(H^+)_R + (OH^-)_R \leftrightarrow (H_2O)_R \tag{2.26}$$

The equilibrium constants of these processes form a set of equations sufficient to determine dependence $g = f([NH_3]_G, \alpha, E, W)$ in an ideal system. However, real systems are substantially nonideal and there is no practical way to determine independently the activity coefficients of their components. We consider in explicit form two obvious reasons for nonideality: (1) formation of hydrates in the ion exchanger phase; and (2) variation of the free energy of ion exchange with loading of the ion exchanger with NH_4^+ expressed as its mole fraction X. The other reasons are described in the previous section.

The second statement of the model, Equation (2.23), is understood in our case to be dissolution of ammonia gas only in free water in the ion exchange phase. The quantity may be calculated from Equation (2.18). In this assumption, Henry's law is formulated as:

$$[NH_3]_R = K_H \times [NH_3]_G \times X_W \tag{2.27}$$

where $[NH_3]_R$ is the concentration of NH_3 in the internal solution; the mole fraction of the free water X_W is calculated from isopiestic data using Equation (2.18); the constant K_H is assumed to be the same as that for equilibrium of an aqueous solution of ammonia and water.

It was calculated as described in Reference[45] and equaled 7.763×10^{-5} (mol/kg)/ (mg/m^3). It follows from such a formulation that the ion exchanger does not absorb ammonia via an ion exchange mechanism in the absence of free water ($X_W = 0$). The ammonia dissociation constant assumed to be the same as that for aqueous media ($K_D = 1.75 \times 10^{-5}$). We applied it to calculate the concentration of NH_4^+ in the internal solution to use the equation for the hydrolysis reaction (2.25) equilibrium coefficient:

$$k_{hyd} = [RH] \times [NH_4^+]_R \times [OH^-]_R / ([RNH_4^+] \times [H_2O]_R) = $$
$$(1 - X) [NH_4^+] [OH^-]/(X \times [H_2O]) \tag{2.28}$$

where $X = [RNH_4]/E$ and $1 - X = [RH]/E$. The coefficient depends on the degree of hydrolysis $(1 - X)$. It has been shown [43,44] that the dependence of the ion exchange equilibrium coefficient $I^+ - H^+$, k, at neutralization of the ion exchanger with alkalis can be described with sufficient accuracy by:

$$pk = pK + \Delta pk \times (X - 1/2) \qquad (2.29)$$

where the equilibrium coefficient k is defined as:

$$k = X \times [H^+]/((1 - X) \times [NH_4^+]) \qquad (2.30)$$

K is the thermodynamic constant of the ion exchange independent of the degree of ion exchange and Δpk is also a constant equal to $pk_{x=1} - pk_{x=0}$. Combining Equations (2.28) through (2.30) with the value for the ionic product of water $[H^+] \times [OH^-] = 1 \times 10^{-14}$ we obtain:

$$pk_{hyd} = 14 - pK - \Delta pk \times (X - 1/2) \qquad (2.31)$$

and

$$X = [NH_3]_R \times K_D/(k_{hyd} + K_D \times [NH_3]_R) \qquad (2.32)$$

We assume that most ammonia absorbed by the ion exchanger from the gas is present in the ion exchanger in the form of ammonium ions bound to the fixed anion—neglecting molecular sorption. The value of $[NH_3]_R$ can be found from Equation (2.27). After its substitution into Equation (2.32), we obtain dependence of the ammonia sorption on its concentration in the gas phase and the relative air humidity:

$$X = K_H \times X_W \times [NH_3]_G \times K_D/(k_{hyd} + K_D \times K_H \times X_W \times [NH_3]_G) \qquad (2.33)$$

In summary, we suggest the following pathway for an a priori calculation of ammonia sorption by an ion exchanger:

1. Obtaining certain initial information: (a) the constant of Henry's law; (b) dissociation constant of the sorbate (ammonia in our example); (c) exchange capacity and acidity parameters of ion exchanger computed from potentiometric titration curves[43,44]; and (d) isopiestic curve of the ion exchanger in acid form
2. Fitting the isopiestic curve to Equation (2.19) and finding parameters q and K
3. Finding the mole fraction of free water XW using Equation (2.18)
4. Calculation of the relative mole fraction of ammonia in the ion exchanger using Equation (2.33) with parameters found in steps 1 through 3 above.

Following this pattern, we calculated the dependence of ammonia sorption on its concentration at fixed air humidity and of the relative air humidity at a fixed concentration on the two ion exchangers. The computed data are in reasonable agreement with the experimental data (Figure 2.6), even if no fitting parameters except those (curve 1) are obtained from the titration curves of the ion exchangers by KOH (instead of NH_4OH). Using more realistic lower values of $\Delta pk = 0.7$ brings the computed data to closer agreement with the experimental ones (curve 2). The proposed

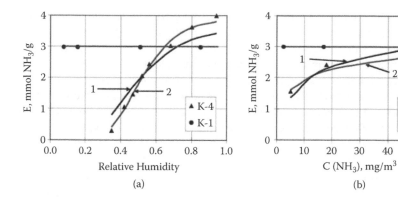

FIGURE 2.6 Ammonia sorption of ion exchangers Fiban K-1 and Fiban K-4 based on relative air humidity at ammonia concentration $[NH_3]_G = 17.5$ mg/m^3. (a). Relative air humidity $\alpha = 0.48$ for Fiban K-1 and $\alpha = 0.55$ for Fiban K-4 (b). The symbols represent experimental data. The curves were computed from Equation (2.33) with the following parameters: Fiban K-1: q = 2, K = 500, pK = 0, Δpk = 0, E = 3.0 m-eq/g. Fiban K-4: (1) q = ½, K = 5, λ = 3.5, pK = 4.93, Δpk = 2.0, E = 4.2 m-eq/g. (2) q = ½, K = 5, λ = 3.5, pK = 4.93, Δpk = 0.7, E = 4.2 m-eq/g.

model correctly reflects the differences in the sorption properties of weak and strong acid ion exchangers. Comparing Figures 2.5 and 2.6, one can see that the ammonia sorption started when free water appeared in the ion exchanger, that is, at $\alpha > 0.4$ for Fiban K-4 and $\alpha > 0.05$ for Fiban K-1.

Both materials have found practical application in removal of ammonia from the air. The model suggested allowed correct prediction of the interval of relative air humidity in which these materials can be used. Especially important was calculation of the working ranges of the filters for removal of traces of ammonia from outdoor air containing usually less than 100 µg/m^3 of NH$_3$, down to levels suitable for high performance clean rooms (1 to 5 µg/m^3). Direct experimental determination of these values is extremely difficult because their working period is usually 1 to 2 years.

2.4 EXPERIMENTAL DATA FOR REMOVAL OF TOXIC GASES FROM AIR

2.4.1 AMMONIA

2.4.1.1 Sorption of Ammonia by Granular Ion Exchangers

The sorption of ammonia by hydrogen forms of cation exchangers based on the reaction

$$RH + NH_3 \leftrightarrow R\text{-}NH_4^+ \tag{2.34}$$

was a subject of the first patents for gas treatment processes using ion exchangers[46,47]. The humidification of a gas or ion exchanger was noted as a condition of efficient sorption. In some works, simultaneous purification of air from ammonia and carbon dioxide was described by mixtures of OH⁻ as a form of a strong base anion, and H⁺ as a strong acid cation exchanger; regeneration was performed by the acid and base after separation of the resins in the filtering layer.

Studies of ammonia sorption by sulfostyrene resins in H⁺ form in static experiments using the McBain balances[48] showed that the rate of sorption strongly depends on the presence of water in the system. In the absence of water, equilibrium in the system was established after 4 to 6 hours. Equilibrium occurred after only 15 minutes when the resin contained 10% water. The sorption isotherms in dry media had typical Langmuir shape with saturation at P_{NH3} = 100 mm Hg. At relative humidity of 25%, the ammonia sorption approached a stoichiometric level at P_{NH3} = 20 mm Hg.

The same process was studied under dynamic conditions[49,50]. At an ammonia concentration of 20 g/m^3 in dry air, an almost immediate penetration of NH_3 through the filtering layer was observed. The dynamic activity of the ion exchanger rose with increasing relative humidity and reached a plateau at 25% (mass) water in the ion exchanger, corresponding to complete saturation of the sulfonic groups of resin with ammonium ions.

The sorption of ammonia by salt forms (Li^+, Mg^{2+}, Ca^{2+}, Al^{3+}, Cu^{2+}, Zn^{2+}, Cd^{2+}, Ni^{2+},Co^{2+}) in sulfostyrene resin[51–53]. The heavy metal cations form complexes with ammonia at low concentrations in the gas phase in the presence of water. Their coordination numbers of four or higher allow at least doubling of the sorption capacity per mass unit of resin matrix. Washing the resins saturated with ammonia in water leads to partial (Cu^{2+}, Zn^{2+}, Cd^{2+}, Ni^{2+}) or complete (Mg^{2+} form) regeneration of ion exchangers and restores their sorption efficiency.

Sorption of ammonia by different ionic forms of macroporous sulfostyrene resins in the absence of water was studied in several works[14, 54–60]. Equilibrium sorption capacity was usually less than that corresponding to formation of four-coordinated ammonia complexes in the ion exchanger. Ammonia sorption by different ionic forms increases in the following order: $Na^+ < H^+ < Zn^{2+} < Cu^{2+} < Co^{2+}$. The presence of the H⁺ form in this series was unexpected.

Although ion exchangers in the form of complexing ions have a higher equilibrium capacity for ammonia, their sorption is slow and their working capacity at dynamic conditions is too low to be of interest in practical air purification applications. Nevertheless, extraction of ammonia from the gas media at high concentrations may be of practical interest. Macroporous ion exchange resins in H⁺ form found application in chemical filters used for deep air purification from ammonia in the air of clean rooms of electronic industries[61]. These resins are used as components of composite filtering media in combination with polyurethane foams.

2.4.1.2   Sorption of Ammonia by Fibrous Ion Exchangers

Sorption of ammonia by fibrous ion exchangers was described[62–70]. The ion exchangers used contained different functional groups (sulfonic, carboxylic, or phosphonic acid) and different polymeric matrices (polypropylene with grafted sulfonated polystyrene, cellulose, polyacrylonitrile). The filtering layers of fibrous ion exchange materials exhibited much higher dynamic sorption activity compared to the fixed beds of their granular analogues and more suitable physical forms for gaseous filters.

The sorption efficiency of these materials increases with increasing acid strength of the functional groups, relative air humidity, and concentration of ammonia in the air. It follows from the cited works that further progress in air purification technologies is connected with use of ion exchange textiles. Air purification processes had not been studied systematically and relationships of the most important factors controlling the process of air purification from ammonia were not established.

Starting in 2003, we performed extensive experimental, theoretical, and technological studies of the implementation of the ion exchange technology to purify air by the removal of ammonia. The studies covered such topics as ammonia removal from industrial and agricultural exhausts (galvanic production, food processing, chemical industries, cattle farms, butchering, etc.); removal of traces of ammonia from the air of clean rooms of semiconductor industries; and purification of breathable air for personal protection (chemical respirators and escape masks). The level of air contamination with ammonia usually does not exceed 300 ppm.

Research was performed on Fiban fibrous ion exchangers containing sulfonic or carboxylic acid functional groups and obtained by radiochemical or chemical modification of industrially produced chemical fibers, mainly polypropylene or polyacrylonitrile. A special interest surrounded ion exchange fibers impregnated by nonvolatile acids in which they mainly act as carriers of the active component binding ammonia. The materials described were produced on a small scale at our institute. The effects of the following factors on the removal of ammonia from the air under dynamic conditions are described in this section:

1. Acidic strength of the functional group of the sorbent
2. Concentration of ammonia in the air
3. Relative air humidity
4. Temperature
5. Thickness of the filtering layer

The influences of ammonia concentration and relative air humidity on the sorption of ammonia from air flow by the Fiban K1 strong acid fibrous ion exchanger and the Fiban K-4 weak base fiber are illustrated by Figures 2.7 through 2.11.

The simplest behavior is observed for the sulfonic ion exchanger: the sorption is little affected by ammonia concentration and air relative humidity within the studied range of these parameters. In the breakthrough points on the sorption curves at different concentrations of ammonia, the sorption is practically equal [see the marked points $(M_1 = 2.50$ mmol NH_3/g; $M_2 = 2.35$ mmol $NH_3/g)$ in Figure 2.8]. Markedly different regularities were observed for the carboxylic acid fiber. In this case, sorption strongly depends on the ammonia concentration and relative air humidity.

At an ammonia concentration in the air of 18 mg/m^3, efficient air purification under accepted conditions is possible only if the relative air humidity exceeds 0.52. At this level, the ion exchanger in equilibrium with the initial gas solution absorbs 2 mmol NH_3/g (~40% of its full capacity). The time before NH_3 breakthrough (protection action time) increases with the α in parallel with increasing full dynamic capacity of the ion exchanger, reaching about 80% of total capacity at $\alpha = 0.94$. At fixed air humidity $\alpha = 0.55$, the protection action time decreases with increasing

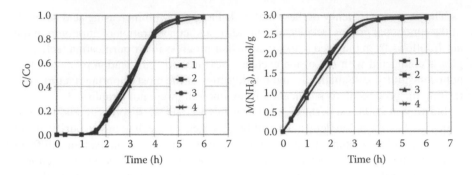

FIGURE 2.7 Breakthrough and sorption curves of ammonia on sulfonic cation exchanger Fiban K-1 (H+ form) at various relative air humidity levels. Experimental conditions: temperature = 25°C; velocity of air filtration = 0.09 m/s; NH_3 concentration = 17 mg/m³; thickness of filtering layer = 3 mm. Relative humidity: 1 = 7.5%; 2 = 15.5%; 3 = 51%; 4 = 85%.

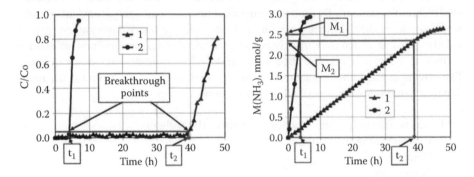

FIGURE 2.8 Breakthrough and sorption curves of ammonia on sulfonic cation exchanger Fiban K-1 (H+ form) at various ammonia concentrations. Experimental conditions: temperature = 25°C; velocity of air filtration = 0.08 m/s; relative air humidity = 48%; thickness of filtering layer = 6 mm. NH_3 concentrations: 1 = 2.5 mg/m³; 2 = 17 mg/m³.

concentration of NH_3 and the full dynamic capacity decreases. This means that weak acid ion exchangers are inefficient for removing trace amounts of ammonia from the gas phase.

A moderate variation of temperature at a constant relative humidity does not affect the breakthrough curve. In our experiments at 15 and 25°C at the same relative humidity ($\alpha = 0.56$), the sorption curves coincided in spite of large differences in concentration of the water vapors (7.18 g/m³ at 15°C and 12.90 g/m³ at 25°C). The effect of filtering layer thickness on the breakthrough curves is illustrated by Figures 2.12 and 2.13.

The dependencies of protection time τ corresponding to the breakthrough capacities on the filtering layer thickness L can be used to calculate the most important technical characteristic of the filtering layer—the length of the mass transfer unit[71]. The parameters needed for calculating this quantity were found from dependences

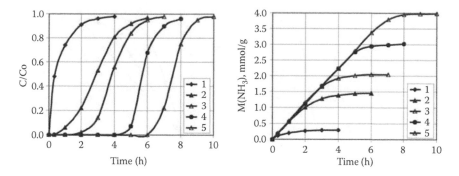

FIGURE 2.9 Breakthrough and sorption curves of ammonia on carboxylic acid cation exchanger Fiban K-4 (H+ form) at various relative air humidity levels. Initial experimental conditions: temperature = 25°C; velocity of air filtration = 0.08 m/s; NH_3 concentration = 18 mg/m³; thickness of filtering layer = 6 mm. Relative humidity: 1 = 35%; 2 = 47%; 3 = 52%; 4 = 68%; 5 = 94%.

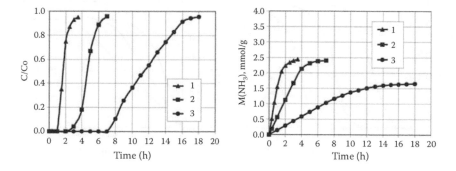

FIGURE 2.10 Breakthrough and sorption curves of ammonia on carboxylic cation exchanger Fiban K-4 (H+ form) at various ammonia concentrations. Experimental conditions: temperature = 25°C; velocity of air filtration = 0.08 m/s; relative air humidity = 55%; thickness of filtering layer = 6 mm. NH_3 concentrations: 1 = 51 mg/m³; 2 = 18 mg/m³; 3 = 5 mg/m³.

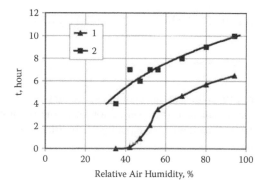

FIGURE 2.11 Dependence of protection action time on relative air humidity for Fiban K-4. Curve 1: (C = 0.05 C_0). Curve 2: time when equilibrium achieved (C = C_0).

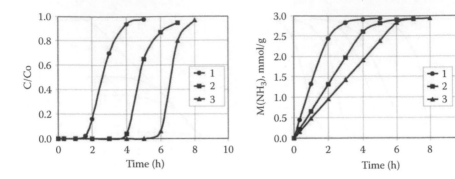

FIGURE 2.12 Breakthrough and sorption curves of ammonia on sulfonic cation exchanger Fiban K-1 (H⁺ form) at various filtering layer thicknesses. Experimental conditions: temperature = 25°C; velocity of air filtration = 0.09 m/s; relative air humidity = 50%; NH_3 concentration = 17 mg/m³. Filter thicknesses: 1 = 3 mm; 2 = 6 mm; 3 = 9 mm.

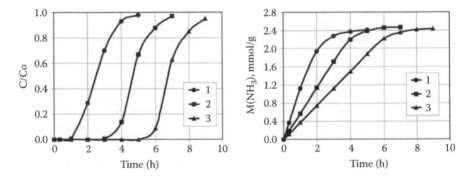

FIGURE 2.13 Breakthrough and sorption curves of ammonia on carboxylic cation exchanger Fiban K-4 (H⁺ form) at various filtering layer thicknesses. Experimental conditions: temperature = 25°C; velocity of air filtration = 0.08 m/s; relative air humidity = 55%; NH_3 concentration = 18 mg/m³. Filter thicknesses 1 = 3 mm; 2 = 6 mm; 3 = 9 mm.

$\tau = f(L)$ using the equations of Shilov (2.35) and Maykel (2.36) as illustrated by Figures 2.14 and 2.15.

$$\tau = K \cdot L - \tau_0 = L/u - \tau_0 = K \times (L - h) \qquad (2.35)$$

where τ_0 is the time required for formation of the stationary sorption front in the filter medium, K is the coefficient of protection action, h is the quantity characterizing value of unused (to breakthrough time) capacity of the filter, and L_0 is the length of the mass transfer unit.

$$f = h/L_0 \qquad (2.36)$$

where f is the factor of symmetry of the breakthrough curve (S_{ABC}/S_{ABCD} in Figure 2.15).

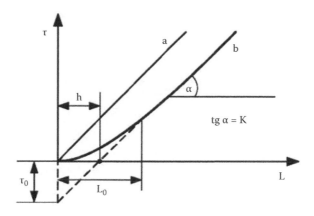

FIGURE 2.14 Determination of parameters of Shilov´s equation.

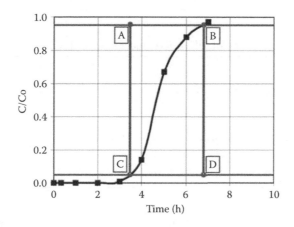

FIGURE 2.15 Determination of parameters of Maykel's equation.

The experimental dependences of $\tau = f(L)$ for fibrous ion exchangers are presented in Figure 2.16 and the parameters of Equations (2.35) and (2.36) appear in are given in Table 2.4.

The length of the mass transfer unit is the lowest for the strong acid cation exchanger. It is at least one power of magnitude smaller than lengths for the granular ion exchangers of the same chemical nature. It follows from the data that the minimal thickness of the filtering layer under the conditions described is approximately 3 mm for Fiban K-1 and 7 mm for Fiban K-4 (double the length of the mass transfer unit). Both materials can achieve efficient air purification, reducing ammonia well below the admissible concentration and odor threshold.

2.4.1.3 Ammonia Sorption by Impregnated Fibers

Ion exchangers absorb large amounts of acids from aqueous solutions[72,73]. The amount of molecularly absorbed acid per gram of fiber on a base of polyacrylonitrile can be comparable to (Figure 2.17) or higher than its exchange capacity.

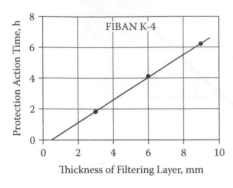
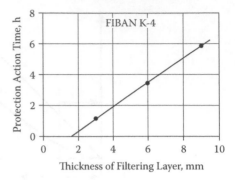

FIGURE 2.16 Experimental dependences $\tau = f(L)$ for Fiban K-1 and Fiban K-4 ion exchangers.

TABLE 2.4
Dynamic Characteristics of Fiban K-1 and Fiban K-4

Ion Exchanger	τ (h)	L (mm)	K (h/mm)	U (mm/h)	τ_0 (h)	H	f	L_0 (mm)
Fiban K-1	1.75	3	0.71	1.41	0.33	0.47	0.28	1.66
	4.00	6						
	6.00	9						
Fiban K-4	1.14	3	0.79	1.27	1.24	1.58	0.44	3.62
	3.43	6						
	5.86	9						

Especially high molecular sorption of inorganic acids was observed for amino–carboxylic polyampholytes containing $R\text{-}CONH(CH_2)_2NH(CH_2)_2NH_2$ and $R\text{-}COOH$ groups. The molecularly absorbed acids can interact with ammonia gas, forming relative ammonium salts encapsulated inside the fibers. Technologies to produce fibrous sorbents via impregnation of ion exchange canvases with concentrated solutions of nonvolatile (sulfuric and phosphoric) acids have been developed and such materials have found practical application in chemical filters. The materials are surface dry and can be handled and processed in the same way as ordinary fibrous ion exchange materials using precautions against direct contact of the filtering materials and corrosive supports with human skin. The ion exchange fibers act as acid carriers and interact with the ammonia gas. Their sorption properties were studied in the same program described for Fiban K-1 and Fiban K-4 ion exchangers.

The materials impregnated with sulfuric acid behave similarly to the Fiban K-1 sulfonic acid cation exchanger. They were prepared from different initial polyampholytes with cationic capacities of 1 to 3 m-eq/g and anionic capacities of 2 to 5 m-eq/g. After equilibrating with solutions of sulfuric acid and drying in the ambient air, they were tested for ammonia sorption as described above. It appeared that differences in the individual sorption properties of the parent fibers almost disappear if the concentration

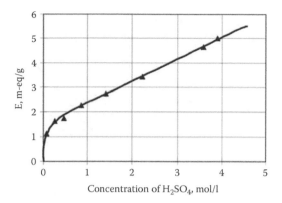

FIGURE 2.17 Sorption of H_2SO_4 by SO_4^{2-} form of amino–carboxylic ion exchanger. E_a = 7.27 m-eq/g. E_b = 1.1 m-eq/g.

of equilibrium acid is sufficiently high. The properties of such sorbents are illustrated in Figures 2.18 through 2.20. The initial experiment conditions were: temperature of 25°C; air filtration velocity of 0.08 m/s; relative air humidity of 50%; NH_3 concentration of 13 mg/m³; and filtering layer thickness of 6 mm, The breakthroughs and total capacities of the materials were similar to those of Fiban K-1. The sorption was independent over wide ranges of ammonia concentration and air humidity.

One drawback of these materials is the need for strict observation of the rules of safety when the materials are mounted in the filters. Materials that are less corrosive and less selective to ammonia were prepared by impregnation of different fibers with solutions of phosphoric and citric acids. Both acids in our range of ammonia concentration and humidity interacted with ammonia only at the first step of dissociation. The capacity for ammonia markedly decreased with decreasing relative air humidity and its concentration in the air (Figure 2.21). Initial experiment conditions were: temperature of 25°C; air filtration velocity of 0.08 m/s; relative air humidity of 43%; and filtering layer thickness of 5 mm.

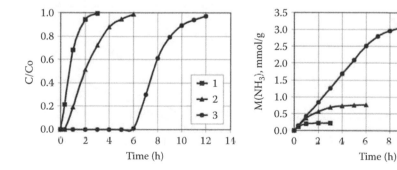

FIGURE 2.18 Breakthrough and sorption curves of ammonia impregnated with sulfuric acid ion exchanger Fiban AK-22-B. E_a = 2.8 m-eq/g. E_b = 2.06 m-eq/g. (1) C (H_2SO_4) = 0.0009 N; (2) C (H_2SO_4) = 0.0083 N; (3) C (H_2SO_4) = 0.8913 N.

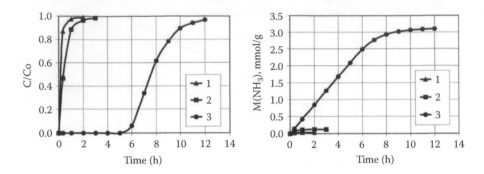

FIGURE 2.19 Breakthrough and sorption curves of ammonia impregnated with sulfuric acid ion exchanger Fiban AK-22-G. $E_a = 1.08$ m-eq/g. $E_b = 4.8$ m-eq/g. (1) C $(H_2SO_4) = 0.0007$ N; (2) C $(H_2SO_4) = 0.0078$ N; (3) C $(H_2SO_4) = 0.9333$ N.

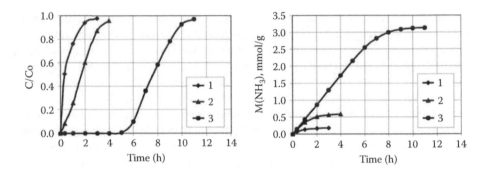

FIGURE 2.20 Breakthrough and sorption curves of ammonia impregnated with sulfuric acid ion exchanger Fiban AK-22-1. $E_a = 1.0$ m-eq/g. $E_b = 4.0$ m-eq/g. (1) C $(H_2SO_4) = 0.0009$ N; (2) C $(H_2SO_4) = 0.0129$ N; (3) C $(H_2SO_4) = 1.0233$ N.

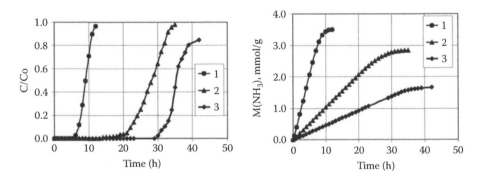

FIGURE 2.21 Breakthrough and sorption curves of ammonia impregnated with phosphoric acid ion exchanger Fiban AK-22-B. $E_a = 2.8$ m-eq/g. $E_b = 2.06$ m-eq/g. (1) C $(NH_3) = 18.6$ mg/m^3; (2) C $(NH_3) = 4.7$ mg/m^3; (3) C $(NH_3) = 2.0$ mg/m^3.

2.4.2 CARBON DIOXIDE

An understanding of carbon dioxide sorption from atmospheric air is necessary for examining its influence on the sorption of toxic gases commonly present in the air in much lower concentrations than CO_2. This understanding is particular important in relation to weak acids with dissociation constants lower than or comparable to that of carbonic acid. For example, hydrogen sulfide and cyanic acid concentrations in the air at the MPC level (for working zones) are 10 mg/m^3 and 0.3 mg/m^3, respectively, while concentration of CO_2 in the ambient air is ~300 mg/m^3. This means that the OH$^-$ form of the anion exchanger in the flow of contaminated air will be rapidly saturated with carbonate and bicarbonate and the weaker alkaline impurities will not be absorbed.

In some cases it is important to remove small amounts of CO_2 from the air used in fuel cells. The removal of carbon dioxide from breathable air is an important element of life support systems intended for long use[74–76]. In such cases, the concentration of CO_2 can be as high as 10 g/m^3.

Most research focusing on carbon dioxide sorption on ion exchangers involved granular ion exchange resins and was performed in the USSR in connection with its space program in the late 1970s. Unfortunately, the research was not completed because interest in the topic faded. The studies were not as systematic as those conducted for ammonia and sulfur dioxide described elsewhere in this chapter.

Equations (2.37) through (2.41) describe equilibrium in a system containing an anion exchanger, water vapor, and carbon dioxide. In an aqueous medium containing carbon dioxide, the following equilibria control the efficiency of sorption by the ion exchanger:

$$(CO_2)_{gas} \leftrightarrow (CO_2)_{\text{ion exchanger}} \qquad (2.37)$$

In an ion exchanger with weak base groups:

$$R_3N + H_2O \leftrightarrow R_3NH^+OH^- \qquad (2.38)$$

$$R_3N + H_2O + CO_2 \leftrightarrow R_3NH^+HCO_3^- + OH^- \qquad (2.39)$$

In a strong base ion exchange:

$$2R_4N^+OH^- + CO_2 \leftrightarrow (R_4N^+)_2CO_3^{2-} \qquad (2.40)$$

$$(R_4N^+)_2CO_3^{2-} + H_2O + CO_2 \leftrightarrow 2\,R_4N^+HCO_3^- \qquad (2.41)$$

A number of published papers and patents cover different methods of carbon dioxide extraction from the air.

Strong base anion exchangers in OH$^-$ and CO_3^{2-} forms were used for CO_2 absorption[77–84]. Application of the OH$^-$ forms makes sense only for removal of trace amounts of CO_2 with disposable filters. Regeneration of the resins to initial OH$^-$ form is possible only by treatment with the alkali solution—a step that is inconvenient from a practical view and causes strong osmotic shocks on the resin beads during the drying and wetting cycles. Using a carbonate form initially and exploiting a carbonate–bicarbonate cycle [Equation (2.41)] is suitable for purification of breathable air. The HCO_3^- form can be regenerated by heating, vacuum, or steam at temperatures above 100°C.

Weak base anion exchangers were also recommended for this purpose[80, 85–87]. Regeneration with water steam or heat in combination with vacuum returned the ion exchanger to the free amine form. Ion exchangers on a base of porous styrene and divinylbenzene matrix with primary and secondary amino groups were used. Macroporous anion exchangers can be used for purification of dry air[81,83]. Different schemes of ion exchange carbon dioxide removal to purify breathable air in cabins of spacecrafts and similar closed systems were suggested[76,88,89].

We now discuss the main principles of carbon dioxide sorption on different ion exchangers established by systematic studies in our laboratory from 1976 to 1980[81–84]. It was conventionally accepted that breathable air may not contain more than 1% (volume) of carbon dioxide. The 1% was used as the starting concentration in our studies of the influences of different factors on sorption. Initially, experiments concerned choices of suitable ion exchangers. Different exchangers were tested under the same conditions.

A column 10 mm in diameter was filled with 5.5 ± 0.5 g of an ion exchange resin in OH⁻ form to a height of 100 ± 5 mm. The resin bead diameter was 0.4 to 1 mm. The air containing $1.01 \pm 0.02\%$ (volume) CO_2 with relative humidity $\alpha = 1$ was passed through the column at a linear velocity of 0.75 m/min at 22 ± 2°C. The outlet air was analyzed. The breakthrough CO_2 curves are presented in Figure 2.22. Table 2.5 compares the data on sorption of carbon dioxide anions

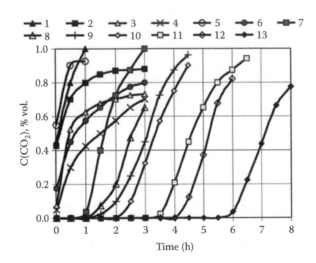

FIGURE 2.22 Breakthrough curves of CO_2 on different anion exchangers. 1 = AN-1; 2 = AN-2Fn; 3 = AN-31g; 4 = EDE-10P; 5 = Dowex-3; 6 = IR-45; 7 = AN-18; 8 = ASD-4 × 1; 9 = ASD-4 × 2; 10 = Dowex-1 × 2; 11 = IRA-401; 12 = IRA-400; 13 = AV-17 × 8. Characteristics of ion exchangers appear in Table 2.5.

TABLE 2.5

Characteristics of Anion Exchangers and Sorption Data for CO_2, CO_3^{2-}, and HCO_3 OH^- form in Various Media

Anion Exchanger	Polymer Matrices and Functional Groups	H_2O (mass %)	E (m-eq/g)	Sorption (mmol CO_2/g Dry Anion Exchanger)				$DA_{0.5}$* (mass %)
				1 M Na_2CO_3 Solution	1 M $NaHCO_3$ Solution	100% CO_2 Gas	1.01% CO_2 Gas	
AN-1	Melamine, formaldehyde; $-NH_2$, $=NH$, $\equiv N$	51.1	5.39	0	0	0.02	–	<0.001
AN-2Fn	Phenol, formaldehyde, polyethilenepolyamine; $=NH$, $\equiv N$	37.7	8.04	0.91	1.57	2.09	0.96	0.001
AN-31g	Epichlorhydrine, ammonia, polyethylenepolyamine; $=NH$, $\equiv N$	55.7	10.70	1.08	1.58	1.83	1.75	0.27
EDE-10P	Epichlorhydrine, polyethylenepolyamine; $=NH$, $\equiv N$, $-NR_3$	59.4	9.43	0.78	1.04	1.17	1.09	0.63
Dowex-3	Styrene, divinylbenzene; $-NH_2$, $=NH$, $\equiv N$	30.8	5.10	0.42	1.20	1.14	0.48	<0.001
IR-45	Styrene, divinylbenzene, polyethylenepolyamine; $-NH_2$, $=NH$, $\equiv N$	35.4	5.79	0.42	1.23	1.06	0.62	0.30
AN-18	Styrene, divinylbenzene; $\equiv N$	41.3	4.35	0.48	0.80	0.88	0.60	1.76
ASD-4 × 1	Styrene, divinylbenzene (1%), triethanolamine	58.9	4.20	0.98	1.17	1.18	–	2.07
ASD-4 × 2	Styrene, divinylbenzene (2%), triethanolamine	46.6	3.46	0.88	1.13	1.40	1.27	3.08
Dowex-1 × 2	Styrene, divinylbenzene (2%); $-N(CH_3)_3$	77.1	5.01	1.82	2.59	2.90	2.80	3.00
IRA-401	Styrene, divinylbenzene (4%); $-N(CH_3)_3$	65.5	4.15	1.71	2.54	2.80	2.67	4.40
IRA-400	Styrene, divinylbenzene (8%); $-N(CH_3)_3$	58.2	4.45	1.87	2.77	2.90	2.86	5.40
AV-17 × 8	Styrene, divinylbenzene (8%); $-N(CH_3)_3$	57.4	4.19	1.81	2.94	3.07	3.03	7.20

* $DA_{0.5}$ = dynamic activity at $C = 0.5 \times C_0$ for air containing 1.01% CO_2.

from different media and by different anion exchangers. The results are summarized as follows:

1. Weak base anion exchangers in equilibrium with wet carbon dioxide or its 1% gaseous solution absorb significant amounts of carbon dioxide, but their dynamic activity is too low for purifying breathable air. They can be used for extracting CO_2 from more concentrated gaseous mixtures under conditions of long contact of resin and gas.
2. The strong base resins with quaternary ammonium groups on the base of styrene and divinylbenzene have high breakthrough capacities on CO_2 and can be used to purify breathable air. Their sorption capacity is almost independent of CO_2 concentration in the range 1 to 100%. The difference in the dynamic activities of IRA-400 and AV-17 (chemical analogues) may be the result of different size distributions of the resin beads—a parameter not controlled in the experiment.

The strong base anion exchangers absorb significantly less carbon dioxide than should be the case if their fixed anionic groups are saturated with HCO_3^- anions based on a comparison of the sorption values and exchange capacities. The anion exchanger in equilibrium with the CO_2 gaseous solution always contains CO_3^{2-}, HCO_3^-, and OH^- counterions. This was confirmed by infrared (IR) spectroscopy[90]. In strong base anion exchangers in equilibrium with 1% CO_2 at 22°C, the HCO_3^- form predominates.

For comparison, Table 2.5 shows the values of carbon-containing species in the resin phase after recalculation for CO_2 content for the resins equilibrated with sodium carbonate or bicarbonate solutions. Due to the competition with OH^- anions that suppressed ionization of the weak base groups, the equilibrium sorption of carbon anions was always lower from a solution than from CO_2 gas. Clearly, based on these studies, the purification of breathable air from CO_2 is possible only with strong base anion exchangers and a carbonate–bicarbonate cycle. Therefore, our further studies focused on that type of resin.

The influence of bead size, thickness of the filtering layer, and air flow rate on the efficiency of air purification was studied using the AV-17 ion exchanger. Three resin fractions were separated by screening. The average bead diameters (determined microscopically) in the swollen state were 1.19, 0.76, and 0.43 mm. Only spherical particles separated from the commercial product by rolling from an inclined plane were used. The exchange capacity and water uptake of the samples used in experiments were 3.82 ± 0.07 m-eq/g and 41.5% of water in the swollen resin, respectively.

The sorption experiments were carried out at 22°C in a column with a diameter of 32 mm and a height of 150 mm (100 and 700 mm in some cases). The CO_2 concentration was 0.30 to 1.25 ± 0.03 %; $\alpha = 1.00$; and the flow rate was 0.5 to 1.25 m/min. The column was filled with a carbonate form of the resin obtained by equilibration with 1 M solution of Na_2CO_3 and the excess rinsed with water.

Figure 2.23 presents the breakthrough curves for the small and large bead fractions of the resin at different air flow rates. Using the resin with smaller beads and lower air

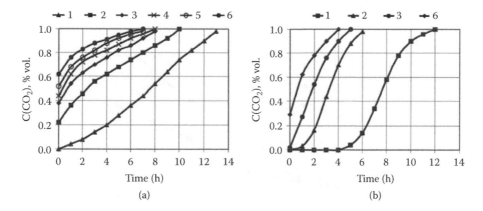

FIGURE 2.23 Breakthrough of CO_2 through column with carbonate form of anion exchanger AV-17. Conditions of the experiment are detailed in text. (a) fraction of resin with average bead size 1.19 mm. (b) fraction of resin with average bead size 0.43 mm. Air flow velocities (m/min): 1 = 0.5; 2 = 1.0; 3 = 2.0; 4 = 3.0; 5 = 4.0; 6 = 5.0.

flow rate aided the protective action of the ion exchange filter. A large difference in the effectiveness of the resin with smaller and larger beads indicates that internal diffusion is a decisive factor limiting the overall rate of air purification from CO_2. The effect of air flow velocity on the dynamic sorption activity at $C/C_0 = 0.5$ ($DA_{0.5}$) and the pressure losses on the column for different resin fractions are shown in Figures 2.24 and 2.25. The dynamic capacity became constant at an air flow rate of 2 m/min on the column filled with the largest beads only when its height reached 700 mm. Under these conditions, full dynamic capacity ($DA_{0.95}$) did not depend on the concentration of CO_2 at $C(CO_2) > 0.8\%$ and decreased quickly (Figure 2.26). This is a good pre-condition for chemical-free regeneration of an ion exchanger by vacuum.

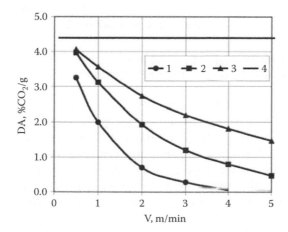

FIGURE 2.24 $DA_{0.5}$ dependence on air flow velocity. Fractions with average bead diameters (mm): 1 = 1.19; 2 = 0.76; 3 = 0.43; 4 = $DA_{0.95}$ for all fractions.

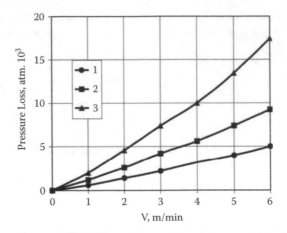

FIGURE 2.25 Pressure loss on column (150 mm) as function of air flow velocity. Fractions with average bead diameters (mm): 1 = 1.19; 2 = 0.76; 3 = 0.43.

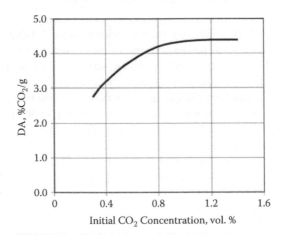

FIGURE 2.26 Dynamic activity of ion exchange column in carbonate form dependent on initial concentration of CO_2 in the air. The experiment conditions are detailed in text.

The variation in cross linkage of the ion exchanger affects sorption activity in a somewhat complicated manner because of overlaps of several opposing tendencies: increasing cross-linkage retards accessibility of CO_2 molecules to the fixed cationic groups; it shortens their diffusion path to the center of the resin bead because of decreasing swelling; the exchange capacity calculated per dry mass of the material decreases, while the capacity per wet mass (or column volume) increased. As a result, one can observe nonmonotonous dependence of sorption versus resin cross-linkage as seen from Figure 2.27.

The maximal equilibrium sorption per mass of wet ion exchanger was observed for resins in OH^- or CO_3^{2-} form with 6 to 12% divinylbenzene (DVB): 8 to 9 and 4 to 4.5%, respectively.

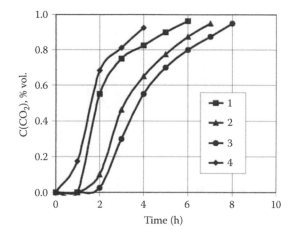

FIGURE 2.27 Breakthrough CO_2 sorption curves for AV-17 strong base anion exchanger in OH^- form with divinylbenzene (1 = 2%; 2 = 6%; 3 = 10%; 4 = 20%). Filtering thickness layer = 100 mm; fraction = 0.40 to 1.0 mm; air flow velocity = 1.55 m/min; initial CO_2 concentration = 1.04% (volume).

In spite of rather high sorption capacities of gel type anion exchangers, the properties of their filtering layers may not be considered to completely satisfy the requirements for purifying breathable air in closed systems. The CO_2 sorption breakthrough curves have diffuse fronts due to poor kinetics of the sorption process even at the low flow rates and with the long filtering layers used in the experiments described. The filtering layer has high pressure that drops rapidly, increasing with decreasing bead size and increasing the airflow rate.

Significantly better properties were revealed by AV-17P, a macroporous anion exchanger of the same type having a specific surface area of 46 m²/g. Its breakthrough curves are very steep (Figure 2.28), indicating excellent kinetics of CO_2 sorption even at very low relative air humidity.

The dry anion exchanger in CO_3^{2-} form at $\alpha = 0$ did not absorb carbon dioxide under the conditions of our experiment. The effect of increasing air humidity was pronounced: a significant protective action time appeared at $\alpha = 0.04$. It rapidly increased and passed through a flat maximum at $\alpha = 0.56$. At relative humidity $\alpha = 1$, the dynamic sorption capacity dropped dramatically because of poor kinetics of the process caused by filling the transport pores in the resin beads with liquid water. Despite that, macroporous strong base anion exchangers are better suited than gel type resins for breathable air purification systems.

2.4.3 SULFUR DIOXIDE

Sulfur dioxide (SO_2) gas is one of the worst pollutants of the atmosphere. Its annual emission to the atmosphere probably approaches a hundred million tons per year. The main sources of emissions are fossil fuel combustion, particularly coal burning power plants; industrial processes such as wood pulping and paper manufacture;

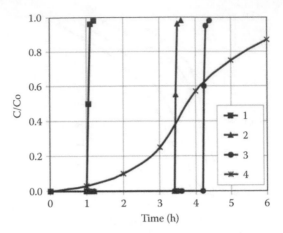

FIGURE 2.28 Breakthrough CO_2 sorption curves for strong base anion exchanger AV-17P in CO_3^{2-} form at various air relative humidity levels: 1 = 4%; 2 = 10%; 3 = 56%; 4 = 100%. Filtering thickness layer = 100 mm; fraction = 0.40 to 1.0 mm; air flow velocity = 0.5 m/min; initial CO_2 concentration = 1.04% (volume).

petroleum and metal refining; and smelting, particularly from sulfide-containing lead, silver, and zinc ores. SO_2 is one of six pollutants mandated for inclusion in the U.S. Environmental Protection Agency (EPA) air contamination criteria by the National Ambient Air Quality Standards.

According to the U.S. regulations, the maximum concentration of SO_2 in the ambient air may not exceed 0.30 ppm (80 $\mu g/m^3$). Technologies for cleaning furnace and vent gases are known and used[91,92]. The main method is neutralization of SO_2 by lime suspension followed by oxidation by atmospheric oxygen. Many attempts to apply ion exchange resins for removing SO_2 from furnace gases[93–100] have been reported. It is clear that the problem is easily solved from a chemical view, but ion exchange technologies cannot compete economically with traditional technologies in handling the huge load of SO_2 emissions.

Several works on the sorption of SO_2 from dry atmosphere or by vacuum using the McBain balances under static conditions allowed calculation of the thermodynamic functions of the process and showed the possibility of using ion exchangers for purification of dry gases[101–108]. These works do not have direct applicability to air purification and will not be discussed here. We restrict our explanations to cases in which the application of ion exchangers is feasible from economic and technical views and consider ion exchanger sorption of SO_2 from air containing it in concentrations not exceeding a few hundred milligrams per cubic meter. Such processes are important for purification of SO_2-contaminated indoor and outdoor air from relatively small sources of air pollution, such as hydrometallurgy, metal refining and smelting, wood pulping, paper manufacture; indoor air cleaning in areas affected by fires; and removal of SO_2 from clean room air of electronic, optical, and fine mechanical operations.

Most experimental works aiming to define the effects of different factors of SO_2 sorption by ion exchangers involved commercial granular ion exchangers and concentrations about three powers of magnitude higher than those defined in our work[97, 109-113]. Sorption of SO_2 by fibrous ion exchangers was studied[114-118]. We concluded from these works that practical applications for solving the problems cited above require fibrous ion exchangers. The next section of this chapter presents the main results of our studies of SO_2 sorption on fibrous strong and weak base anion exchangers already in use for air purification.

2.4.3.1 SO₂ Sorption by Fibrous Anion Exchangers

The sorption of sulfur dioxide was systematically studied on two fibrous ion exchangers in HCO_3^- ionic form: Fiban A-5 (weak base fiber) and Fiban A-6 (strong base fiber). We chose this ionic form, because in practical air purification the OH^- form rapidly converts to a mixed carbonate–bicarbonate form due to its reaction with atmospheric CO_2 whose concentration is ~300 mg/m³. The bicarbonate form rapidly reaches equilibrium with atmospheric CO_2 and its interaction with SO_2 in a stationary regime can be expressed by the following equilibria:

$$2R_4N^+OH^- + CO_2 \leftrightarrow (R_4N^+)_2CO_3^{2-} \tag{2.42}$$

$$(R_4N^+)_2CO_3^{2-} + H_2O + CO_2 \leftrightarrow 2\,R_4N^+HCO_3^- \tag{2.43}$$

$$2R_4N^+OH^- + SO_2 \leftrightarrow (R_4N^+)_2SO_3^{2-} \tag{2.44}$$

$$(R_4N^+)_2SO_3^{2-} + H_2O + SO_2 \leftrightarrow 2\,R_4N^+HSO_3^- \tag{2.45}$$

R_4 denotes four radicals of the functional group. It may be convenient to interpret this process as the ion exchange of monocarbonate and bicarbonate ions with monosulfite and bisulfite in a system containing hydroxy anions. The relative equilibrium equations can be derived from Equations (2.42) through (2.45). If the functional group is a weak base, at least one radical is H and the following additional equilibrium equation is needed:

$$R_3HN + H_2O \leftrightarrow R_3HNH^+OH^- \tag{2.46}$$

In this case hydrolysis of all ionic forms is important and may be represented by equations such as:

$$R_3HN^+An^- + H_2O \leftrightarrow R_3HNH^+OH^- + H^+ + An^- \tag{2.47}$$

In the presence of oxygen, intensive oxidation of SO_2 to SO_3 catalyzed by the anion exchanger occurs:

$$SO_2 + \tfrac{1}{2}\,O_2 \rightarrow SO_3 \tag{2.48}$$

It is followed by formation of sulfuric acid, whose sorption by the ion exchanger is also described by equilibrium type [Equations (2.44) and (2.45)]. We can see from this brief discussion that SO_2 sorption from air is not symmetric to the process of NH_3 sorption by cation exchangers. Experimental data and elaborated theories are insufficient for quantitatively describing SO_2 sorption. In this section we can only qualitatively describe the influence of different factors on SO_2 removal from air by anion exchangers.

The scheme for studying SO_2 sorption was the same as that for investigating ammonia sorption as described earlier. The influence of relative humidity on the SO_2 breakthrough curves is shown by Figures 2.29 and 2.30. It is surprising that the Fiban A-5 weak base anion exchanger revealed a better ability to remove SO_2 from the air at equal relative humidity than the Fiban A-6 strong base at a concentration of 32 to 37 mg/m³. The dependence shapes of breakthroughs and full capacities from α are also different (Figures 2.31 and 2.32).

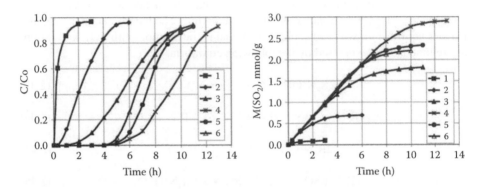

FIGURE 2.29 Breakthrough and sorption curves of sulfur dioxide on anion exchanger Fiban A-5 (HCO_3^- form) at various relative air humidity levels. Experimental conditions: temperature = 25°C; velocity of air filtration = 0.08 m/s; SO_2 concentration = 37 mg/m³; thickness of filtering layer = 6 mm. Relative air humidity α: 1 = 0.15; 2 = 0.22; 3 = 0.28; 4 = 0.49; 5 = 0.68; 6 = 0.91.

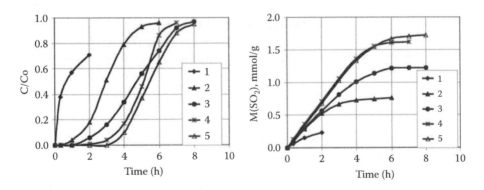

FIGURE 2.30 Breakthrough and sorption curves of sulfur dioxide on anion exchanger Fiban A-6 (HCO_3^- form) at various relative air humidity levels. Experimental conditions: temperature = 25°C; velocity of air filtration = 0.08 m/s; SO_2 concentration = 32 mg/m³; thickness of filtering layer = 6 mm. Relative air humidity α: 1 = 0.28; 2 = 0.33; 3 = 0.33; 4 = 0.48; 5 = 0.92.

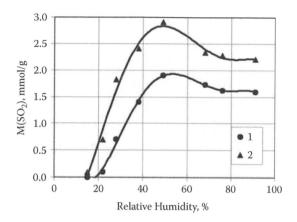

FIGURE 2.31 Dependence of breakthrough (curve 1) and full (curve 2) sorption capacity of Fiban A-5 on relative air humidity.

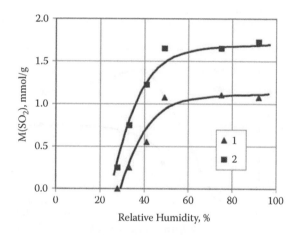

FIGURE 2.32 Dependence of breakthrough (curve 1) and full (curve 2) sorption capacity of Fiban A-6 on relative air humidity.

The maxima in the curves for Fiban A-5 at $\alpha = 0.5$ are probably explained by increasing hydrolysis of ionic forms of the weak base fiber according to equations such as (2.47). The hydrolysis should be much weaker for a strong base fiber. In this case, we observed a plateau in the curve sorption due to relative humidity. The higher sorption ability of the weak base Fiban A-5 may arise from its large hydrophilicity compared to strong base Fiban A-6. As seen from Figure 2.33, water sorption in the whole α range in the first case is higher. It is important to note that Fiban A-5 contains more free water than Fiban A-6. This may be the reason the protective action of the Fiban A-5 filtering layer appears at lower relative humidity than that for Fiban A-6 (Figure 2.34).

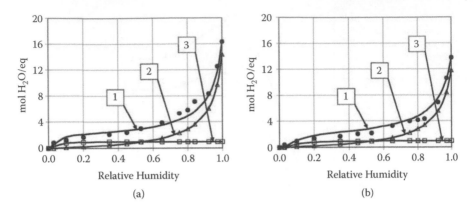

FIGURE 2.33 Isopiestic curves of ion exchangers Fiban A-5 (a) and Fiban A-6 (b). The symbols were calculated from experimental data. Curve 1: theoretical curve computed from Equation (2.19) with the following parameters: (a) q = 2; K = 450; λ = 0. (b) q = 2; K = 100; λ = 0. Curves 2 and 3 are based on the amounts of free [Equation (2.18)] and hydrate [Equation (2.14)] water in α, respectively.

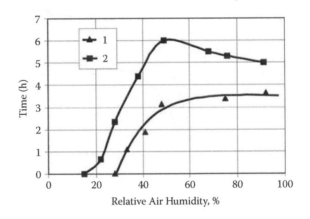

FIGURE 2.34 Dependence of protection action time of Fiban A-5 (curve 1) and Fiban A-6 (curve 2) on relative air humidity.

Another reason for the higher sorption ability of Fiban A-5 may be catalytic oxidation of $SO_2 \rightarrow SO_3$ [Equation (2.48)] with the subsequent formation of H_2SO_4 if the conversion on the ternary amino groups of Fiban A-5 is faster than on the quaternary ammonium groups of Fiban A-6. The rate of overall SO_2 sorption is controlled at this step and the difference between weak and strong base ion exchangers is not significant at the studied concentration of SO_2 in the air. That is because difference in degree of neutralization of the weak and strong base groups by strong H_2SO_4 can be seen only at very low SO_2 concentrations.

Another factor controlling the process rate may be diffusion of the sorbate molecules to the middle of the fiber. We performed an experiment to determine the importance of the diffusion step. Air passing through the filtering layer of Fiban A-5 was stopped at the breakthrough point for 12 hours, then allowed to continue to complete saturation of the sample. After the rest period, the breakthrough time increased from 5 to 6 hours (breakthrough capacities were 1.6 and 1.8 mmol/g, respectively). The total dynamic capacity increased from 2.2 to 2.9 mmol/g. This experiment shows the importance of the diffusion step in the rate of the reaction.

Figures 2.35 and 2.36 illustrate the influence of SO_2 concentration on breakthrough curves. The breakthrough capacity of the weak base fiber increased with increasing concentration of SO_2, while the protective action time decreased. With the strong base fiber, the breakthrough capacity passed through a flat maximum and remained almost constant The dependences of total sorption on concentration (Figure 2.37) are well described by the Langmuir equation with coefficients given in Table 2.6.

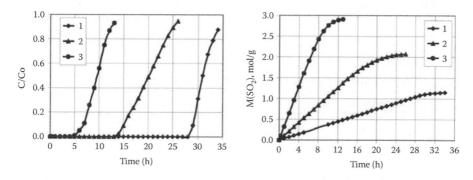

FIGURE 2.35 Breakthrough and sorption curves of sulfur dioxide on anion exchanger Fiban A-5 (HCO_3^- form) at various sulfur dioxide concentrations. Experimental conditions: temperature = 25°C; velocity of air filtration = 0.08 m/s; relative air humidity = 50%; thickness of filtering layer = 6 mm. $C(SO_2)$: 1 = 2.8 mg/m³; 2 = 10.5 mg/m³; 3 = 37 mg/m³.

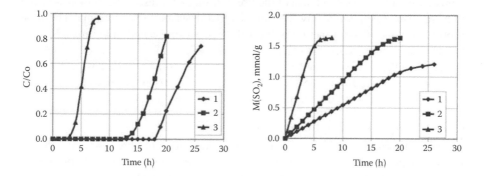

FIGURE 2.36 Breakthrough and sorption curves of sulfur dioxide on anion exchanger Fiban A-6 (HCO_3^- form) at various sulfur dioxide concentrations. Experimental conditions: temperature = 25°C; velocity of air filtration = 0.08 m/s; relative air humidity = 50%; thickness of filtering layer = 6 mm. $C(SO_2)$: 1 = 3.1 mg/m³; 2 = 11 mg/m³; 3 = 33 mg/m³.

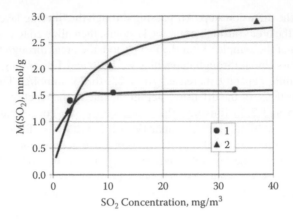

FIGURE 2.37   Dependence of equilibrium SO_2 sorption on concentration for Fiban A-6 (1) and Fiban A-5 (2). Points calculated from Langmuir equation with parameters in Table 2.6.

TABLE 2.6
Coefficients of Langmuir Equation

Ion Exchanger	a_∞, mmol/g	K	Equation
Fiban A-5	3.1	0.224	$a = 3.1 \times 0.224 \times c / (1 + 0.224 \times c)$
Fiban A-6	1.6	2.115	$a = 1.6 \times 2.115 \times c / (1 + 2.115 \times c)$

The equation was used for extrapolation of the sorption values at the lower concentrations of interest for air purification in clean rooms of electronic industries (below 1 mg/m³). At SO_2 concentrations above 10 mg/m³, the Fiban A-5 weak base was more efficient than the Fiban A-6 strong base. At lower concentrations, the opposite occurred: Fiban A-5 lost its sorption ability dramatically and Fiban A-6 retained its relatively high level. Dependence of the breakthrough curves on the thickness of the filtering layer was qualitatively the same as that described for ammonia. The time of protection action as a function of layer thickness is presented in Figure 2.38. The data characterizing the filtering layer appear in Table 2.7. These results indicate that the length of the mass transfer unit is always below 2 mm. These data were obtained for relative air humidity of 0.50 according to the clean room requirements of electronic industries that are the main users of such filtering materials.

2.4.3.2   Sorption of SO_2 by Impregnated Fibers

An efficient sorbent of SO_2 was obtained by impregnation of fibrous aminocarboxylic polyampholytes with sodium carbonate[119]. As an example, we describe properties of one such sorbent prepared from the Fiban AK-22 ion exchanger with a cationic capacity of 2.80 m-eq/g and an anionic capacity of 2.06 m-eq/g. The water uptake varied from 35 to 70% at pH from 1 to 12. After contact with sodium carbonate followed by washing with water, the material practically did not absorb SO_2 from air under the conditions of our experiment (Figure 2.39, curve 1).

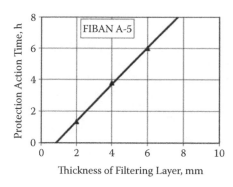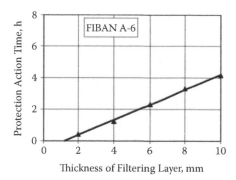

FIGURE 2.38 Experimental dependences $\tau = f(L)$ for Fiban A-5 and Fiban A-6 ion exchangers.

TABLE 2.7
Dynamic Characteristics of Fiban A-5 and Fiban A-6 Ion Exchangers

Ion Exchanger	τ (h)	L (mm)	K (h/mm)	u (mm/h)	τ_0 (h)	H	f	L_0 (mm)
Fiban A-5	1.33	2	1.17	0.86	0.97	0.83	0.48	1.72
	3.78	4						
	6.00	6						
Fiban A-6	0.60	2	0.60	1.70	0.38	0.64	0.42	1.55
	2.11	4						
	3.30	6						
	4.52	8						
	5.30	10						

* Definitions of the symbols are given in Section 2.4.1.2.

After equilibrating this fiber with 1 M sodium carbonate, removing the excess solution, and drying, the material revealed rather poor sorption activity (Figure 2.39, curve 2). The sodium carbonate partially crystallized on the surfaces of the fibers. To prevent crystallization and improve sorption activity, a fibrous material should contain nonvolatile components to decelerate precipitate formation and attract water from relatively dry media used in clean rooms. The best result among several substances tested was achieved with glycerol (Figure 2.39, curve 3). The nonwoven fabric impregnated with a sodium carbonate–glycerol solution had a clean dry surface and revealed excellent sorption to SO_2. It has practical applications in chemical filters for air purification systems in clean rooms.

2.4.4 HYDROGEN CHLORIDE

Air purification with HCl has created problems for chemical industries, metal finishing workshops, and thermal treatment of garbage. Much HCl becomes polyvinylchloride. The heavily contaminated air is purified by wet scrubbing, but removal of trace amounts

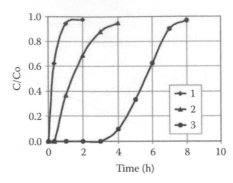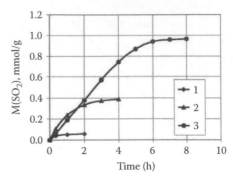

FIGURE 2.39 Breakthrough and sorption curves of sulfur dioxide on ion exchanger Fiban AK-22 impregnated with sodium carbonate. Experimental conditions: temperature = 25°C; velocity of air filtration = 0.08 m/s; relative air humidity = 0.50; thickness of filtering layer = 6 mm; SO_2 concentration = 33 mg/m³. Solutions used for the treatment of ion exchanger: 1 = sodium carbonate, followed by washing with water; 2 = sodium carbonate; 3 = sodium carbonate with glycerol.

of HCl (< 100 mg/m³) remains a problem. Based on its strong acidic properties, the best solution for HCl may be sorption by anion exchangers. Analysis of the literature indicates that air purification to remove trace amounts of HCl is poorly studied.

Sorption of HCl by the AV-17 strong base resin in OH⁻ form from the air at concentration ranges of 50 to 5,000 mg/m³ was studied[120]. The experimental conditions were: thickness of sorption bed = 200 mm; bead size = 0.5 to 1 mm; flow rate = 0.2 m/s; temperature = 20 to 25°C; $\alpha = 0.7$ to 0.9. It was established that the ion exchanger could absorb up to 8 m-eq/g of HCl while its exchange capacity was 4 m-eq/g. The same phenomenon was observed on a fibrous weak base anion exchanger[2].

HCl sorption depended strongly on air humidity. It started at $\alpha = 0.15$ and reached maximum at $\alpha = 0.5$. Superequivalent sorption is possible at the expense of formation of HCl[72,73]. At low relative humidity, HCl sorption is determined by the permeability of the polymeric matrix of the ion exchanger and is minimally sensitive to the type of functional group[121]. Similar data were obtained for fibrous[122,123] and granular ion exchangers[124–126] under static conditions using McBain balances. Systematic experimental data on the effects of important factors controlling HCl sorption by fibrous anion exchangers were recently obtained in our laboratory.

Four fibrous ion exchangers with differing chemical natures (Table 2.8) were studied. The fibers were in carbonate form. The initial concentrations of HCl in the air were 57 and 80 mg/m³. The breakthrough curves of HCl at different humidity levels are presented in Figures 2.40 and 2.41. The horizontal lines in the figures denote the exchange capacities on different anion exchange groups. In all cases, the sorption curve cross these values, i.e., superequivalent sorption occurs. No dependence peculiarities g = g(τ) were observed at the cross-section point. This means that sorption occurs in parallel according to two mechanisms: (1) the exchange of carbonate on Cl⁻ and/or the addition of HCl to the amino group; and (2) the association of ionic pairs H⁺Cl⁻ in the ion exchanger[127]. The interaction of HCl with the anion exchanger can be formulated as follows:

TABLE 2.8
Main Properties of Ion Exchangers Used in Studies of HCl Sorption

Ion Exchanger /Main Acidity Parameters of Anionic Groups	Functional Groups	E_b (m-eq/g)	Water Uptake (g H_2O/g)
Fiban A-1/ $pK_b = -1.0$	$-N^+(CH_3)_3$	2.5	0.62
Fiban A-5/ $pK_b = 4.7$	$-N(CH_3)_2; =NH$ $-COOH$	3.31 ($-NR_2$) 1.02 ($-COOH$)	1.13
Fiban A-6/ $pK_{b1} = 2.2; pK_{b2} = 5.0$	$-[N(CH_3)_2C_3H_5O]^+$ $-N(CH_3)_2$	1.96 ($-N^+R_3$) 0.65 ($-NR_2$)	1.16
Fiban AK-22-G/ $pK_{b1} = 4.0;$ $pK_{b2} = 6.0; pK_{b3} = 9.0$	$=NH; -NH_2$ $-COOH$	1.5 (pK_{b1}); 1.1 (pK_{b2}); 1.3 (pK_{b3}); 0.53 ($-COOH$)	0.62

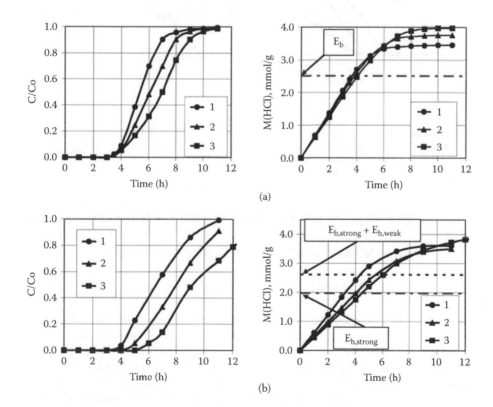

FIGURE 2.40 Breakthrough and sorption curves of HCl on Fiban A-1 (a) and Fiban A-6 (b) anion exchangers. α: 1 = 35%; 2 = 50%; 3 = 60%. Experimental conditions: initial concentration of HCl = 57 mg/m³; thickness of filtering layer = 9.5 ± 0.5 mm; air flow velocity = 0.11 m/s; temperature = 18°C.

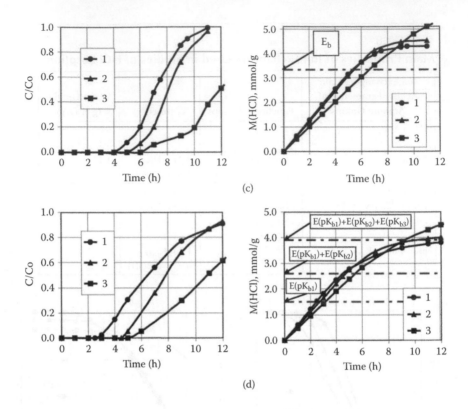

FIGURE 2.41 Breakthrough and sorption curves of HCl on Fiban A-5 (c) and Fiban AK-22-G (d) ion exchangers. α: 1 = 35%; 2 = 50%; 3 = 60%. Experimental conditions are the same as those in Figure 2.40.

$$\left(R_i \equiv N^+\right)_2 CO_3^{2-} + 2HCl \xrightarrow{H_2O} 2R_i \equiv N^+Cl^- + H_2O + CO_2 \qquad (2.49)$$

$$R_i \equiv N^+HCO_3^- + HCl \xrightarrow{H_2O} R_i \equiv N^+Cl^- + H_2O + CO_2 \qquad (2.50)$$

$$R_i \equiv N^+Cl^- + nHCl \leftrightarrow R_i \equiv N^+Cl^- \cdot n\,HCl \qquad (2.51)$$

where R_i is a hydrocarbon radical of the functional group or H atom. Equation 2.51 corresponds to superequivalent sorption (g_s) that can be determined precisely only for the monofunctional Fiban A-1 strong base fiber by washing it from the ion exchanger, contacting the HCl gas, and removing excess HCl with water. The HCl bound to the fixed anionic groups was washed away with KNO_3 and the Cl^- content g_e was determined. The sum of $g_s + g_e$ coincided with the HCl sorption calculated from the breakthrough curve. The separation of HCl for equivalent and superequivalent sorption for weak base groups is less certain due to hydrolysis of the sorption complex during the washing of the ion exchanger with water. Table 2.9 shows the values of equivalent and superequivalent sorption at different stages of the process.

TABLE 2.9

Equivalent and Superequivalent Sorption of HCl

Ion Exchanger	$\alpha = 35\%$			$\alpha = 50\%$			$\alpha = 60\%$		
	τ	g_s	g_e	τ	g_s	g_e	τ	g_s	g_e
Fiban A-1	3.8	0.08	2.07	4.2	0.11	2.15	4.5	0.23	2.23
	5.0	0.46	2.38	5.0	0.60	2.40	5.0	0.75	2.38
	7.0	0.92	2.40	7.0	1.00	2.45	7.0	1.22	2.45
	9.0	1.02	2.42	9.0	1.15	2.50	9.0	1.35	2.47
	11.0	1.11	2.43	11.0	1.30	2.50	11.0	1.50	2.50
	$\tau_{p.a.} = 3.7$			$\tau_{p.a.} = 4.0$			$\tau_{p.a.} = 4.0$		
Fiban A-5	5.0	0.57	2.28	6.0	0.84	2.53	-	-	-
	7.0	1.41	2.33	7.0	1.60	2.45	7.0	1.29	2.50
	9.0	1.72	2.40	9.0	1.90	2.50	9.0	2.22	2.47
	11.0	1.94	2.40	11.0	2.14	2.45	11.0	2.61	2.47
	$\tau_{p.a.} = 4.5$			$\tau_{p.a.} = 5.5$			$\tau_{p.a.} = 7.0$		
Fiban A-6	4.0	0.69	1.60	5.0	1.28	1.63	6.0	1.59	1.67
	7.0	1.27	1.75	7.0	2.07	1.68	7.0	2.25	1.70
	9.0	1.60	1.70	9.0	2.78	1.67	9.0	2.81	1.73
	11.0	1.80	1.70	11.0	2.86	1.67	11.0	3.06	1.67
	$\tau_{p.a.} = 4.1$			$\tau_{p.a.} = 5.0$			$\tau_{p.a.} = 6.0$		
Fiban AK-22-G	5.0	0.54	2.37	5.0	0.33	2.53	6.0	0.80	2.63
	7.0	0.78	2.50	7.0	0.92	2.62	7.0	1.25	2.70
	9.0	1.00	2.60	9.0	1.13	2.73	9.0	1.93	2.80
	11.0	1.08	2.67	11.0	1.30	2.63	11.0	2.47	2.80
	$\tau_{p.a.} = 3.1$			$\tau_{p.a.} = 5.0$			$\tau_{p.a.} = 6.0$		

τ = time after beginning of process (hours). g_e = equivalent sorption (mmol/g). g_s = superequivalent sorption (mmol/g). $\tau_{p.a}$ = protection action time; hours to $C/C_0 = 0.05$.

At the breakthrough point of the Fiban A-1 strong base anion exchanger, almost 100% of its exchange capacity is used. The other 45% of HCl absorbed because of the formation of associates. Fiban A-6 contains quaternary ammonium groups with $pK_b = 2.2$ [(three units lower than Fiban A-1 ($pK_b = -1$)]. It also contains secondary amino groups with $pK_b = 5.0$. At the breakthrough point, it absorbs about 88% HCl to its total exchange capacity and 117% to the capacity on the strong base groups. The value of equivalent sorption (defined as the acid not removed by water) is 84% of the capacity on strong acid groups. The amount of absorbed HCl is much greater than the total anion exchange capacity.

Fiban A-5 contains only weak base (predominantly ternary amino) groups with $pK_b = 4.7$. After its washing with water, only 2.4 mmol/g of Cl⁻ is left in the fiber— less than its exchange capacity. It can be assumed that the carboxylic acid groups in the fiber block some of the amino groups due to formation of hydrogen bonds R_3N ××× HOOCR. The HCl sorption occurs predominantly on the free amino groups whose amount equals the difference in the anionic and cationic capacities of the fiber (2.29 mmol/g). This agrees well with the experimental value of chemically bound Cl⁻. The superequivalent sorption of HCl in this case reaches 3 mmol/g fiber.

The Fiban AK-22-G ion exchanger exhibits the highest anionic capacity and lowest sorption activity. It may be due to the weak basicity of the functional groups (pK_b = 9.0) and/or may be connected to a low amount of free water in the ion exchanger, as calculated by Equation (2.19). Figure 2.42 presents the isopiestic curves and amounts of hydrate and free water for the Fiban A-5 and AK-22-G weak base anion exchangers studied. Table 2.10 compares the amounts of hydrate and free water in the fibers at the relative air humidity at equilibrium. Both values are lower for Fiban AK-22-G.

A large superequivalent sorption of HCl even at low concentrations mixed with air is a good precondition to regenerate ion exchangers with water or low concentrated solutions of HCl. By recycling water in a regeneration bath, it was possible to obtain 6 to 7% HCl from air containing 100 mg/m³ of HCl[2].

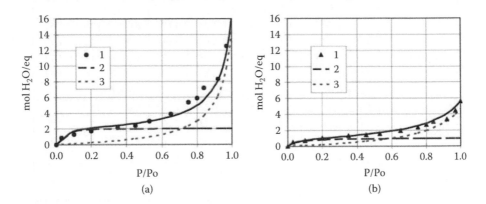

FIGURE 2.42 Isopiestic curves of ion exchangers Fiban A-5 (a) and Fiban AK-22-G (b). The points represent experimental data; the lines are theoretical, computed from Equation (2.19) with the following parameters: Fiban A-5: q = 2; K = 450; λ = 0. Fiban AK-22-G: q = 1; K = 20; λ = 0. Curve 1: W = f (α). Curve 2: hydrate water. Curve 3: free water.

TABLE 2.10
Free and Hydrate Water Dependent on Relative Humidity

Ion Exchanger	α	Hydrate Water (mol/eq)	Free Water (mol/eq)
Fiban A-5	0.35	1.964	0.486
	0.50	1.982	0.878
	0.60	1.988	1.277
Fiban AK-22-G	0.35	0.875	0.406
	0.50	0.909	0.701
	0.60	0.923	0.977

2.4.5 OTHER GASES

Nitrogen oxides—These compounds are the main toxic and chemically active contaminants of atmosphere. The main sources of air contamination with nitrogen oxides are heat power stations, automobile engines, chemical plants, galvanic enterprises, and garbage processing operations. N_xO_y are represented by N_2O, NO, N_2O_3, NO_2 (N_2O_4), and N_2O_5. NO is formed by high temperature burning of organic fuel at the expense of both atmospheric nitrogen (thermal NO) and nitrogen-containing components of the fuel (fuel NO). The process of building a nitrogen oxide pool in the atmosphere is represented by:

$$N_2 + O_2 \rightarrow 2NO \tag{2.52}$$

Note that 90 to 99% of total N_xO_y is released at the outlets of burning chambers. At ambient conditions, NO is oxidized quickly by atmospheric oxygen to NO_2:

$$2NO + O_2 \rightarrow 2NO_2 \tag{2.53}$$

In its turn, NO_2 polymerizes forming the dimer:

$$2NO_2 \leftrightarrow N_2O_4 \tag{2.54}$$

The $\Delta H°$ dimerization is 57.3 KJ/mol. At normal pressure and temperature, 31% of the nitrogen oxides are represented by NO_2; at 100°C, the percentage is 88, and at 140°C it reaches almost 100%. NO has no acid–base properties and may not be removed by acidic or basic sorbents or chemicals. Nevertheless it can be removed from the air because of its spontaneous conversion to NO_2. The presence of the other oxides can be ignored because under ambient conditions they are inert (N_2O) or decompose according to reactions:

$$N_2O_3 \nrightarrow NO + NO_2 \tag{2.55}$$

$$2N_2O_5 \nrightarrow 4 NO_2 + O_2 \tag{2.56}$$

For decontamination of large-scale emissions of nitrogen oxides, selective noncatalytic and catalytic oxidation technologies are used. With noncatalytic technology, furnace gases (containing mainly NO) are mixed with ammonia or carbamide[128]; the reaction occurs at 900 to 1150°C. Catalytic oxidation on platinum or oxide catalysts needs a temperature of 250 to 400°C[129]. This technology is more expensive but better suited to treatment of smaller emissions including those from internal combustion engines (in which carbon monoxide can serve as a reductant).

In many cases, decontamination of relatively small emissions of nitrogen oxides at ambient temperatures is required. Such emitters include galvanic and metal etching workshops and chemical factories. The main component of the oxides is NO_2, and the most sensible solutions are medium and small size air treatment plants based on chemosorption. Anion exchangers in OH^- form were tested for removing nitrogen oxides from exhaust gases of nitric acid production containing high concentrations of N_xO_y[130]. Both strong and weak base anion exchangers revealed high sorption capacities but the degree of retention was about 60%. The other 40% passed the filtering layer as NO. This is in agreement with the stochiometry of NO_2 conversion to nitric acid because three moles of NO_2 convert into two moles of nitric acid and one mole converts in NO.

The retention of NO_2 in OH^- form in the AV-17 strong base anion exchanger from air flow at concentrations of 0.1 to 10 g/m^3 was studied[131]. The thickness of the filtering layer was 10 to 15 cm, linear flow rate was 0.106 m/s, and relative air humidity was below 90%. The dynamic activity of the column was very low (~2 mg/g) and independent of NO_2 concentration. The equilibrium capacity (at complete saturation) was very high and more than twice the exchange capacity of the resin (8 mmol/g).

A special case is removal of trace amounts of nitrogen oxides in the ambient atmospheres of clean rooms. The most convenient technique is using fibrous chemosorbents such as Vion[132] or Fiban[133]. Nevertheless, removal of nitrogen oxides from the air was never complete. It seems clear that alkaline absorbents are not suitable for fine air purification to remove nitrogen oxides because the process is slow and not effective for NO sorption. We know of no acceptable methods for air purification from nitrogen oxides at ambient temperature at their low concentrations (< 100 ppm).

We assume that the filtering material for this purpose can be a fibrous ion exchange material impregnated with a reaction solution that eliminates NO_2 by non-acid–base reactions. We tested two types of reaction solutions removing nitrogen dioxide from air at the expense of chemical reaction. The first is reduction of NO_2 with formation of non-volatile substances:

$$Na_2SO_3 + 2NO_2 + NaOH \rightarrow 2NaNO_2 + Na_2SO_4 + H_2O \quad (2.57)$$

The efficiency of this process decreases quickly because of oxidation of the sulfite to sulfate by atmospheric oxygen. A similar behavior was observed with the other reductants. The reduction of sulfite may have been suppressed by the addition of nonvolatile polyalcohols, among which glycerol was the most efficient. Figure 2.43 illustrates the efficiency of adding glycerol to the sulfite solution at 20°C with an initial NO_2 concentration of 16 mg/m^3.

Solution 4 is suitable for use in regenerable filters for removal of trace amounts of nitrogen dioxide from the air. Among different ion exchange fibers impregnated with a sodium sulfite–glycerol solution and additional antioxidants, a nonwoven material suitable for deep air purification from NO_2 (Fiban N-1) was obtained. See Table 2.11.

The second reaction used to prepare a highly efficient material for fine air purification from nitrogen oxides was diazitation of primary amines by nitrites in the presence of moderately strong acids used in the syntheses of diazo compounds:

$$ArNH_2 + NaNO_2 + 2HX \rightarrow ArN_2^+X^- + NaX + 2H_2O \quad (2.58)$$

The Fiban N-2 fibrous ion exchange material used for this reaction was stable to oxidation and highly efficient for removing trace amounts of NO_2 from the air.

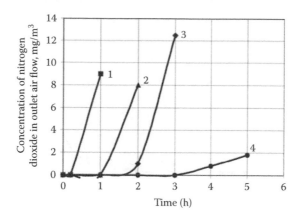

FIGURE 2.43 Breakthrough curves of NO$_2$ (air flow passed through water solutions). Curve numbers correspond to numbers in Table 2.11.

TABLE 2.11
Compositions of Absorption Solutions

Number of Curves in Figure 2.43	Concentration of Molar Na$_2$SO$_3$,	Concentration of Glycerol (Vol. %)
1	0.5	0
2	0.5	1.6
3	0.5	3.8
4	0.5	14.6

The breakthrough curves of NO$_2$ obtained under the same conditions evidenced high efficiency of the new NO$_2$-removing fibrous materials (Figure 2.44). The breakthrough capacity (5% of initial concentration) depends on the relative air humidity. At NO$_2$ concentration of 10 mg/m^3 it increases from 2.5% at $\alpha = 0.48$ to 4% at $\alpha = 0.72$. Such solutions can be used in frame or slit type chemical filters as regeneration or impregnation solutions in disposable cartridges.

Carbamide is a somewhat efficient reductant of nitrogen oxides at low temperatures, especially when NO$_2$ and NO are present in equimolar ratios[134]. Carbamide is used in aqueous solutions in packed bed devices. We tested it as a liquid phase in frame and slit chemical filters[133] with a nonwoven anion exchange fabric acting as a carrier of the active carbamide. The degree of air purification at NO$_2$ concentrations of 10 to 50 mg/m^3 was almost independent of concentration and somewhat high (70 to 80%) for this contaminant.

Hydrogen sulfide, mercaptans, and carbon bisulfide—Air purification by removing hydrogen sulfide in concentrations exceeding the MPC from working zones (10 mg/m^3 for working zones) in industries such as natural gas extraction and processing, chemical manufacturing, animal skin processing, cattle farms and others is now a reality. A separate issue is the removal of trace amounts of H$_2$S present in the air (10 to 100 µg/m^3) of clean rooms in the precise mechanical, electronics,

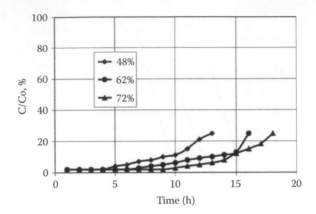

FIGURE 2.44 Breakthrough curve of NO_2 in dependence on relative air humidity for Fiban N-2. $C_0(NO_2) = 10$ mg/m³; velocity of air filtration = 0.06 m/s.

and pharmaceutical industries, all of which require air with low relative humidity (<50%).

A number of publications and patents[135–142] describe sorption of hydrogen sulfide by OH⁻ forms of anion exchangers. Despite the positive aspects of using this process for removal of H_2S from gases, it is not useful for practical air purification because of unfavorable competition with carbon dioxide, whose concentration in ambient air is about 300 mg/m³ and greatly exceeds the MPC level of H_2S whose concentration in contaminated air is usually below 100 mg/m³. Furthermore, H_2S has a dissociation constant at the first step of three powers of magnitude lower than carbonic acid. It appears that only the Amberlist XN-1007 and A-27 ion exchange resins[138] can be used for this purpose because they selectively absorb H_2S in the presence of CO_2. Cation exchanger ions that form nonsoluble sulfides (Cu^{2+}, Ni^{2+}, and Zn^{2+}) in principle can be used for selective H_2S sorption[140,143], but the results reported do not show their practical use for air purification.

We suppose that the problem can be solved with the help of catalytic conversion of H_2S to elemental sulfur using fibrous ion exchangers as carriers of the catalyst. The Fe (III)–ethylenediamine tetraacetate (EDTA) complex is an efficient catalyst of this process[144]. Our work[145–147] suggested the removal of H_2S from the air via catalytic oxidation of H_2S to elemental sulfur by atmospheric oxygen in the presence of Fe(III) complexed with EDTA in alkaline solutions. The main advantage of this system is selective oxidation of H_2S to S^0 without formation of noticeable amounts of side products such as thiosulfates, sulfates, and sulfites. This process can be formulated as follows:

$$H_2S_{gas} \rightarrow H_2S_{solution} \tag{2.59}$$

$$H_2S + OH^- = HS^- + H_2O \tag{2.60}$$

$$2Fe^{3+} + EDTA + HS^- = S^0 + H^+ + 2Fe^{2+} + EDTA \tag{2.61}$$

$$2Fe^{2+} + EDTA + H_2O + 1/2\ O_2 = 2Fe^{3+} + EDTA + 2OH^- \tag{2.62}$$

The dry catalyst in these processes is inactive.

Clean rooms demand materials that can remove H_2S from air at ~50% relative humidity. For this purpose, a special fibrous catalyst was prepared by adding different humidifying agents to a fiber. Figure 2.45 illustrates the optimal composition of the catalytic fiber for H_2S removal from the air.

In pilot plant experiments with slit type contacts (tangential air flow in the slit between two planes with ion exchange fabric), the catalytic solution was used to wet the packed ion exchange bed. The degree of conversion of H_2S to sulfur was 70 to 90% if the residence time of the purified air in the mass exchange unit was 0.35 second. This is in good agreement with existing data[148]. It appeared that the degree of the air purification that depends on the time of contact with the catalytic solution τ is described by the following empirical equation:

$$C/C_0 = e^{-10 \times \tau} \tag{2.63}$$

The working time before regeneration or changing the impregnation solution is much longer than that for sorption processes. The catalytic system reduces its activity when access to the active component is mechanically blocked with colloid particles of sulfur.

Anion exchangers also absorb mercaptans and can be used for deodorizing air[149–151]. The filtering layer can be a few centimeters thick and should be wet with sodium chloride solution. Mercaptans in the ion exchanger may form disulfide or convert into relative alcohols. In cation exchangers containing heavy metal ions such as copper or silver, they form relative mercaptides.

Gaseous wastes of viscose production contaminate the air with carbon bisulfide and hydrogen sulfide. The air can be purified of these compounds by a strong base anion exchanger sprayed with an alkali solution[152, 153]. Carbon bisulfide can form with functional groups of some anion exchangers of the xanthogenate type and can be easily destroyed by solutions of diluted hydrogen chloride in the regeneration process[154].

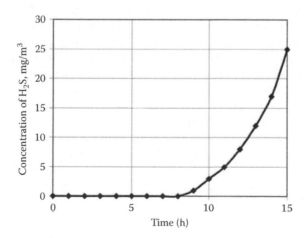

FIGURE 2.45 Breakthrough curve of H_2S on fibrous catalyst (Fiban AK-22 parent material). $C_0(H_2S) = 60$ mg/m³; thickness of single catalyst layer = 3 mm; catalyst contained 0.18 mmol Fe/g.

Chlorine—Disproportionate Cl_2 in air–aqueous media forms chloride ions:

$$Cl_2 + H_2O \rightarrow HOCl + HCl \qquad (2.64)$$

$$2HOCl \rightarrow 2HCl + O_2 \qquad (2.65)$$

The anion exchanger in alkaline form catalyzes these processes and finally converts to Cl^- form. Sorption of chlorine and bromine from the air by strong base anion exchangers has been studied[155, 156]. The protective action time of the ion exchange column depended strongly on relative humidity. A mechanism of the sorption process was suggested. The principles of chlorine sorption have been described[157–160]. The dynamic activity of the AV-17 strong base anion exchanger as a function of water content in the resin, chlorine concentration, and thickness of the filtering layer was also studied[157].

Removal of chlorine from the air by ion exchange fibers on bases of cellulose was studied[158] and the results led to development of a universal respirator to prevent dust and gas inhalation. The rate of sorption by a strong base anion exchanger from air was studied in relation to air humidity[159]. The same resin was studied for chlorine removal from the air in a fluidized bed. The influences of chlorine concentration and air flow rate were studied and the basic calculations for an air purifying plant were developed.

Bromine—An anion exchanger in bromide form readily absorbed bromine from the air and formed polybromide anions[161–163]. The effects of resin swelling, bromine concentration, and initial ionic form on sorption have also been studied.

Iodine—I is easily absorbed by anion exchangers in OH^- and halogenide ionic forms with the formation of polyiodine complexes. Because the removal of radioactive iodine from the air and gaseous exhausts of nuclear power stations is vital, iodine removal was studied extensively and described in several papers[164–169] and patents[170–173]. Especially efficient for iodine sorption were anion exchangers containing ternary amino groups and quaternary ammonium groups. The sorption of iodine can reach 2 g/g of ion exchanger. The removal of iodine from the anion exchanger after sorption can be achieved easily by a solution of reductants like hydrazine, sodium sulfite, or thiosulfate. Sorption of iodine by different ionic forms of fibrous polyacrylonitrile anion exchangers is especially efficient[167]. The sorption of iodine is fast and the amount absorbed greatly exceeds the exchange capacity of the ion exchanger.

Hydrazine—The application of strong acid cation exchangers to remove hydrazine from air was described[174,175]. The influences of main parameters controlling the dynamics of hydrazine sorption under different conditions (concentration of hydrazine, air flow rate, thickness of absorbing layer, type and form of ion exchanger) allowed construction of hydrazine filters with varying purifying capacities. The KU-2 industrial sulfonic cation exchanger exhibited high performance in purifying air from hydrazine.

Ozone—A method of air purification from ozone was developed on the basis of ozone reaction with iodide anion exchangers causing the formation of elemental iodine which is strongly held by the anion exchange resin[176]. Another method was based on reaction of hydrazine with ozone, leading to the formation of elemental nitrogen and oxygen[177]. The purifying air is passed through a sulfonic cation exchanger in the form of hydrazonium ions. The ozone interacts with the hydrazonium sorption complex with formation of hydrogen from the sulfonic groups and O_2, N_2 gases, and water according to the reaction:

$$R\text{-}SO_3^-N_2H_5^+ + 2O_3 \rightarrow R\text{-}SO_3H + N_2 + 2O_2 + 2H_2O \qquad (2.66)$$

The process is practically stoichiometric. It is significantly more efficient than the one based on the iodide form of anion exchanger.

Acetic acid—The air purification to remove the vapors of acetic acid was studied[178]. The dynamics of sorption was studied using the AV-17 strong base anion exchanger in the hydroxonium form and the carboxylic acid cation exchanger in sodium form. Both resins readily absorbed acetic acid (26 and 46 mass %). The sorption is high and independent of the relative humidity at $\alpha = 0.45 \div 0.95$, but rapidly decreases at lower relative humidity. After sorption, the resins can be partially regenerated with water, yielding 5% acetic acid solution.

Mercury—The vapor of mercury can be removed from the air by sodium forms of melamine–formaldehyde or melamine–carbamide–formaldehyde resins ion exchangers[179], as well as by sulfonic cation exchangers[180]. A method to remove mercury vapors and aerosols by fibrous ion exchangers in the forms of oxidative counterions was suggested[181]. Ions such as Fe^{3+} and MnO_4^- can serve as the oxidants. The AV-17 strong base anion exchanger in polyiodine ionic form was an efficient Hg sorbent[182]. Mercury vapor was suggested to absorb by the cation exchanger in the form of a divalent mercury counterion (Hg^{2+})[183]. The ion exchanger after sorption can be regenerated by nitric acid solutions. The Hg^{2+} is reduced to Hg^+ and the capacity of the sorbent is equal to the contents of the divalent Hg in the resin (~30% mass for KU-2). The regeneration also can be achieved by passing the Hg^{2+} solution through the filtering layer.

Hydrogen fluoride and silicon tetrafluoride—Air purification by removing hydrogen fluoride and silicon tetrafluoride is applicable mainly to phosphorous fertilizer and microelectronics production. These gases are often present together in the air. A number of works focused on using anion exchangers to remove them from the air. In a wet atmosphere, SiF_4 hydrolyzes and forms HF. Pure HF is readily absorbed from the air by anion exchangers. Its sorption mechanisms were studied extensively and described[184–186].

Strong base ion exchangers absorb HF in amounts that largely exceeded their exchange capacities due to formation of associates $(HF)_n$ in the ion phase, similar to the sorption of HCl discussed above. The Fiban AK-22-G weak base ion exchanger appeared a very efficient sorbent for purifying air of HF[2]. Its shallow filtering layers (8 mm) were able to purify air containing 20 to 800 mg HF/m³ at a flow rate of 8–10 m/min at relative humidity of 0.1 to 0.9. The breakthrough capacity of the ion exchanger was 160 mg/g, independent of concentration and relative humidity.

The latter phenomenon is a peculiarity of HF. The equilibrium capacity at 800 mg/m³ concentration reached almost 16 mmol HF/g. It can be explained by the strong solvating properties of HF. The superequivalently absorbed HF can be removed easily from the ion exchanger by water. Several fluor-containing substances (HF, SiF_4,

F_2) may be present simultaneously in the air. Several methods to purify the air by removal of HF and its accompanying substances were described[187,188]. The removal of HF from the air by carboxylic acid cation exchange fibers in salt forms was investigated[189]. Fluorides of the cations were formed in the phase of ion exchange and the carboxylate groups converted to hydrogen form. Simultaneous sorption of hydrogen fluoride and silicon tetrafluoride by a strong base anion exchanger was described[190–194].

2.5 APPLICATION OF ION EXCHANGERS FOR AIR PURIFICATION

2.5.1 CONSTRUCTION OF GASEOUS ION EXCHANGE FILTERS

Air purification with granular ion exchangers can be carried out on fixed, moving, or fluidized filtering beds as illustrated by Figure 2.46[195–197]. A stationary layer can provide deep air purification of the air flows, but the application of fixed bed devices in gaseous processes has serious limitations:

1. High aerodynamic resistance (1,500 to 4,000 Pa) at the minimal possible thickness of the filtering layer (100 to 200 mm) and filtration rate (0.1 to 0.3 m/s)
2. The need for removal of dust
3. A large change of resin bead size during sorption (regeneration, drying and wetting) cycles—especially significant for weakly dissociating resins during their transformation from nonionized hydrogen and hydroxonium forms to ionized salt forms

The efficiency of gas purification can be greatly improved if the sorption layer is sprayed with an acid or alkali solution to remove bases and acids, respectively. This can be done in cocurrent or countercurrent mode[198]. The spray converts the ion exchanger for continuous absorption to the active packed bed. At the same time, the resistance of the filtering layer to gas flow sharply increases and the problem of

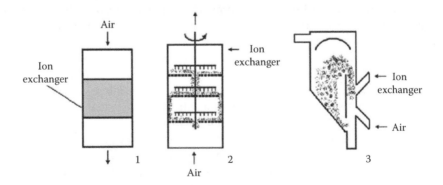

FIGURE 2.46 Variants of using granular ion exchangers in air purification systems. 1. Stationary layer. 2. Moving bed. 3. Fluidized bed.

generation of aerosol droplets in the purified air appears. The apparatus becomes more complex and the energy consumption increases.

The application of fluidized beds allows good reduction of aerodynamic resistance with a simultaneous increase of the filtration rate. In fluidized beds, dry and hot gaseous flows can be treated and the necessary air humidity can be controlled by spraying water into the gas flow. Conversely, fluidized beds impose strict requirements on resin quality related to their uniformity and resistance to mechanical and osmotic shocks. Fluidized bed technology is applicable to purification of gas flows containing 1 to 10 g/m^3 of the removed substance. Under this condition, the process is very efficient. Even in a one-step process, the saturation of the sorbent approaches equilibrium and the degree of gas purification reaches 95 to 99%. The regeneration of the ion exchanger is carried out in the usual column apparatus with a high aspect ratio of the ion exchanger bed (h:d > 10).

Systems utilizing continuous electrochemical regeneration in the cells formed by ion exchange membranes with ion exchange resins have been described[199]. In practice, all the described techniques of air purification are possible, but to date they have not found wide practical application.

Air filters utilizing nonwoven ion exchange canvases have practical applications. Their filtering elements use ion exchange fibers in thin (5 to 10 mm) layers to form large filter areas in relatively small volumes (30 m^2/m^3). The low linear air flow across the filtering layer (0.05 to 0.15 m/s) causes a low pressure drop on the filtering unit in combination with its high filtering capacity (~5,000m^3/h for a unit with a 1 m^3 filtering chamber). Such filters are suitable for removing active impurities in concentrations <300 mg/m^3.

Ion exchange filters for ventilation systems are rectangular chambers containing sets of parallel frames formed by layers of nonwoven ion exchange fabric. The frames form a system of slits with altering inlets and outlets as shown in Figure 2.47[117,118], connected with the feed and outlet pipes of the ventilation system, respectively. A regeneration system allows periodic or continuous functioning of the filter.

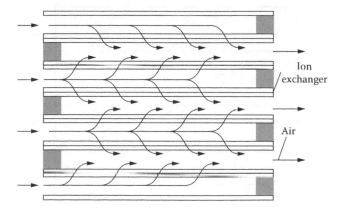

Ion
exchanger

Air

FIGURE 2.47 Placing ion exchange canvas in filtering chamber of frame type (RIF) ion exchange filter, top view.

Filtering devices with parallel slits formed by sorption-active materials and open from the inlet and outlet sides can be used when extremely low aerodynamic resistance of the filter is required and when the air contains dust[200]. Figure 2.48 illustrates the principles of this type of filtering unit. The air to be purified passes through the slits between the plates of the absorbent of the fibrous ion exchanger. The regenerating solution is delivered periodically or continuously to the tops of the plates and flows down inside the fibrous sorbent material. The filtering air and regenerating solution can move in crossflow or counterflow mode.

Filters for continuous gas purification utilizing moving bands of ion exchange material were described[201]. The filter has two chambers in which sorption and regeneration occur. The latter is achieved by moving the ion exchange band through the regeneration solution. The construction is complex and imposes severe requirements on mechanical properties of the filtering material. We did not find publications citing practical use of these systems.

A rotor type apparatus for continuous air purification was suggested[202,203]. The filtering unit is a cylinder with a wall of ion exchange fabric; the wall has openings. The cylinder is positioned horizontally and its lower part is placed into a bath with a regenerating solution. It has an axial tube used as an inlet for the air. The cylinder is sealed at the other end. Air enters the cylinder through the axial tube and exits as purified air through the wall into a collector. The cylinder rotates around the axial tube and the ion exchange material is continuously impregnated with the regenerating solution. The layer of ion exchanger moves relative to the air flow and regeneration solution. The construction restricts the filtration area and implies a high pressure loss on the filter. These systems are not useful for large scale air purification.

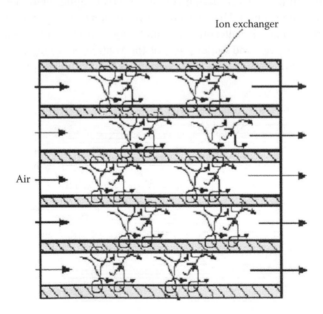

FIGURE 2.48 Top view of filtering chamber of ion exchange filter with parallel slits (FC type).

Ion exchange disposable filters are used for purification of air containing trace amounts of contaminants. The service time can be as long as several months or years and their regeneration at the point of use is not practical. Examples include purification of the air of clean rooms or protecting the air in residences from tobacco smoke and other contaminants. A number of patents and publications suggested systems and filtering elements of different types and shapes[204–207].

The most practical construction for a filtering element made from a nonwoven ion exchange fabric is a pleated plate containing an ion exchanger and supporting rigid polymeric air-permeable material with a filter area of ~1.5 m². The plates are assembled in a filtering element installed in a conventional clean room. Such filters are produced by M+W Zander and Purofil and are used to purify atmospheric air of trace amounts of ammonia, sulfur dioxide, and other impurities. Different types of Fiban materials are used in such filters (Figure 2.49). Strong acid and strong base ion exchange resins are also applied to filtering elements of this type. They are integrated into air-permeable polyurethane foam or glued between the layers of inert fibrous materials.

2.5.2 FILTERING ON ION EXCHANGE FABRICS

Ion exchange filters of types RIF and CF have practical uses for purifying industrial exhaust air containing ammonia and acid gases. These filters are intended for integration into existing ventilation systems in industries and are produced in Belarus (RIF and FC), Russia (IVF), and Japan (EPIX). We will use the examples of Fiban filter plants to illustrate the construction and operation of these filters. Figure 2.50 is a simplified schematic drawing.

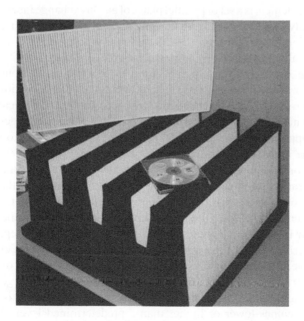

FIGURE 2.49   Filtering element and filter pack with Fiban material used in clean rooms of electronics industries. (Courtesy of M+W Zander Products GmbH, Stuttgart, Germany.)

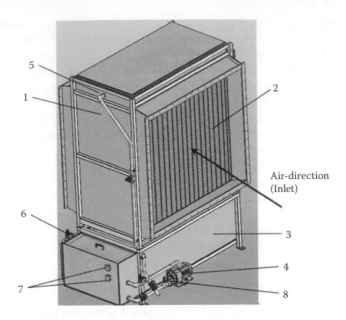

FIGURE 2.50   RIF type Fiban filtering plant. 1. Rectangular housing unit. 2. Chamber. 3. Regeneration solution tank. 4. Pump. 5. Upper chamber. 6. Water supply system. 7. Liquid level sensors. 8. pH sensor.

The filtering element is a set of parallel plates of an ion exchange fabric (Figures 2.47 and 2.48) rigidly joined in a rectangular unit (2) and placed in a chamber (1). The air to be purified passes through the inlet diffuser to the outlet. The regeneration solution is contained in a tank (3). For alkaline contaminants such as ammonia, the solution can be 3 to 7% sulfuric or phosphoric acid; a sodium carbonate solution is usually used for acidic contaminants. When contaminant concentration in the outlet approaches a predetermined level detected by a gas sensor or chemical method, the pump (4) is switched on for a short time and transfers part of the regeneration solution into the upper chamber (5). The pump capacity must allow it to work for 1 to 3 minutes. The upper chamber construction ensures that the regenerating solution slowly flows along the filtering material down to the regeneration tank (3).

The regeneration solution impregnating the filter material (capillary liquid and absorbed solution) and the functional groups of the ion exchanger interact with the contaminant, removing it from the air. When the concentration of the contaminant in the outlet air increases to a predetermined level, regeneration is repeated. During the working period, some of the water from the regeneration system evaporates and should be replaced from the water supply system (6) after the signal of a liquid level sensor (7) in the tank. The concentration of the acid or base in the regeneration tank decreases during the working period. It is measured by a glass electrode pH sensor (8). When it becomes lower or higher than a predetermined level, the regenerant solution is pumped to a receiver and a fresh regenerating solution is pumped into the tank (3). Figure 2.51 is a photograph of the filtering plant.

FIGURE 2.51 Filter of RIF-10 000. (Courtesy of IMT Ltd., Minsk, Belarus.)

The filters can include additional equipment such as sensors at the outlet and inlet, a purifying air humidifier, a monitoring system, and other devices. The filtering plant can be designed for fully or partially automatic operations. In the simplest case, when concentration of the contaminant in the air is approximately constant, the working cycle of the filter is programmed so that the regeneration is switched at equal time intervals for equal periods.

The working regime of an ion exchange filter is illustrated with a typical example. The exhaust air of a galvanic workshop contains acidic impurities in concentrations ~50 ppm. The air should be purified to the permitted level before release to the atmosphere. The amount of the air to be purified is 10,000 m³/h.

The RIF-10 filtering plant is suitable for this purpose. Its mass (exclusive of the regeneration solution) is 735 kg. The area needed for installation of the square shaped filtering plant is 2.1 m²; the height is 2.5 m. The filtering chamber has a volume of 2.5 m³ and contains 80 m² of ion exchange canvas 10 mm thick, with a total mass of 80 kg. The regeneration tank contains 600 liters of the regeneration solution (5% Na_2CO_3). The power of the pump is 750 W. During regeneration, it pumps 250 liters of solution per regeneration. The time between regenerations is constant (30 minutes). The pump works 1 minute in the regeneration mode. The filter purifies incoming air to a level <2 ppm. The regenerating solution is replaced when its pH reaches 8. Under these conditions, replacement is required after 120 hours of continuous work. The amount of the purified air between changes of the regenerant is 1.2 million m³. The amount of water added to the regenerating solution is 6 m³. Filters of this type with capacities of 5,000 to 20,000 m³/h have been installed in many industries in Russia, Belarus, Ukraine, and other countries.

REFERENCES

1. IEC. Report on Cincinnati ACS Meeting. *Ind. Eng. Chem.* 47, 5, 1955.
2. Soldatov, V.S., Elinson, I.S. and Shunkevich, A.A. Application of fibrous ion exchangers in air purification from acidic impurities. In *Proceedings of International Symposium on Hydrometallurgy.* Chapman & Hall, London, 1994, 837.
3. Soldatov, V.S. et al. Air pollution control with fibrous ion exchangers. In *Chemistry for the Protection of the Environment.* Plenum, New York, 1996, 2, 107.
4. Turkolmez, S. Application of ion exchange resins for air deodoration. Part I. *Wasser Luft Betrieb* 9, 737, 1965.
5. Turkolmez, S. Application of ion exchange resins for air deodoration. Part II. *Wasser Luft Betrieb* 9, 812, 1965.
6. Gostomczyk, M.A. and Kuropka, J. Investigation of sorption of acid gases by anion exchangers, *Environ. Prot. Eng. (Poland)* 1, 135, 1977.
7. Bogatyriov, V.K. *Ion Exchangers in Mixed Layers (Ionity v smeshannom sloje).* Khimia, Leningrad, 1968, 175.
8. Urano, K. and Mizuuchi, A. Application of ion exchange resins in air purification. *Kobun. High Polym. Jap.* 21, 594, 1972.
9. Hashida, J. Investigation of ion exchange resins, adsorbing gases, *Bull. Osaka Munic. Technol. Res. Inst.* 50, 117, 1974.
10. Lebedev, K.B. et al. *Ion Exchangers in Color Metallurgy (Ionity v tsvetnoj metallurgii).* Khimia, Moscow, 1975, 284.
11. Vulikh, A.I., Aloviajnikov, A.A., and Nikandrov, G.A. A new application of ion exchangers: gas purification. In *Ion Exchange (Ionnij obmen).* Khimia, Moscow, 1981, 214.
12. Ashirov, A. *Ion Exchange Purification of Wastewaters, Solutions and Gases (Ionoobmennaya ochistka stochnykh vod, rastvorov i gazov),* Khimia, Leningrad, 1983, 295.
13. Chikin, G.A. and Miagkoj, O.N. Ion exchangers in gas sorption technologies. In *Ion Exchange Methods of Substance Purification (Ionoobmennye metody ochistki veschestv).* Voronezh University, 1984, 326.
14. Vulikh, A.I., Bogatyriov, V.A. and Aloviajnikov, A.A. Application of ion exchange resins for sorption and purification of gases, *Zh. Vsesojuz. Khimich. obschestva im. Mendeleeva* 15, 425, 1970.
15. Shramban, B.I. Investigation of the process of sorption of hydrogen fluoride from the gas phase by anion exchanger AV-17. Dissertation, MITCT, Moscow, 1972.
16. Soldatov, V.S. Syntheses and the main properties of Fiban fibrous ion exchangers, *Solvent Extract. Ion Exch.* 26, 513, 2008.
17. Soldatov, V. et. al. *New Materials and Technologies for Environmental Engineering. Part I. Structure and Syntheses of Ion Exchange Fibers,* Monographs of Polish Academy of Sciences, Lublin, 2004, 21, 130.
18. Zagorodni, A.A. *Ion Exchange Materials: Properties and Applications.* Elsevier, Oxford, 2007, 477.
19. Soldatov, V.S. et al. Sorption of water vapor by salt forms of fibrous anion exchanger Fiban A-1. *Russ. J. Appl. Chem. (Zh. Prikladnoj Khimii)* 63, 2285, 1990.
20. Clarence, R.C., Lake, J., and Leighton, S.M. U.S. Patent 3,275,549, 1966.
21. Radl, V. and Krejkar, E. Cation exchangers as drying agents for gases and liquids. *Chem. Prum. (Poland)* 12, 579, 1962.
22. Wymore, C.E. Sulfonic type cation exchange resins as desiccants. *Ind. Eng. Chem. Prod. Dev.* 1, 173, 1962.
23. Shamilov, T.O. et al. Gas drying by fibrous anion exchangers. *Chem. Industr. (Khimicheskaja promyshlennost)* 3, 181, 1980.

24. Miagkoj, O.N. et al. Gas drying by fibrous anion exchangers. In *Application of Ion Exchange Materials (Primenenie ionoobmennykh materialov)*. Voronezh University, 1981, 50.
25. Glueckauf, E. and Kitt, G.P. A theoretical treatment of cation exchangers. III. The hydration of cations in polystyrene sulphonates. *Proc. Roy. Soc.* 228, 322, 1955.
26. Gregor, H.P. et al. Studies on ion-exchange resins. V. Water vapor sorption. *J. Colloid. Sci.* 7, 511, 1952.
27. Boyd, G.E. and Soldano, B.A. Osmotic free energies of ion exchangers. *Z. Electrochem.* 57, 162, 1953.
28. Gregor, H.P. and Frederick, M. Thermodynamic properties of ion exchange resins: Free energy of swelling as related to ion selectivities. *Ann. N.Y. Acad. Sci.* 57, 87, 1953.
29. Van Krevelen, D.W. *Properties of Polymers: Correlations with Chemical Structure.* Translated from Russian. Khimia, Moscow, 1976, 416.
30. Chalych, A.E., *Diffusion in Polymeric Systems (Diffuzia v polimernykh sistemakh).* Khimia, Moscow, 1987, 312.
31. Soldatov, V.S. et al. Quantitative description of water sorption by salt forms of fibrous anion exchanger Fiban A-1. *Russ. J. Appl. Chem. (Zh. Prikladnoj Khimii)* 63, 2291, 1990.
32. Gantman, A.I. and Veshev, S.A. Water uptake and swelling of ion exchangers. *Russ. J. Phys. Chem. (Zh. Fizicheskoj Khimii)* 59, 2615, 1985.
33. Sosinovich, Z. et al., Investigation of the hydration of strong anion exchangers using the model of stepwise hydration. *Dokl. Akad. Nauk. BSSR* 22, 920, 1978.
34. Katz, B.M., Kutarov, V.V. and Kutovaja, L.M. Kinetics of water vapor sorption by anion exchange fiber on the base of cellulose and polyacrylonitrile. *Russ. J. Appl. Chem. (Zh. Prikladnoj Khimii)* 64, 1713, 1991.
35. Brunauer, S. *Adsorption of Gases and Vapors*, Vol. 1. Humphrey Milford, London, 1943, 512.
36. Jovanovich, D.S. Physical adsorption of gases. *Kolloid Z. Z. Polym.* 235, 1, 1203, 1969.
37. White, H.J. and Eyring, H. Adsorption of water by swelling high polymeric materials. *Text. Res. J.* 17, 523, 1947.
38. Sosinovich, Z. et al. Thermodynamics of water sorption on Dowex 1 of different cross-linking and ionic forms. In *Ion Exchange and Solvent Extraction*, Marcel Dekker, New York, 1985, 9, 303.
39. Popovski, D. and Mitrevski, V. Method for generating water sorption isotherm models. *Electron. J. Environ. Agric. Food Chem.*, 4, 945, 2005.
40. Soldatov, V.S. and Kosandrovich, E.G. A new equation of isopiestic curve for polyelectrolytes. *Dokl. NAN Bel.* 49, 66, 2005.
41. Soldatov, V.S. and Kosandrovich, E.G. Chemical equilibria between the ion exchanger and gas phase. In *Recent Advances in Ion Exchange Theory and Practice. Proceedings of IEX.* Fitzwilliam College, Cambridge, July 9–11, 2008, 103.
42. Kirgintsev, A.N., Lukjanov, A.V., and Vulikh, A.I. Isotherms of water adsorption by KU-2 resin in different ionic forms. In *Ion Exchangers and Ion Exchange (Ionity i ionnyj obmen)*, Voronezh University, 1966, 34.
43. Soldatov, V.S. Quantitative presentation of potentiometric titration curves of ion exchangers. *Ind. Eng. Chem. Res.*, 34, 2605, 1995.
44. Soldatov, V.S. Potentiometric titration of ion exchangers, *React. Funct. Polymers* 38, 112, 1998.
45. Kosandrovich, E.G. Sorption of ammonia and sulfur dioxide by fibrous ion exchangers. Dissertation, Minsk, 2005.
46. McRae, W.A. and Rigopulos, P.N. U.S. Patent 3,330,750, 1967.

47. Vulikh, A.I., Zagorskaya, M.K., and Bogatiriov, V.L. USSR Aut. Certificate 222340, 1968.
48. Krejkar, E. Sorpce plynneho amoniaku na katexech. *Chem. Prum. (Poland)*, 13, 110, 1963.
49. Vulikh, A.I. et al. Sorption of ammonia and chlorine by ion exchange resins in dynamic conditions. *Dokl. AN SSSR* 160, 1072, 1965.
50. Vulikh, A.I., Zagorskaya, M.K., and Varlamova, L.V. Sorption of ammonia from the gas phase by carboxylic acid cation exchangers. *Ind. Sanitary Gas Purif. (Promyshlennaya i sanitarnaya ochistka gasov)*, 6, 14, 1984.
51. Vulikh, A.I. et al. Sorption of ammonia from the gas phase by cation exchangers in form of complex forming ions. In *Proceedings of Conference on Ion Exchange*, Moscow, 1979, 208.
52. Vulikh, A.I. and Zagorskaya, M.K. Sorption of ammonia by salts forms of cation exchangers. In *Proceedings of Sixth Conference on Synthesis, Analysis, and Studies of Pure Compounds of Rare Metals*, Novosibirsk, 1968, 14.
53. Jiro, O. et al. Synthesis of a new type adsorbent for the removal of toxic gas by radiation-induced graft polymerization. *Int. J. Rad. Appl. Instrum. C* 35, 113, 1990.
54. Lazarev, M.Y. and Katz, B.M. Sorption of HCl and NH$_3$ by sulphonic cation exchanger KU-23 in different ionic forms. *Theory Pract. Sorption Proc. (Teorija i praktika sorbtsionnykh protsessov)*, 18, 74, 1986.
55. Kamata, S. and Fashiro, M. Sorption of ammonia by ion exchangers. *J. Chem. Soc. Jap. Ind. Chem. Sect.*, 73, 19, 1970.
56. Seredin, B.I. and Nikolaev, N.I. Features of diffusion and interaction of ammonia with copper ions in the cation exchanger, *Russ. J. Phys. Chem. (Zh. phizicheskoy khimii)* 46, 99, 1972.
57. Katz, B.M. et al. Dynamics and mechanism of ammonia sorption by sulphonic cation exchanger KU-23 in different ionic forms. In *Application of Ion Exchange Materials (Primenenie ionoobmennykh materialov)*, Voronezh University, Voronezh, 1981, 28.
58. Varenko, V.P. et al. Modified ion exchange resin–chemical absorbent of ammonia and methyleneamine. *Russ. J. Appl. Chem. (Zh. Prikladnoj Khimii)* 52, 2199, 1979.
59. Katz, B.M., Olontsev, V.F., and Lazarev, M. Purification of air–gas mixtures by sulphonic cation exchanger KU-23. *Russ. J. Appl. Chem. (Zh. Prikladnoj Khimii)* 56, 1886, 1983.
60. Vdovina, G.P. et al. Studies of ammonia sorption by ion exchangers from air–gas mixtures. *Russ. J. Appl. Chem. (Zh. Prikladnoj Khimii)* 55, 2580, 1982.
61. Inoue, Hiroshi et al. Japan Patent 1,621,244, 2006.
62. Zverev, M.P. Chemosorption fibers: Vion, a prospective material for the environment protection. *Industr. Energ. (Promyshlennaya energetika)*, 2, 12, 1982.
63. Kurilenko, O.D. et al. Ion exchange fibrous materials on the base of cellulose in gas purification. *Vestnik AN USSR* 7, 37, 1975.
64. Vulikh, A.I. et al. Purification of vent gases by fibrous ion exchange sorbents. *Color Metall. (Tsvetnye metally)*, 4, 38, 1981.
65. Ermolenko, I.N. and Lubliner, I.P. Sorption of ammonia by fibrous ion exchangers at dynamic conditions. *Russ. J. Appl. Chem. (Zh. Prikladnoj Khimii)* 45, 748, 1972.
66. Nizovtseva, O.P. and Shunkevich, A.A. Sorption of ammonia by fibrous cation exchangers in H$^+$-form. *Vestik AN BSSR* 5, 62, 1978.
67. Malinovskij, E.K. et al. Effect of exchange capacity of carboxylic acid fiber Vion KN-1 on its sorption and physical mechanical properties. *Russ. J. Appl. Chem. (Zh. Prikladnoj Khimii)* 63, 64, 1990.
68. Soldatov, V.S. Chemically active textile materials for filtration and purification of gases and liquids. *Proceedings of Seventh World Filtration Congress*, Budapest, May 1996, 1, 213.

69. Soldatov, V.S. et al. Air pollution control with fibrous ion exchangers. In *Environmental Science Research*. Plenum, London, 1996, 51, 55.
70. Soldatov, V.S. New fibrous ion exchangers for purification of liquids and gases. In *Studies in Environmental Science*. Elsevier, Amsterdam, 1984, 23, 353.
71. Harjula, R. and Lehto, J. Memorandum of International Workshop on Uniform and Reliable Nomenclature, Formulations and Experimentation for Ion Exchange, Helsinki, May 30–June 1, 1994. *React. Funct. Polymers* 27, 147, 1995.
72. Polhovski, E.M. and Soldatov, V.S. Non-exchange sorption of electrolytes by ion exchangers. II: Sorption of sulfuric acid and lithium sulfate by Dowex 1 × 8 resin. *React. Funct. Polymers* 60, 49, 2004.
73. Polhovski, E.M. and Soldatov, V.S. Sorption of inorganic acids by fibrous ion exchanger on the base of hydrozinated nitrogen. *Vestik NAN Bel. Ser. Khim. Nauk.* 5, 37, 2003.
74. Perry, L.J. and Le Van, M.D. Air purification in closed environments: Overview of spacecraft systems. *www.natick.army.mil/soldier/jocotas/ColPro_Papers/Perry-LeVan.pdf*, 12.05.2009, 1.
75. Brown, D.L., Glass, W., and Greatorex, J.L. Performance of an electrochemical device for simultaneous carbon dioxide removal and oxygen generation. *Chem. Eng. Progr. Sympos. Ser.* 62, 50, 1966.
76. Kester, F.L. U.S. Patent 3,659,400, 1972.
77. Vulikh, A.I. and Arkhipov, S.V. Sorption of carbon dioxide by carbonate form of anion exchanger AV-17. *Russ. J. Appl. Chem. (Zh. Prikladnoj Khimii)* 1, 216, 1968.
78. Arkhipov, S.V. and Vulikh, A.I. Equilibrium and mechanism of reaction in system anion exchanger AV-17 in carbonate form: Carbonic acid. In *Thermodynamics of Ion Exchange (Termodinamika Ionnogo Obmena)*. Nauka i Technika, Minsk, 1968, 207.
79. Miagkoj, O.N., Krutskih, A.S., and Meleshko, V.P. Reagent-free regeneration of anion exchangers used for carbon dioxide sorption. In *Theory and Practice of Sorption Processes (Teoriya i praktika sorbtsionnykh protsessov)*. Voronezh University, 1975, 10, 106.
80. Chizhevskaya, A.B. et al. Characteristics of anion exchangers on the ability to absorb carbonate ions and carbon dioxide. *Russ. J. Appl. Chem. (Zh. Prikladnoj Khimii)* 50, 1776, 1977.
81. Elinson, I.S., Chizhcvskaya, A.B., and Soldatov, V.S. Dynamics of carbon dioxide sorption by anion exchanger AV-17 × 8 in carbonate form. *Russ. J. Appl. Chem. (Zh. Prikladnoj Khimii)* 50, 2228, 1977.
82. Elinson, I.S. et al. Sorption of carbon dioxide by anion exchanger AV-17. In *Ion Exchange and Chromatography (Ionnij obmen i khromatografia)*. Voronezh University, 1976, 254.
83. Tsigankova, A.V. et al. Sorption of carbon dioxide by strong base anion exchangers with different cross-linkage. *Russ. J. Appl. Chem. (Zh. Prikladnoj Khimii)* 53, 1, 74, 1980.
84. Elinson, I.S. et al. Influence of the air humidity on sorption of carbon dioxide by anion exchangers. *Russ. J. Appl. Chem. (Zh. Prikladnoj Khimii)* 53, 1237, 1980.
85. United Kingdom Patent 1,296,888, 1969.
86. Beliakova, L.D. et al. USSR Aut. Certificate 361620, 1971.
87. Beliakova, L.D. et al. Adsorption of carbon dioxide by macroporous anion exchangers. *Dokl. AN USSR* 213, 1311, 1973.
88. Tepper, F. et al., Regenerable carbon dioxide sorbents. *Aero. Med.* 40, 3, 291, 1969.
89. Smart, R.C. and Derrick, W.S. Research on sorption of carbon dioxide: properties of ion exchange resins. *Anesthesiology* 18, 216, 1957.
90. Vulikh, A.I. and Arkhipov, S.V. Sorption of carbonic acid by anion exchanger AV-17 in carbonate form. *Russ. J. Appl. Chem. (Zh. Prikladnoj Khimii)* 1, 216, 1968.

91. Malin, K.M. *Handbook of Sulfuric Acid (Spravochnik sernokislotchika)*, 2nd ed. Khimia, Moscow, 1971, 744.
92. Strauss, W., *Industrial Gas Cleaning*. Translated from English. Khimia, Moscow, 1981, 616.
93. Vulikh, A.I. and Nikandrov, G.A. USSR Aut. Certificate 331807, 1972.
94. Krejcar, E., Czech Patent 107,940, 1963.
95. Glowiak, B. and Gostomczyk, A. Investigation of sulfur dioxide sorption on anion exchangers. *Wiad. Chem. (Poland)* 5, 399, 1968.
96. Glowiak, B. and Gostomczyk, A. Sorption of sulfur dioxide by anion exchange resins. In *Proceedings of Second International Clean Air Congress*. Academic Press, New York, 1971, 771.
97. Glowiak, B. and Gostomczyk, A. Investigation of effectiveness of SO_2 sorption by ion exchange resins. I. Laboratory studies. *Staub-Reinhalt Luft* 33, 29, 1973.
98. Glowiak, B. and Gostomczyk, A. The investigation of effectiveness of SO_2 sorption by ion exchange resins. II. Pilot plant studies. *Staub-Reinhalt Luft* 33, 387, 1973.
99. Beliakova, L.D. et al. Adsorption of sulfur dioxide by macroporous vinylpyridine anion exchangers. *Coll. J. (Kolloidnyj zhurnal)* 38, 1060, 1976.
100. Smola, V.I. and Keltsev, N.V. Sorption of sulfur dioxide by ion exchange resins. In *Atmosphere Protection from Sulfur Dioxide (Zaschita atmosfery ot dvuokisi sery)*. Metallurgia, Moscow, 1976, 223.
101. Cole, R. and Shulman, H.L. Adsorbing sulfur dioxide on dry ion exchange resins for reducing air pollution. *Industr. Eng. Chem.* 52, 859, 1960.
102. Layton, L. and Youngquist, G.R. Sorption of sulfur dioxide by ion exchange resins. *Industr. Eng. Chem. Fundament.* 8, 317, 1969.
103. Youngquist, G.R. and Garg, S.R. Sorption of sulfur dioxide by microreticular ion exchange resins. *Ind. Eng. Chem. Process Des. and Develop.* 11, 259, 1972.
104. Hashida, I. and Nishimura, M. Adsorption of sulfur dioxide from the gas phase on m-aminophenol condensation resins. *J. Chem. Soc. Jap., Chem. Industr. Chem.* 1, 179, 1972.
105. Avgul, N.N. et al. The heat of adsorption of sulfur dioxide by macroporous weak base anion exchangers and determination of their porous structure. *Coll. J. (Kolloidnyj zhurnal)* 39, 339, 1977.
106. Hashida, I. and Nishimura, M. Adsorption of sulfur dioxide by porous polyvinylpyridine resins. *J. Chem. Soc. Jap. Chem. Industr. Chem.* 6, 1195, 1973.
107. Avgul, N.N. et al. The heat of adsorption of sulfur dioxide by macroporous weak base anion exchangers. *Coll. J. (Kolloidnyj zhurnal)* 38, 129, 1976.
108. Hashida, I. and Nishimura, M. Adsorption and desorption of sulfur dioxide by macroreticular strong base anion exchanger *Sci. Ind.* 50, 131, 1973.
109. Pinaev, A.V. and Muromtseva, L.S. Sorption of sulfur dioxide by synthetic resins. *Russ. J. Appl. Chem. (Zh. Prikladnoj Khimii)* 41, 2092, 1968.
110. Nikandrov, G.A., Vulikh, A.I., and Zagorskaya, M.K. Sorption of sulfur dioxide by anion exchangers. In *Ion Exchange and Chromatography (Ionnyi obmen i chromatografia)*. Voronezh University, 1971, 15, 218.
111. Subbotin, A.I. and Tkachenko, V.I. Adsorption of sulfur dioxide by ion exchangers KB-4 and KNP-1B. *Chem. Ind. (Khimicheskaya promyshlennost)* 9, 681, 1976.
112. Beasley, G.H. and Cartier, P.G. U.S. Patent 3,945,811, 1976.
113. Vulikh, A.I. et al. Sorption of SO_2 by fixed layers of granular anion exchangers from wet gaseous mixtures. *Russ. J. Appl. Chem. (Zh. Prikladnoj Khimii)* 55, 1297, 1982.
114. Varlamova, L.V., Ksenzenko, V.I., and Aloviajnikov, A.A. Dynamics of sulfur dioxide sorption on fibrous cation exchanger Vion KN-1. *Ind. Sanitary Gas Purif. (Promyshlennaya i sanitarnaya ochistka gazov)* 6, 11, 1984.

115. Bergen, R.L. Removal of SO_2 from the gas mixture by the fibers containing polymeric amines. In *Technology for the Future to Control Industrial and Urban Wastes*. Special Centennial Symposium, University of Missouri. Rolla, 1971, 73.

116. *Fibers with Special Properties (Volokna s osobymi svojstvami)*. Khimia, Leningrad, 1980, 240.

117. Soldatov, V.S., Shunkevich, A.A., and Sergeev, G.I. Synthesis, structure and properties of new fibrous ion exchangers. *React. Polymers* 7, 159, 1988.

118. Shunkevich, A.A., Sergeev, G.I., and Elinson, I.S. Fibrous ion exchangers in protection of the atmosphere. *Zh. vsesoyuznogo khimicheskogo obschestva im. Mendeleeva* 35, 64, 1990.

119. Kosandrovich, E.G. and Soldatov, V.S. Sorption of sulfur dioxide from the air by fibrous polyampholyte. In *Proceedings of Third Belarus Conference on Scientific Technical Problems of Production of Chemical Fibers in Belarus*. Mogilev, 2006, 75.

120. Vulikh, A.I. and Aloviajnikov, A.A. Sanitary air purification by ion exchange sorbents. *Proc. GINTSVETMET* 47, 194, 1979.

121. Barash, A.N., Zverev, M.P., and Kalianova, N.F. Exploitation properties of chemosorption fiber Vion AN-1. *Chem. Fibers (Khimicheskie volokna)* 3, 37, 1987.

122. Zverev, M.P., Barash, A.N., and Grebennikov, S.F. Processes of mass exchange in gas sorption by fibers Vion AN-1 and Vion AS-1. *Chem. Fibers (Khimicheskie volokna)* 6, 9, 1988.

123. Baskin, Z.L. et al., Dynamics of the sorption of hydrogen chloride and ammonia on the carboxyl group-containing fiber Vion KN-1. *Fibre Chem.* 19, 15, 1988.

124. Katz, B.M., Lazarev, M.Y., and Malinovskij, E.K. Effect of the capacity and porosity of ion exchanger AN-221 on sorption of SiF_4 and HCl. *Russ. J. Appl. Chem. (Zh. Prikladnoj Khimii)* 53, 1175, 1980.

125. Katz, B.M., Lazarev, M.Y. Swelling of weak base macroporous anion exchangers at sorption of HCl gas and water vapors. *Russ. J. Appl. Chem. (Zh. Prikladnoj Khimii)* 55, 1971, 1982.

126. Katz, B.M. et al., Sorption of HCl by weak base macroporous anion exchanger AN-511. *Russ. J. Appl. Chem. (Zh. Prikladnoj Khimii)* 55, 2093, 1982.

127. Hashida, I. and Nishimura, M. Adsorption of hydrogen chloride on porous resins with different functional groups. *J. Chem. Soc. Jap.* 3, 569, 1975.

128. Khodakov, Y. S. *Nitrogen Oxides and Heat Energetics: Problem Solving (Oksidy azota i teploenergetika: problemy i reshenia)*. EST-M, Moscow, 2002, 184.

129. Ismagilov, Z.R., Kerzhentsev, M.A., and Susharina, T.L. Catalytic methods of reducing nitrogen oxides emissions at burning fuels. *Russ. Chem. Rev. (Uspekhi khimii)* 59, 1676, 1990.

130. Nabieva, Z.M., Merenkov, K.V., and Yusipov, M.M. Nitrogen oxide sorption from exhaust by ion exchange adsorbents. In *Proceedings: Ion Exchange Materials and Their Applications (Ionoobmennye materially i ikh primenenie)*. Alma-Ata, 1968, 232.

131. Vulikh, A.I. et al. Sorption of nitrogen dioxide by the alkaline form of anion exchanger AV-17. In *Ion Exchange and Chromatography (Ionnij Obmen i Khromatografia)*. Voronezh University, 1976, 256.

132. Ganzha, G.F. et al., Sanitary air purification from nitrogen oxides by fibrous ion exchange materials. *Industr. Sanitary Gas Purif. (Promyshlennaya i sanitarnaya ochistka gazov)* 3, 14, 1978.

133. Soldatov, V.S. et al. Air purification from nitrogen dioxide by aqueous solution of carbamide on ion exchange fibrous carrier. *Chem. Technol. (Khimicheskaya tekhnologia)* 5, 34, 2006.

134. Zaitsev, V.A. et al. The furnace gases of heat electrical power stations, *Chem. Industr. (Khimicheskaya promyshlennost)* 3, 39, 1993.

135. Vulikh, A.I., Lukjanova, G.V., and Bogatirjov, V.L. USSR Aut. Certificate 234595, 1969.
136. Alovjainikov, A.A., Vulikh, A.I., and Riabikina, L.G. The sorption of hydrogen sulphide by ion exchangers. *Theory Pract. Sorption Proc. (Teorija i praktika sorbtsionnykh protsessov)* 13, 99, 1980.
137. Shimko, I.F. et al. Purification of low-concentrated vent gases of viscose productions from carbon bisulfide and hydrogen sulfide. *Chem. Fibers (Khimich. Volok.)* 6, 6, 1984.
138. Rohm & Haas Co. Ion exchange resins for removal hydrogen sulfide from air flows. *Sulphur* 77, 45, 1968.
139. Grebennikov, S.F. et al. Investigation of the gas exhaust of viscose productions from hydrogen sulfide, *Russ. J. Appl. Chem. (Zh. Prikladnoj Khimii)* 55, 354, 1982.
140. Lobanova, G.A. et al. Gas purification by metal forms of sulfonic cation exchanger KU-2. *Russ. J. Appl. Chem. (Zh. Prikladnoj Khimii)* 55, 2333, 1982.
141. Riemer, H. Switzerland Patent 563,799, 1975.
142. Pollio, F. and Kunin, R. Macroreticular ion exchange resins as hydrogen sulfide sorbents. *Ind. Eng. Chem. Prod. Res. Dev.* 7, 62, 1968.
143. Alovjainikov, A.A., Vulikh, A.I., and Riabikina L.G. Sorption of hydrogen sulfide by ion exchangers. *Theory Pract. Sorption Proc. (Teorija i praktika sorbtsionnykh protsessov)* 13, 99, 1980.
144. Nejaglov, A.A. et al. Kinetics and mechanism of the liquid phase oxidation of hydrogen sulfide by chelate complex of trivalent iron (Fe^{3+}-EDTA). *Kinet. Catal. (Kinetika i kataliz.)* 32, 548, 1991.
145. Potapova, L.L. et al., Oxidation of hydrogen sulfide on fibrous anion exchanger Fiban A-1 with coordination saturated complex Fe-EDTA. *Dokl. Akad. Nauk. Bel.* 42, 54, 1998.
146. Potapova, L.L. et al. Oxidation of hydrogen sulfide on fibrous anion exchangers with fixed Fe-EDTA complexes. *Russ. J. Appl. Chem. (Zh. Prikladnoj Khimii)* 73, 780, 2000.
147. Soldatov, V.S., Kashinsky, A.V., and Martinovich, V.I. Catalytic air purification from hydrogen sulfide by ion exchange fiber Fiban. *Chem. Technol. (Khimicheskaya tekhnologia)* 6, 2009, 55.
148. Gritsenko, A.I., Galanina, I.A., and Zinovieva, L.M. *Gas Purification from Sulfur Compounds at Exploitation of Gas Fields (Ochistka gazov of sernistykh soedinenij i ekspluatatia gazovykh mestorozhdenij).* Nedra, Moscow, 1985, 270.
149. Hasida, I. and Nisimura, M. Adsorption of mercaptans on ion exchange resins from the gas phase. *J. Chem. Soc. Jap. Industr. Chem. Sec,* 71, 1939, 1968.
150. Rutkovcki, J.D. Removal of mercaptans from waste gases. In *Deodoration of Waste Gases: Proceedings of Institute of Environment Protection of Wroclav Polytechnical Institute.* Poland, 1975, 28, 192.
151. Croydon, C.F., Studies on application of selected anion exchangers for mercaptan vapor removal from air/waste gases. In *Proceedings of Second World Filtration Congress,* London, 1979, 4.
152. Kuropka, J. and Gostomczyk, M.A. Investigations on kinetics of carbon disulfide sorption on anion exchangers. *Environ. Protect. Eng.* 4, 87, 1978.
153. Tsaplina, L.A. and Davankov, A.B. USSR Aut. Certificate 141856, 1961.
154. Tsaplina, L.A. and Davankov, A.B. Adsorption of hydrogen disulfide by anion exchange resins. *Russ. J. Appl. Chem. (Zh. Prikladnoj Khimii)* 30, 609, 1966.
155. Ksenzenko, V.I., Nikonova, I.N., and Zakgeim, A.Y. Investigation of mechanism of sorption of elemental haloids by anion exchanger. In *Ion Exchange and Chromatography (Ionnij Obmen i Khromatografia).* Voronezh University, 1982, 83.

156. Ksenzenko, V.I., Zilberg, G.A., and Galushenko, G.M. Air purification from small admixtures of chlorine and bromine on ion exchanger AV-17. In *Ion Exchange and Chromatography (Ionnij Obmen i Khromatografia)*. Voronezh University, 1982, 253.

157. Vulikh, A.I. et al. Sorption of ammonia and chlorine by ion exchange resins in dynamic conditions. *Dokl. AN SSSR*, 160, 1072, 1965.

158. Asaulova, T.A. et al. Air purification from chlorine and chlorine compounds by ion exchange fibrous materials. In *Ion Exchange and Chromatography (Ionnij Obmen i Khromatografia)*. Voronezh University, 1976, 253.

159. Galushenko, G.M. et al. Sorption of chlorine by ion exchanger AV-17 × 8. *Trans. Moscow Inst. Fine Chem. Technol. (Trudy Moskovskogo Instituta Tonkoj Khimicheskoj Tekhnologii)* 4, 74, 1974.

160. Chernova, S.P. and Ksenzenko V.I. On calculation of process of absorption of gases by solid sorbents. *Chem. Industr. (Khimicheskaya promyshlennost)* 5, 48, 1967.

161. Ksenzenko, V.I., Nikonova, I.N., and Chernova, S.P. The mechanism and kinetics of bromine sorption by anion exchanger AV-17. In *Ion Exchangers in Economics (Ionity v narodnom khoziajstve)*. Niintekhim, Moscow, 1973, 74.

162. Ksenzenko, V.I. and Nikonova, I.N. On the character of bromine sorption by ion exchanger AV-17. *Trans. Moscow Inst. Fine Chem. Technol. (Trudy Moskovskogo Instituta Tonkoj Khimicheskoj Tekhnologii)* Moscow, 1973, 122.

163. Irving, Y. and Wilson, P.D. The absorption of polybromide on an anion exchange resin. *J. Inorg. Nucl. Chem.* 26, 2235, 1964.

164. Giona, A.R. et al. Adsorption of iodine vapor by ion exchange resins. I. *Energia Electr.* 39, 419, 1962; II. ibid. 40, 649, 1963; III. ibid. 42, 33, 1965.

165. Nikolaev, A.V. et al. Sorption of iodine by anion exchange resins at dynamic conditions. *Dokl. AN SSSR* 167, 841, 1966.

166. Miagkoj, O.N., Meleshko, V.P., and Serdiukova, M.I. Desorption of elemental iodine from anion exchange resins by solutions of reductants. In *Ion Exchange and Chromatography (Ionnij Obmen i Khromatografia)*. Voronezh University, 1971, 241.

167. Serdiukova, M.I. et al. Sorption of iodine on fibrous polyacrylonitrile anion exchangers. In *Ion Exchange and Chromatography (Ionnij Obmen i Khromatografia)*. Voronezh University, 1976, 200.

168. Sugii, A., Ogawa, N., and Ogawa, H. Removal of radioactive iodine from wastes by column with ion exchange resin. *Radioisotopes*, 27, 654, 1978.

169. Giona, A.R. The process of extraction of radio-iodine from gaseous products of used nuclear fuel. *Ingegnere*, 35, 887, 1961.

170. Simidzu, H., Midzuuty, A. and Yokoyama, S. Japan Patent 48-34320, 1973.

171. Simidzu, H., Midzuuty, A. and Pokoyama, F. Japan Patent 48-37672, 1973.

172. Cejnar, F. Chez.SSR Patent 146,453, 1972.

173. Keener, R.L. and Kittle, P.A. U.S. Patent 3,943,229, 1976.

174. Miagkoj, O.N., Serdiukova, M.I., and Perunova, N.A. Sorption of hydrazine vapor by ion exchange materials. *Theory Pract. Sorption Proc. (Teoria i praktika sorbtsionnykh protsessov)* 15, 73, 1982.

175. Vulikh, A.I., Nazarov, V.I., and Zagorskaya, M.K. Sorption of hydrazine from the gas phase by cation exchanger KU-2. In *Applications of Ion Exchange Materials (Primenenije ionoobmennyh materialov)*. Voronezh University, 1981, 97.

176. Miagkoj, O.N., Petrunin, A.N., and Serdiukova, M.I. Deozonation of the air by anion exchangers In *Ion Exchange and Chromatography (Ionnij Obmen i Khromatografia)*. Voronezh University, 1976, 247.

177. Miagkoj, O.N. and Serdiukova, M.I. Destruction of ozone on cation exchanger KU-2 in hydrazonium form. *Chem. Chem. Technol. (Khimia i Khimicheskaya Technologia)* 22, 888, 1979.

178. Subbotin, A.I. and Tkachenko, V.I. Air purification from vapor of acetic acid by ion exchangers. *Plastics (Plasticheskije massy)* 5, 38, 1955.
179. Odanaka, Akira et al. Japan Patent 51-5638, 1976.
180. Miagkoj, O.N. et al. Regeneration of sulfonic ion exchanger used for sorption of vapor and aerosol of mercury. In *Ion Exchange and Chromatography (Ionnij Obmen i Khromatografia)*. Voronezh University, 1976, 260.
181. Skrepnik, V.A, Fedorovskaya, L.F., and Scherbina, K.F. USSR Aut. Certificate 2827000, 1981.
182. Vulikh, A.I., Bogatyriov, V.L., and Dubinina, E.G. USSR Aut. Certificate 1248619/23-26, 1968.
183. Vulikh, A.I. et al. USSR Aut. Certificate 311493, 1970.
184. Zagorskaya, M.K. et al. Distribution of hydrogen fluoride between the anion exchanger and gas phase. In *Ion Exchange and Chromatography (Ionnij Obmen i Khromatografija)*. Voronezh University, 1976, 85.
185. Shramban, B.I., Ksenzenko, V.I., and Zakheim, A.Y. Sorption of hydrogen fluoride by anion exchanger AV-17 × 8 from the gas phase. *Izvest. Vuz. Ser. Khim. Nauk.* 15, 1112, 1972.
186. Legenchenko, I.A. and Mikhailovina, C.K., Sorption of hydrogen fluoride by ion exchange resins. *Russ. J. Phys. Chem. (Zh. Fizicheskoj Khimii)* 49, 1530, 1975.
187. Shramban, B.I. et al. USSR Aut. Certificate 327936, 1972.
188. Vulikh, A.I., Zagorskaya, M.K., and Ksenzenko, V.I. Sorption of hydrogen fluoride from the gas phase. *Dokl. Akad. Nauk. SSSR*, 167, 1059, 1967.
189. Zagorskaya, M.K. et al. The air purification from hydrogen fluoride by carboxylic acid cation exchange fibers in salt forms. In *Ion Exchange and Chromatography (Ionnij Obmen i Khromatografija)*. Voronezh University, 1976, 249.
190. Shramban, B.I., Afonina, N.D., and Pavlukhina, L.D. Sorption of hydrogen fluoride and silicon tetrafluoride by anion exchanger AV-17. *Russ. J. Phys. Chem. (Zh. Fizicheskoj Khimii)* 49, 713, 1975.
191. Katz, B.M. and Malinovskij, E.K. The sorption ability of some ion exchange resins toward silicon tetrafluoride. *Izvest. Vuz. Ser. Khim. Nauk.* 20, 528, 1977.
192. Katz, B.M., Malinovskij, E.K., and Ennan, A.A. Sorption of silicon tetrafluoride by macroporous ion exchange resins. *Russ. J. Appl. Chem. (Zh. Prikladnoj Khimii)* 50, 1980, 1977.
193. Katz, B.M. and Malinovskij, E.K. Kinetics of silicon tetrafluoride sorption by weak base macroporous anion exchangers. *Russ. J. Appl. Chem. (Zh. Prikladnoj Khimii)* 51, 1989, 1978.
194. Pavlukhina, L.D., Shramban, B.I., and Ksenzenko, V.I. On comparative sorption ability of HF and SiF$_4$ molecules on anion exchanger AV-17. In *Ion Exchange and Chromatography (Ionnij Obmen i Khromatografija)*. Voronezh University, 1976, 260.
195. Beskov, V.S. and Safronov, V.S. *General Chemical Technology and Fundamentals of Industrial Ecology (Obschaya khimicheskaya tekhnologiya i osnovy promyshlennoj ekologii)*. Khimia, Moscow, 1999, 472.
196. Seniavin, M.M. *Ion Exchange (Ionnij obmen)*. Nauka, Moscow, 1981, 272.
197. Shulman, H.L. et al. Development of a continuous countercurrent fluid-solids contactor: Improvement of contacting efficiency. *Ind. Eng. Chem. Process Des. Dev.* 7, 493, 1968.
198. Temple, A.R. U.S. Patent 3,407,045, 1968.
199. Wallace, R.A. U.S. Patent 3,727,375, 1973.
200. Khamizov, R.Kh. and Tikhonov, N.A. On the possibility of purification of anodic gases of aluminum production from fluoride and sulfur compounds by the wet filtration on anion exchange fibrous materials. *Sorption Chromatogr. Proc. (Sorbtsionnye i khromatograficheskie protsessy)* 2, 331, 2002.

201. Messinger, H. and Daly, J.F. U.S. Patent 3,498,026, 1970.
202. Miagkoj, O.N., Krutskikh, A.S., and Zarodin, G.S. A rotor type apparatus for continuous ion exchange gas purification. *Appl. Ion Exchange Mater. (Primenenie ionoobmennykh materialov)* 5, 28, 1981.
203. Miagkoj, O.N. et al. A rotor apparatus for continuous ion exchange air purification from acid and base impurities, *Ser. Indust. Energ. Environ. Prot. Energ. Supply (Ser. Promyshlennaya energetika, okhrana okruzhauschej sredy i energosnabzhenie sudov)* 9, 56, 1982.
204. Takashi, K. et al. U.S. Patent 3,804,942, 1974.
205. Chang-S.L. et al. U.S. Patent 0039600A1, 2005.
206. Kinkead, D.A., Rezuke, R.W., and Higley, J.K. U.S. Patent 5,626,820, 1997.
207. Akira, T., Yioko, S., and Takashi, K. U.S. Patent 6,723,151B2, 2004.

3 Applications of Selective Ion Exchange for Perchlorate Removal, Recovery, and Environmental Forensics

Baohua Gu, John Karl Böhlke, Neil C. Sturchio, Paul B. Hatzinger, Andrew Jackson, Abelardo D. Beloso, Jr., Linnea J. Heraty, Yongrong Bian, Xin Jiang, and Gilbert M. Brown

CONTENTS

ABSTRACT

Perchlorate (ClO_4^-) is a widespread contaminant found in drinking water and groundwater that has far-reaching ramifications ranging from public health issues to potential liabilities arising from environmental clean-up requirements. The chapter summarizes recent developments in highly selective and regenerable ion exchange technologies for removing ClO_4^- from contaminated water. The technologies rely on a unique, highly specific resin to trap ClO_4^-. The resin is then regenerated and ClO_4^- is either destroyed or recovered—leading to significant cost reduction and waste minimization. The ability to recover trace quantities of pure ClO_4^- from contaminated media also allows identification of the sources of its contamination through stable isotope ratio analysis of chlorine and oxygen atoms. We provide detailed descriptions of the techniques for extracting, purifying, and crystallizing trace amounts of ClO_4^- and characterizing its isotopic composition for fingerprinting in the environment.

3.1   INTRODUCTION

Perchlorate (ClO_4^-) contamination in groundwater and surface water is an emerging environmental problem in the United States and abroad, with far-reaching ramifications ranging from public health concerns to potentially large impacts on agriculture, defense industries, and environmental remediation. Apart from its obvious anthropogenic sources (e.g., rocket fuels, explosives, and fireworks)[1,2], ClO_4^- is found in nature as a minor component of salt deposits in the hyperarid Atacama Desert in northern Chile[3,4] and in some parts of the United States[5–9].

Various treatment technologies have been developed to remove trace quantities of ClO_4^- from contaminated media, and their pros and cons have been described in detail by Gu and Coates[1]. This review focuses on a selective ion exchange technology that allows both the efficient removal of ClO_4^- from contaminated water and the quantitative recovery of ClO_4^- for disposal or possible reuse[10–13]. Ion exchange is also perhaps the only technology capable of recovering and concentrating trace quantities of ClO_4^- from contaminated environmental media, subsequently allowing forensic evaluation through stable isotope ratio analysis of chlorine and oxygen atoms[4,5,9,14,15]. This chapter intends to:

- Provide a brief overview of recent developments in ion exchange technology for perchlorate removal in contaminated water resources
- Evaluate the efficiency and longevity of ClO_4^--specific, bifunctional ion exchange resin for perchlorate treatment
- Examine the regeneration of spent ion exchange resins and the recovery of ClO_4^- using a tetrachloroferrate-based displacement technique[11,12]
- Describe techniques and procedures to extract, purify, and crystallize trace quantities of ClO_4^- from contaminated environmental media such as groundwater and soils[4,5,9,14,15]
- Describe stable isotope techniques to characterize the isotopic composition of ClO_4^- for source identification and fingerprinting in the environment

3.2   SELECTIVE ION EXCHANGE FOR PERCHLORATE REMOVAL

Ion exchange technology has been used for water treatment for more than half a century because of its simplicity, high capacity and efficiency, and ability to operate at a relatively high flow rate with a small treatment module. Today, the most commonly used ion exchange technologies for ClO_4^- removal from contaminated water are (1) selective but nonregenerable strong base anion exchange resins, (2) nonselective or low selective anion exchange resins with sodium chloride (NaCl) brine regeneration, and (3) highly selective and regenerable strong base anion exchange resins.

In the first case, the spent resin cannot be regenerated using conventional brine washing[12,16,17], so it is either discarded or incinerated after reaching sorption capacity. The resin bed must then be replaced. The change-out time for the resin depends on the feed ClO_4^- concentration and water quality. In the second case, the resin is regenerated by flushing with concentrated NaCl brine solution. However, because of its relatively low selectivity (or low sorption efficiency and capacity for ClO_4^-), the resin indiscriminately removes other common anions such as sulfate (SO_4^{2-}) and nitrate (NO_3^-).

Because these anions exist in contaminated groundwater or surface water usually at orders of magnitude higher concentrations than ClO_4^-, they occupy most of the ion exchange sites on the resin (>99%), resulting in an extremely poor efficiency for ClO_4^- removal[18-20]. The spent resin bed must be regenerated frequently, producing large volumes of secondary brine waste containing ClO_4^-. These factors contribute to relatively high capital and operating costs of conventional ion exchange technologies that are discussed in numerous reports and publications[16-18,21].

Finally, the highly selective and regenerable ion exchange technology for ClO_4^- removal is based primarily on the use of a bifunctional anion exchange resin jointly developed at the Oak Ridge National Laboratory and the University of Tennessee [10,22,23]. The bifunctional resin has been demonstrated in both laboratory and field studies to have high efficiency and longevity in removing ClO_4^- from contaminated groundwater at varying concentrations[10,11,13]. Furthermore, a novel tetrachloroferrate displacement technique was developed to desorb ClO_4^- from the resin bed using a small volume of regenerant solution. This technique allows the resin to be reused and results in minimal secondary waste production and substantially reduced capital cost compared to resin replacement[11,12]. The bifunctional resin and tetrachloroferrate regeneration technology is also the basis for recovering ClO_4^- from contaminated water and soils for environmental forensics studies. The remediation and forensic applications of this technology serve as the focus of the remainder of this chapter.

3.3 RESIN SELECTIVITY AND BIFUNCTIONAL ANION EXCHANGE RESIN

Ion exchange is a process by which ions of interest in a solution phase, such as ClO_4^-, are removed by exchange with ions associated with the resin solid phase. Most anion exchange resins are made from styrenic and acrylic polymers containing positively charged quaternary ammonium surface functional groups (so-called Type I resins) that electrostatically attract negatively charged anions such as ClO_4^- in solution. The resin, as received from the manufacturer, usually contains adsorbed chloride (Cl^-) or hydroxide (OH^-) as counterions.

When exposed to water or other media containing anions (e.g., ClO_4^-), these anions enter the resin and exchange with the Cl^- or OH^-. At equilibrium, the resin and the solution will contain both the original counterion and the anions from the environmental media, although usually not in the same ratio. Depending on their effectiveness and affinity to sorb ClO_4^- (or the anion of interest), the resins can be operationally categorized into selective and nonselective anion exchangers. The selective resin sorbs ClO_4^- strongly and effectively in the presence of competing anions such as SO_4^{2-}, Cl^-, NO_3^-, and bicarbonate (HCO_3^-). These competing anions usually exist in groundwater and surface water at much higher concentrations (orders of magnitude) than ClO_4^-, and non–selective resins sorb these ions indiscriminately or by the mass reaction based on their charges and concentrations. Therefore, divalent SO_4^{2-} anions usually are more strongly sorbed than monovalent anions on nonselective acrylic-based ion exchangers[20]. As an example, a binary exchange reaction between ClO_4^- and Cl^- can be written as:

$$R–Cl^- + ClO_4^- \rightleftharpoons R–ClO_4^- + Cl^-$$

where $R–Cl^-$ and $R–ClO_4^-$ represent Cl^- and ClO_4^- anions associated with the resin. Because both Cl^- and ClO_4^- are monovalent anions, the thermodynamic equilibrium constant (K) is written as:

$$K = \frac{\{R - ClO_4^-\}\{Cl^-\}}{\{R - Cl^-\}\{ClO_4^-\}}$$

where the { } braces refer to the activities of these ionic species. By design, different ion exchangers prefer certain counter ions to other ions in solution. The K value is thus called the *selectivity coefficient*—a measure of the preference of counterions to other anions in solution. A more practical description of this equilibrium behavior is called the *distribution* (or *partitioning*) *coefficient* (K_d), defined as:

$$K_d = \frac{[R - ClO_4^-]}{[ClO_4^-]}$$

where $[R–ClO_4^-]$ and $[ClO_4^-]$ represent the concentrations of sorbed ClO_4^- on the resin bed and free ClO_4^- ions in the solution phase, respectively. A high K_d value thus indicates a high selectivity or efficiency of the resin to sorb ClO_4^- from the solution phase. An advantage of using the distribution coefficient K_d is its simplicity because it can be calculated regardless of the presence of any other anions such as SO_4^{2-}, Cl^-, NO_3^-, CO_3^{2-}, and HCO_3^- in the environment. The disadvantage is that K_d is not a constant and may vary based on factors such as the initial concentration of the target anion, solution ionic composition, nature of the competing anions (e.g., charge-to-size ratio, and hydration energy)[24], surface functional groups, total anion exchange capacity (AEC), and structural properties of resin matrices[18,22,23].

The bifunctional anion exchange resin was designed to maximize sorption selectivity K_d of target anions such as ClO_4^- without sacrificing sorption kinetics. The resin was originally developed to remove trace quantities of radioactive pertechnetate (TcO_4^-) at low microgram-per-liter or nanogram-per-liter concentrations from groundwater at the U.S. Department of Energy's Paducah Gaseous Diffusion Plant site in Kevil, Kentucky[22,23,25].

Because of their relatively low charge-to-size ratios and hydration energies (or large ionic sizes and hydrophobicities), both ClO_4^- and TcO_4^- are strongly adsorbed by the nonpolar polystyrenic ion exchange resins, as compared with matrices containing oxygen molecules, such as polyacrylic resins in which the styrene monomer is replaced by the acrylic monomer in the polymer chain.

Similarly, the so-called Type II anion exchange resins are formed by the replacement of one of the trialkylammonium groups with an ethanol group, resulting in increased hydrophilic or polar characteristics. Thus, polyacrylic and Type II resins usually are much less selective than Type I polystyrenic resins for sorbing ClO_4^- and $TcO_4^{-[18,22,23]}$. By increasing the size of the trialkylammonium groups or the exchange sites on the resin (e.g., from trimethylammonium to trihexylammonium), the hydrophobicity of the resin and the charge separation distance of the exchange sites are increased, favoring the sorption of large anions like ClO_4^- and TcO_4^-. Therefore, as the trialkyl chain length increases, the resin selectivity increases, as illustrated in Table 3.1[10,18]. The 1-week K_d values at equilibrium increased from ~47,000 mL/g to >3,000,000 mL/g when the trialkyl chain lengths increased from trimethyl to trihexyl. However, one drawback of increasing trialkyl chain length is slower adsorption of the target anions to the resin, partly due to a decreased rate of diffusion of these ions from water into the hydrophobic resin matrix.

Table 3.1 shows that although the trihexyl resin had the highest selectivity for ClO_4^- at equilibrium, its 1-hour K_d was the lowest of the resins tested. Overall, a balance must be maintained between ion selectivity and reaction kinetics. A desire to optimize these factors led to the development of the bifunctional anion exchange resin that consists of two trialkylammonium groups, one having long chains (e.g., trihexyl) for higher selectivity and one having shorter chains (e.g., triethyl) for improved reaction kinetics. In comparison with the monofunctional trimethyl and trihexyl resins, the bifunctional resin shows faster reaction kinetics and a greatly

TABLE 3.1

Comparison of ClO_4^- Distribution Coefficient (K_d) Values of Resins with Varying Surface Trialkylammonium Functional Groups

Resin Exchange Site	1-hr K_d (mL/g)	24-hr K_d (mL/g)	168-hr K_d (mL/g)
Trimethylammonium	17,600	41,600	46,900
Tripropylammonium	233,000	657,000	905,000
Trihexylammonium	5,800	2,900,000	>3,300,000
Bifunctional (triethyl- and trihexylammonium)*	165,000	1,877,000	1,842,000

* *Purolite D-3696 resin.*

Note: Equilibrium time varied from 1 to 168 h; initial added ClO_4^- concentration was 10 mg/L. (*Source:* Gu, B. and Brown, G.M. In *Perchlorate Environmental Occurrences, Interactions, and Treatment*, Gu, B. et al., Eds. Springer: New York, 2006, 209–251. With permission.)

increased 1-hour K_d value. Rapid sorption of ClO_4^- to the resin matrix is crucial under the flow conditions typical of drinking water treatment because a short hydraulic residence time (HRT)—usually less than a few minutes—is required for such systems to be economically viable. More details about water treatment for ClO_4^- using ion exchange resins are provided in subsequent sections.

3.4 PERFORMANCE OF BIFUNCTIONAL ANION EXCHANGE RESINS

The performance of ion exchange resins to remove ClO_4^- from contaminated water is usually evaluated in continuous flow-through systems to mimic pump-and-treat field remediation scenarios. Feed water containing ClO_4^- and competing anions is added continuously to a fixed resin bed until saturation. As ion exchange progresses, the adsorbed ClO_4^- concentration at the inlet of the resin bed attains equilibrium with the ClO_4^- concentration in the feed water.

The ClO_4^- further in the column equilibrates between resin and water at an increasingly lower total ClO_4^- (due to adsorption closer to the inlet in the bed) so that the concentration at the outlet of the resin bed can be very low or below the detection limit even though the resin near the inlet of the bed has reached saturation. This fixed bed flow-through approach can remove ClO_4^- to a much lower concentration with less resin than typical batch or fluidized bed contacting techniques because, for the concentration in the effluent solution to approach zero, the ClO_4^- concentration in the resin bed must also approach zero. In a batch type system, this is only possible if the amount of resin is infinite whereas, in a packed bed system, the effluent concentration depends only on the concentration near the outlet—usually at trace level during much of the treatment process (assuming no channeling). As the loading continues, a concentration gradient forms along the resin bed and moves further down until the breakthrough of ClO_4^- is observed[18]. The abilities of various ion exchange resins to remove ClO_4^- from contaminated water have been studied extensively[10,16-18,26], and additional details appear in a recent review by Gu and Brown[18].

Laboratory and field studies have demonstrated that bifunctional resins are highly selective and efficient at removing ClO_4^- from contaminated groundwater[10,11]. In a small-scale field experiment at a contaminated site in California, the resin was found to be capable of treating ~100,000 bed volumes (BVs) of groundwater before a ClO_4^- breakthrough occurred at an initial influent ClO_4^- concentration of ~50 µg/L[10]. The resin bed was operated at ~2 BV/min (HRT ~30 s). The flow rate was appreciably faster than that typically employed during water treatment (~0.1 to 0.5 BV/min). This translates to a treatment period prior to resin regeneration or replacement of about 5 months during operation at 0.5 BV/min, or nearly 2 years when operated at 0.1 BV/min.

In another large-scale field demonstration in California at a much higher influent ClO_4^- concentration (~870 µg/L), the bifunctional resin treated ~38,000 BV of groundwater before the breakthrough of ClO_4^- occurred (Figure 3.1). The resin bed was approximately 450 L and operated at 1 to 1.3 BV/min. More than 18 kg of ClO_4^- were removed by the resin bed during this treatment period. Note that the

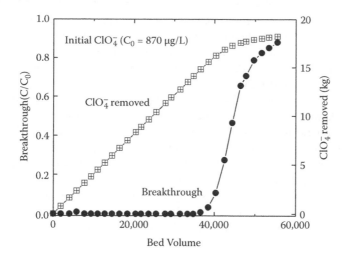

FIGURE 3.1 Perchlorate breakthrough and mass removal using bifunctional anion exchange resin bed containing Purolite A-530E in a field groundwater treatment study in California. Average feed ClO_4^- concentration was 870 µg/L and flow rate was about 150 gpm (or ~1 BV/min). (*Source:* Modified from Gu et al. *Environ. Sci. Technol.* 2007, 41, 6277–6282.)

lower volume of groundwater treated in this field study as compared to the small-scale experiment above was due mainly to the ClO_4^- feed concentration being more than 17 times greater than that in the previous study (50 µg/L).

We also noted that the amount of groundwater treated was reduced only by a factor of about 2.6 (i.e., 100,000 versus 38,000 BV). This is because the amount of groundwater treated (or quantity of ClO_4^- removed) depends on the anion exchange capacity of the resin, its selectivity (K_d) for ClO_4^-, and the influent feed concentration. This relationship is usually expressed as: $Q = K_d \cdot C$, where Q is the amount of ClO_4^- removed by the resin bed and C is the feed ClO_4^- concentration. However, K_d is not a constant here because the sorption isotherms of ClO_4^- on resin are usually nonlinear[10]. A higher K_d value is usually observed at the lower feed ClO_4^- concentration. Therefore, this relationship cannot be simply used to predict the amount of water treatment at different feed concentrations.

The K_d value can also vary with the background ionic compositions and concentrations because these ions compete with ClO_4^- for sorption onto the resin bed. Depending on water quality, the numbers of BVs of water treated may vary significantly. For common anions encountered in groundwater and surface water (e.g., SO_4^{2-}, HCO_3^-, Cl^-, NO_3^-), NO_3^- is generally the most effective competitor with ClO_4^- for sorption onto strong base anion exchange resins because of its relatively low hydration energy ($\Delta G^0 = -314$ kJ/mol). In contrast, SO_4^{2-} is the least effective competitor with ClO_4^- because of its high hydration energy ($\Delta G^0 = -1,103$ kJ/mol)

The bifunctional resin is particularly effective for removal of ClO_4^- at low concentrations because of its high selectivity. As stated earlier, this property allows the removal and recovery of trace quantities of ClO_4^- from various environmental media[10]. At relatively high ClO_4^- concentrations, selectivity is not an issue. By using radioactive TcO_4^- as an analog (because of its chemical similarity to ClO_4^-), we showed that the bifunctional resin is capable of removing TcO_4^- at nanogram-per-liter concentrations from contaminated groundwater containing about six orders of magnitude higher concentrations of competing anions such as NO_3^-, Cl^-, SO_4^{2-}, and HCO_3^- [23].

The field experimental results (Figure 3.2) demonstrated that the resin was able to treat >700,000 BV of contaminated groundwater at a flow rate of ~6 BV/min before a 3% breakthrough of TcO_4^- occurred. These results further indicate the high efficiency of the bifunctional resin to remove ClO_4^- or TcO_4^- from contaminated groundwater or surface water. The technique is now used to remove trace quantities of ClO_4^- from contaminated drinking and other water and also to concentrate and separate ClO_4^- at low microgram-per-liter or even nanogram-per-liter concentration levels such as those found in natural groundwater or soil leachate solutions [4,5,9,27], as described later in this chapter.

3.5   RESIN REGENERATION AND PERCHLORATE RECOVERY

3.5.1   RESIN REGENERATION

Regeneration of spent ion exchange resin (rather than disposal and replacement with fresh resin) is desirable to reduce operational costs during groundwater or drinking water treatment. Because ion exchange systems for ClO_4^- are generally inexpensive to design and install, resin replacement is often the major cost driver for the technology. Depending on the feed ClO_4^- concentrations and water chemistry, the resin replacement time for a typical treatment system may vary from a few weeks to several months.

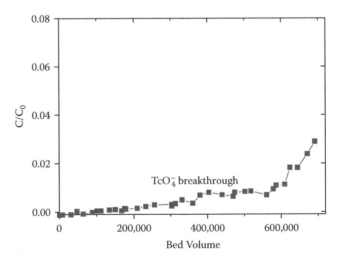

FIGURE 3.2   Treatment of pertechnetate (TcO_4^-)-contaminated groundwater using a bifunctional anion exchange resin at the U.S. Department of Energy's Paducah gaseous diffusion plant in Kevil, Kentucky. Average feed TcO_4^- concentration was about 0.07 µg/L and flow rate was about 6 BV/min. (*Source:* Modified from Gu et al. *Environ. Sci. Technol.* 2000, 34, 1075–1080.)

The conventional approach for resin regeneration is washing by a strong brine solution (e.g., 6 to 12% NaCl) so that the sorbed anions from the treated water (ClO_4^-, NO_3^-, SO_4^{2-}) are replaced by Cl^- ions—saturating the resin with Cl^- as the counterion. However, this technique is applicable only for nonselective polyacrylic and Type II ion exchange resins[16,17,25,28]. Moreover, the inability to easily treat ClO_4^- in the regenerant brine or dispose of the brine has resulted in few recent applications of this approach for ClO_4^-.

The exceptionally high affinity of ClO_4^- on many resins, even on Type I monofunctional strong base anion exchange resins, results in poor regeneration via typical brine techniques and large volumes of concentrated brine solutions are often required[12,16,17,25]. For example, Batista et al.[16] reported that several bed volumes of concentrated brine removed only a small fraction of the sorbed ClO_4^- from a Sybron SR-7 (monofunctional tripropyl anion exchange resin) bed and that heating the ClO_4^--laden resins during regeneration led to only limited improvement.

Other studies have shown that, at 3 M (or 18%) NaCl, even the least selective polyacrylic resin, Purolite A-850, has a K_d value of ~17 mL/g for TcO_4^- sorption[18]. Based on this value, a single bed volume of resin must contact with >10 BV of 3 M NaCl to remove sorbed TcO_4^- ions on the resin. The bifunctional resin is much more selective, and its K_d remains above 50,000 mL/g in 3 M NaCl for ClO_4^- or TcO_4^-. Thus, it is impractical to regenerate this resin with NaCl brine. Until recently, any Type I ion exchange resins used for removing ClO_4^- from contaminated water were incinerated or otherwise discarded after saturation[11].

New techniques have been developed to regenerate bifunctional and other Type I anion exchange resins loaded with ClO_4^- so that the spent resin bed can be recycled or reused in order to reduce costs[11,12,18]. The new method for bifunctional resin involves a mixed solution of $FeCl_3$ and HCl, in which tetrachloroferrate ($FeCl_4^-$) ions form in the presence of excess amounts of Cl^- by equilibrium:

$$FeCl_3[aq] + Cl^- \rightleftharpoons FeCl_4^- \tag{3.1}$$

Like ClO_4^-, $FeCl_4^-$ is a large, poorly hydrated anion and, because of this trait, it can more effectively displace ClO_4^- from the bifunctional resin than can Cl^- or most other counterions. In practice, a mixed solution of 1 M $FeCl_3$ and 4 M HCl was effective in eluting ClO_4^- ions from a spent resin bed[11]. The use of 4 M HCl has another advantage in removing mineral deposits and cleaning the resin so it can be reused. If needed, organic solvents such as methanol can be used to enhance the elution of ClO_4^- and the removal of organic materials from the resin. However, after regeneration, the resin must be washed thoroughly with water. The sorbed $FeCl_4^-$ ion has a desirable chemical property: it decomposes in water or dilute acidic solutions based on the following established chemical equilibria:

$$FeCl_4^- \rightleftharpoons Fe^{3+} + 4\,Cl^- \tag{3.2}$$

$$FeCl_4^- \rightleftharpoons FeCl^{2+} + 3\,Cl^- \tag{3.3}$$

$$FeCl_4^- \rightleftharpoons FeCl_2^+ + 2\ Cl^- \tag{3.4}$$

$$FeCl_4^- \rightleftharpoons FeCl_3 + Cl^- \tag{3.5}$$

By decreasing the Cl^- concentration, the $FeCl_4^-$ anion converts to positively charged or neutral Fe [III] species such as Fe^{3+}, $FeCl^{2+}$, $FeCl_2^+$, and $FeCl_3$ that are readily desorbed from the resin by charge repulsion. The resin is regenerated to its original state with Cl^- as the counterion by charge balance. Because the rinse water is acidic but does not contain ClO_4^-, it can be readily disposed of after proper neutralization by passing the effluent solution through a carbonate gravel filter or simply adding carbonate or bicarbonate to the rinse water.

This novel regeneration methodology offers a cost-effective means to regenerate or recycle strong base anion exchange resins loaded with ClO_4^- because both $FeCl_3$ and HCl are relatively inexpensive and readily available. Both field and laboratory experiments show that the regeneration is highly efficient; almost 100% recovery of the exchange sites can be achieved by rinsing with 1 to 2 BV of the regenerant solution[11,13].

The elution profiles and the amounts of ClO_4^- recovered from repeated regenerations of a resin bed [~570 L] used to treat contaminated groundwater (Figure 3.1) are shown in Figure 3.3[11,26]. During resin regeneration, the effluent regenerant solution was collected every quarter BV and analyzed for ClO_4^- to facilitate determination of an accurate mass balance of recovered ClO_4^-. The sorbed ClO_4^- was rapidly eluted and concentrated in roughly the first bed volume of the regenerant solution (Figure 3.3). The maximum ClO_4^- concentration, usually occurring in the third or fourth quarter of the first BV, reached as much as 115,000 mg/L (or ~1.2 M), representing a concentration factor of more than five orders of magnitude in comparison with the ClO_4^- concentration in the groundwater [~870 μg/L]. The amount of ClO_4^- recovered was about 20 kg for the second and third regeneration cycles, (Figure 3.3)—roughly equal to the amount of ClO_4^- removed during the water treatment phase.

Note that the smaller amount of ClO_4^- (~16 kg) recovered during the first regeneration was due to the use of a smaller resin bed (~450 L) during the initial treatment[26]. Nonetheless, these observations indicate the high efficiency of the tetrachloroferrate displacement technique to desorb ClO_4^- from the spent resin; the regeneration yielded a recovery of 96 to 100% of the sorbed ClO_4^-. Moreover, when continuous flow of groundwater was reinitiated, the regenerated resin removed ClO_4^- with efficiency and capacity similar to those of the virgin resin. Similar results were obtained when evaluating the performance of the bifunctional resin to remove ClO_4^- from contaminated groundwater containing high concentrations of dissolved solids (SO_4^{2-} up to 180 mg/L and Cl^- up to 400 mg/L)[13,26]. Although the total ClO_4^- recovered may vary based on initial loading, the sorbed ClO_4^- can be quantitatively recovered regardless of water quality. This is in contrast to the low efficiency of ClO_4^- recovery from nonselective polyacrylic and Type II ion exchange resins using concentrated NaCl brines, as stated earlier[16,17,25].

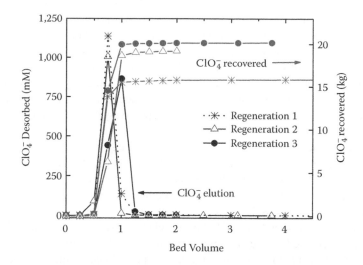

FIGURE 3.3 Elution profiles (left Y axis) and cumulative ClO_4^- recovery (right Y axis) during repeated regeneration of spent resin bed used to remove ClO_4^- from contaminated groundwater. Regenerant solution was 1 *M* $FeCl_3$ and 4 *M* HCl. (*Source:* Modified from Gu et al. *Environ. Sci. Technol.* 2007, 41, 6277–6282.)

A major advantage of using the $FeCl_3$–HCl regeneration technique is that the recovered ClO_4^- is highly concentrated in a small volume of the regenerant solution. The rest of the spent regenerant solution (eluted after the first BV) can be reused so secondary waste is minimized. Furthermore, ClO_4^- in the regenerant solution can be chemically reduced to Cl^- and H_2O at moderate temperature and in the presence of ferrous chloride $(FeCl_2)$[11,29]. While ClO_4^- ions are destroyed, ferrous ions are oxidized to ferric ions so that the process renews the Fe^{3+} ion in the regenerant solution, allowing it to be recycled. In other words, the entire regeneration and ClO_4^- destruction process produces little secondary waste and thus overcomes the major problem associated with conventional throw-away ion exchange and/or NaCl brine regeneration methodologies.

3.5.2 Perchlorate Recovery

Because ClO_4^- is concentrated in a small volume of the $FeCl_3$–HCl regeneration solution, it can be precipitated and then recovered as pure ClO_4^- salts such as $KClO_4$ and $CsClO_4$ because of their relatively low solubility in water ($K_{sp} = 1 \times 10^{-2}$ or 4×10^{-3} at 20°C, respectively). At a concentration of about 1 *M* or 100,000 mg/L ClO_4^-, theoretical calculations indicate that about 98% of ClO_4^- can be precipitated as $KClO_4$ solids, assuming that the final K^+ concentration is maintained at 0.5 *M* or higher.

A slightly higher recovery may be obtained by using Cs^+, but Cs salts (e.g., CsCl) are generally more expensive than K salts. In practice, we found that about 13 g of

$KClO_4$ (or ~9.4 g ClO_4^-) are recovered by mixing 30 mL of saturated KCl solution with 100 mL of the regenerant solution (with 1 M ClO_4^-) [11]. This represents a recovery of ~94% of ClO_4^- from the spent regenerant. A higher initial ClO_4^- concentration gives a higher yield of precipitated ClO_4^-. In other words, less ClO_4^- will be left in the solution phase at equilibrium because $KClO_4$ has a solubility of about 20 g/L in pure water at 20°C. For example, at initial ClO_4^- concentrations of 10,000 and 5,000 mg/L, recoveries of 62.5 and 57.4% of ClO_4^-, respectively, were obtained by precipitation with KCl at 4°C.

Other anions such as NO_3^- and SO_4^{2-} that can also be sorbed by the anion exchange resin and may thus exist in the spent regenerant solution usually do not form precipitates with K or Cs because of their high solubility in water. However, small quantities of NO_3^-, SO_4^{2-}, Fe^{III}, and/or organics may coprecipitate with $KClO_4$ or $CsClO_4$, affecting the purity of recovered $KClO_4$ or $CsClO_4$ salts. Purity is not a significant concern if the recovered ClO_4^- is to be disposed of or recycled for use in pyrotechnics[1]. However, purity of the recovered salt is critically important if stable isotope analysis is to be performed on the sample for forensics or other purposes because impurities can affect both the accuracy and precision of this technique[4,5,9,14,15].

The potential for such impurities increases when ClO_4^- is recovered from natural environments where it exists at very low concentrations compared to other species such as NO_3^-, Cl^-, and SO_4^{2-} from surface evaporite deposits. In particular, trace amounts of ions such as perrhenate (ReO_4^-), pertechnetate, periodate (IO_4^-), permanganate (MnO_4^-), and chlorate (ClO_3^-) may present problems at comparable concentrations because these ions are also concentrated on the resin bed and eluted by the $FeCl_3$–HCl solution. Thus, the identification and removal of these impurities in recovered ClO_4^- is critical to the stable isotope analysis for ClO_4^- fingerprinting, as described below.

In summary, the ability to recover ClO_4^- from an ion exchange resin in a small volume of regenerant solution is beneficial for several reasons. First, secondary waste requiring treatment or disposal is substantially reduced compared to conventional brine regeneration. In addition, the ClO_4^- recovered during environmental remediation can potentially be recycled for use in pyrotechnics and other materials. Finally, the technique developed to remove and recover ClO_4^- for remediation also provides a means to collect and purify samples for perchlorate fingerprinting via stable isotope analysis of Cl and O. This technique continues to provide new information about sources of ClO_4^- contamination in the environment, mechanisms of formation of natural ClO_4^-, and bacterial fractionation of Cl and O isotopes in the biodegradation of ClO_4^-. [4,5,9,27] More details on isotopic analysis of ClO_4^- are provided in the next section.

3.6   PERCHLORATE FINGERPRINTING AND ENVIRONMENTAL FORENSICS

The recent development of techniques for analyzing stable isotope ratios of O and Cl in ClO_4^- led to applications for distinguishing sources of ClO_4^- in the environment[4,5,9,27]. Perchlorate can be derived from both anthropogenic and natural sources. Synthetic ClO_4^- is widely used as an oxidant in energetic materials such as rocket propellants, explosives, and pyrotechnics. Natural ClO_4^- can also form in the atmosphere and is found in relatively high concentrations (~0.2 wt%) in nitrate-rich salt deposits from the Atacama Desert in Chile[3,4] and at low concentrations in the High Plains Region of western Texas and eastern New Mexico[30]. Chilean nitrate also has been widely used in fertilizer production for well over a century. As a consequence, many groundwater and surface water supplies in the U.S. and abroad show elevated concentrations of ClO_4^- in the microgram-per-liter to milligram-per-liter range. It has also been found in a wide variety of food products and in human urine and milk[31–38].

Stable isotope ratio analysis of ClO_4^- offers a unique tool to identify contamination sources and also provide a fundamental understanding of ClO_4^- formation processes in nature. Studies have shown that natural ClO_4^- from the Atacama Desert contains excess ^{17}O, indicating an atmospheric origin from photochemical reactions involving ozone (O_3)[4]. This finding is consistent with O isotope studies of NO_3^- and SO_4^{2-} from the Atacama Desert and indicates atmospheric sources for those anions as well[39]. Anthropogenic ClO_4^- has a Cl isotope ratio similar to that of its Cl source, and an O isotope ratio related to that of the water used for its production by electrolysis[5,40]. Thus, isotope ratio analysis can in principle distinguish natural and synthetic ClO_4^- and also different varieties of synthetic ClO_4^-, provided that the water and Cl used for its manufacture have distinct isotope signatures.

However, analysis of ClO_4^- isotope ratios requires milligram quantities of pure ClO_4^- salts, but most environmental samples from sources such as natural water, soils, and plant materials contain very low concentrations of ClO_4^-, usually nanogram-per-kilogram to microgram-per-kilogram quantities, depending on the matrix. For example, at a concentration of 1 µg/L, thousands of liters of water must be extracted to ensure a sufficient quantity of ClO_4^- to be purified, crystallized, and analyzed for stable isotopes. This assumes an extraction efficiency of 100%, although this is rarely achieved. Furthermore, most environmental samples often contain several orders of magnitude higher concentrations of unwanted ions, dissolved and suspended solids, and organic impurities that make the extraction and purification of a small quantity of ClO_4^- a rather formidable task. Here we describe a well established methodology to extract, purify, and analyze ClO_4^- isotope ratios for fingerprinting in the environment[4,5,9,27]. The general procedures are:

- Leaching or extraction of ClO_4^- from environmental media such as soil, mineral deposits, or groundwater
- Collection and concentration of ClO_4^- on an ion exchange resin column
- Extraction and recovery of ClO_4^- from the resin column
- Purification and crystallization of ClO_4 and verification of sample purity
- Analysis of O isotopes by isotope ratio mass spectrometry (IRMS)
- Analysis of Cl isotopes by IRMS

3.6.1 PERCHLORATE EXTRACTION AND CONCENTRATION
FROM ENVIRONMENTAL MEDIA

Trace amounts of ClO_4^- in groundwater or drinking water (microgram-per-liter concentrations) are first concentrated onto packed glass or PVC columns with Purolite A-530E, a ClO_4^--specific bifunctional anion exchange resin, as described earlier. Note that other ClO_4^-–specific ion exchange resins such as Amberlite PWA-2 may also be used to concentrate ClO_4^-, but the recovery of ClO_4^- during regeneration with them was found to be very low (<10%)[18] because of their different resin backbones and surface exchange sites. Water is pumped directly from a well or faucet to the inlet port of the resin column to trap ClO_4^-.

Because ClO_4^- has a low sorption affinity for most tubing materials and metal or plastic connections, selection of these materials for sampling is not critical. The size of the resin column and the flow rate may vary, depending on the concentration of ClO_4^- in water and site-specific conditions. For example, for a 100-mL PVC column, a maximum flow rate of 2 L/min is recommended for processing water containing ~1 µg/L ClO_4^-. At this flow rate, ~3 mg of ClO_4^- will be trapped on the resin column in 24 h. Assuming no loss of ClO_4^- during sample collection and purification, this quantity is sufficient for Cl and O isotope analysis. However, due to potential losses, particularly during purification, it is desirable to collect ~10 mg of ClO_4^- for stable isotope analysis. This quantity also allows for multiple analyses of a single sample.

If sample collection time is not a constraint, a slower flow rate is recommended for better capture of ClO_4^- from the contaminated water. This is especially true if the water contains a relatively high concentration of NO_3^- (>50 mg/L) because of its competitive sorption with ClO_4^- onto the resin bed, as previously noted. The breakthrough of ClO_4^- can occur at a high flow rate as a result of short contact time and potential channeling. For water containing a relatively high concentration of ClO_4^- (>10 µg/L), a smaller resin column (20 mL) can be used to minimize the amount of regenerant solution processed and the loss of ClO_4^- during recovery and crystallization (described below).

In general, a sufficient but not excessive amount of the resin should be used with the goal of capturing all ClO_4^- in the water. It is therefore recommended that the effluent water be analyzed for ClO_4^- to assess capture efficiency. The resin (1 mL) has sufficient exchange capacity to sorb 10 mg ClO_4^- under ideal conditions. Also, depending on water quality, a cartridge prefilter (e.g., sediment filter) may be used to remove any fines or suspended solids from the water. However, an activated carbon filter should not be used because such filters can also sorb substantial quantities of ClO_4^- and thus decrease the recovery on the resin column[1].

For collection and purification of ClO_4^- from soils or mineral deposits, samples are generally homogenized and subsequently leached with ClO_4^--free water. This process is effective for recovering ClO_4^- because of its high solubility and low affinity to most minerals[41]. With the exception of soils containing high amounts of iron or aluminum oxyhydroxides at relatively low pH conditions, little or no ClO_4^- retention should occur in soils. The leachate containing ClO_4^- may be filtered to remove particulates and subsequently concentrated onto a packed resin column as described above. Upon completion of the sampling and preconcentration, the resin column can be preserved by saturating the resin with 1 M HCl and storing at 4°C prior to ClO_4^- extraction and analysis. This procedure is necessary, especially if ClO_4^- is collected from an anaerobic environment where microbial degradation of ClO_4^- may occur[42,43].

3.6.2 PERCHLORATE RECOVERY, PURIFICATION, AND VALIDATION

The resin columns with trapped ClO_4^- extracted from water or soil and/or mineral leachates are subsequently transferred to the laboratory for removal of unwanted impurities via resin cleaning followed by recovery and purification of adsorbed ClO_4^-. The cleaning or prewash step is important because field resin columns usually contain large quantities of dissolved organics or humic materials, other anions such as NO_3^-, SO_4^{2-}, etc., and suspended solids such as clays and oxides.

As previously noted, NO_3^-, SO_4^2 and dissolved organics are major competing ions sorbed by the resin during ClO_4^- collection because their concentrations are usually orders of magnitude higher than that of ClO_4^- in water. The prewash step is usually accomplished by redispersing the resin in 4 M HCl and repacking the resin into a preparative chromatographic glass column, followed by washing with 3 to 5 BV of the HCl solution. The use of 4 M HCl effectively removes most of sorbed SO_4^{2-}, NO_3^-, carbonates, and some organics, but the process does not remove significant amounts of ClO_4^- because it is sorbed much more strongly than the other anions as described earlier. Figure 3.4 illustrates a typical elution profile in which SO_4^{2-} ions are eluted quickly with only 1 BV of the HCl solution. NO_3^- is more strongly sorbed

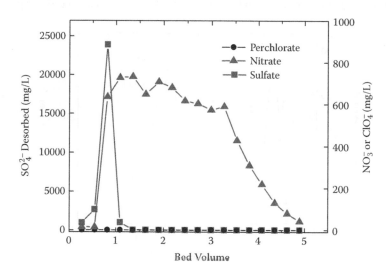

FIGURE 3.4 Typical elution profiles of SO_4^{2-}, NO_3^-, and ClO_4^- by leaching with 4 M HCl from a field resin column used to concentrate ClO_4^- from the Amargosa Valley in southern Nevada. The eluted ClO_4^- concentration is negligible.

than SO_4^{2-} and desorbed slowly. Small amounts of NO_3^- are observed in the effluent even after leaching with about 5 BV of the 4 M HCl solution.

The elution of NO_3^- from the resin is more rapid when the resin is leached directly with a solution of 1 M FeCl$_3$ and 4 M HCl[2]; in this case, the NO_3^- is eluted within 1 BV followed by ClO_4^-, because FeCl$_4^-$ ions formed in the FeCl$_3$ and HCl solution are much more effective than Cl$^-$ in desorbing NO_3^- and ClO_4^- (Figure 3.5). This elution sequence ($SO_4^{2-} \to NO_3^- \to ClO_4^-$) is consistent with the fact that SO_4^{2-} is least strongly sorbed by Type I polystyrenic anion exchange resins among the three anions, whereas ClO_4^- is the most strongly sorbed[19]. Note that the elution of organics is not monitored but can be clearly observed by the color of the initial effluent solution.

Organic solvents such as methanol may be used along with HCl to increase the removal efficiency of sorbed organics. Most sorbed or trapped carbonates are also desorbed or degassed under acidic pH and can readily be identified by gas bubbles initially in the column.

Following the acid prewash, the resin column is eluted with a combination of 1 M FeCl$_3$ and 4 M HCl to desorb and recover ClO_4^-. The effluent solution is usually collected using a fraction collector at 0.1 to 0.2 BV intervals and analyzed for ClO_4^- concentrations by ion chromatography (IC). Because the eluted ClO_4^- is concentrated in less than 1 BV of effluent (Figure 3.5), only a small fraction of the solution with concentrated ClO_4^- is finally collected in order to minimize the amount of effluent solution to be processed during ClO_4^- recovery and crystallization. The IC analysis is necessary to determine which fractions are to be collected and to quantify the total ClO_4^- eluted from the column for mass balance determination at the end of the recovery and purification processes.

Two different approaches can be used to recover ClO_4^- from the FeCl$_3$–HCl effluent solution[4,14,27] and both have advantages and disadvantages. The first approach[4]

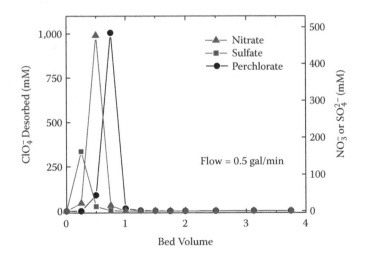

FIGURE 3.5 Typical elution profiles of SO_4^{2-}, NO_3^- and ClO_4^- from the resin bed (without pre-washing with 4 M HCl) by desorption with a mixture of 1 M FeCl$_3$ and 4 M HCl. (*Source:* Modified from Gu et al. *Environ. Sci. Technol.* 2007, 41, 6277–6282.)

involves the neutralization of the acidic effluent solution with NaOH to a pH above 9 and below 10, in which ferric ions form $Fe(OH)_3$ precipitates and can thus be separated from the solution containing ClO_4^-. The clear supernatant solution is concentrated, for example, by using rotary evaporation under vacuum [Savant Speed Vac]. During this process, salts such as NaCl and precipitates are removed periodically, and the clear solution with concentrated ClO_4^- is recovered.

The sample can also be filtered, depending on the size and quantity of precipitate that varies based on the impurities in each sample. Based on the total ClO_4^- recovered from the resin, the concentration ratio (initial to final volumes) may vary from 2 to >100 or until a sufficiently high concentration of ClO_4^- (>3000 mg/L) is obtained in the concentrated solution. This high concentration is necessary to crystallize ClO_4^- salts with the addition of Cs^+ or K^+ at a later stage. However, the concentration step (especially at high concentration ratios) can result in substantial loss of ClO_4^-. The loss can be minimized by redissolving the salts and repeating the extraction, elution, and recovery using a smaller resin column (usually <5 mL).

The second approach is to use a cation exchange resin [H^+ form] such as Bio-Rad AG-50 W × 12 resin [100 to 200 mesh] to remove ferric ions in the $FeCl_3$–HCl solution[14,27]. This procedure is followed by heating and evaporation of water and HCl until a sufficiently high concentration of ClO_4^- is obtained, as noted earlier. This approach has the advantage of avoiding the precipitation of abundant Fe hydroxides and NaCl as encountered in the first approach. However, one potential disadvantage of using cation exchange for Fe removal is the loss of some ClO_4^- due to its sorption onto the cation exchange resin matrix, despite the fact that both ClO_4^- and resin exchange sites are negatively charged. The selection of cation exchange resins is thus important because of differences in their compositional and surface properties; some resins (e.g., the Fe^{3+}-specific Purolite S–957 resin, H^+ form) may cause >35% loss of ClO_4^-.

Finally, the purified and concentrated ClO_4^- in solution is crystallized by the addition of ions such as Cs^+ or K^+, although the crystallization of $CsClO_4$ is usually preferred because of its slightly lower solubility than that of $KClO_4$ salt in water at ~4°C. During this process, saturated CsCl or CsOH solutions are added to cause supersaturation and precipitation of $CsClO_4$. In general, a higher initial ClO_4^- concentration results in a higher recovery of the $CsClO_4$ salt, as stated earlier.

For example, at a concentration of 100,000 mg/L ClO_4^-, theoretical calculations suggest that about 98% of ClO_4^- should be precipitated as $KClO_4$ if a K salt is used, and more than 99% of ClO_4^- should precipitate as $CsClO_4$ salt. In reality, however, a lower recovery may be obtained, particularly when handling a small volume (< 1 mL) and low concentrations of ClO_4^- in the recovered solution. Recovery of about 94% of $KClO_4$ at a ClO_4^- concentration of 100,000 mg/L was reported [11]. At 5,000 mg/L ClO_4^-, only about 40 to 60% of ClO_4^- could be obtained because of difficulties in completely recovering milligram quantities of ClO_4^- from a small volume of concentrated solution. Potential isotope fractionation during the concentration, recovery, and crystallization of ClO_4^- has been evaluated and usually found within 0 to 1 ‰ ($\overline{O}^{18}O$) with the above procedures.

As stated earlier, the purity of recovered ClO_4^- salts is critical in the analysis of the O and Cl isotope ratios. Therefore, during the recovery of the crystalline

ClO_4^- salt, which is usually performed by filtration, the wet crystals are washed with a few drops of 90 to 95% methanol in water to ensure that their surfaces are free from other salts. To ensure that the recovered ClO_4^- (e.g., $CsClO_4$) is of the highest purity for IRMS, the $CsClO_4$ solids are generally examined for purity using techniques such as nondestructive Raman spectroscopy.

The sample is first inspected under a Raman imaging microscope, and Raman spectra are collected from at least two or three individual salt crystals. $CsClO_4$ crystals usually exhibit needle-shaped morphologies, although crystal sizes may vary depending on the concentration of ClO_4^- in the final solution before crystallization. Cubic or spherical crystals generally indicate impurities of NO_3^- or other salts and can be easily detected by micro Raman spectroscopy with characteristic Raman shifts around 1050 cm^{-1} for NO_3^-, 980 to 985 cm^{-1} for SO_4^{2-}, and 970 or 332 cm^{-1} for perrhenate (ReO_4^-). See Figure 3.6.

As shown in Figure 3.6a, $CsClO_4$ solids have three strong characteristic Raman bands at 937, 627, and 460 cm^{-1}, respectively, with the strongest symmetrical stretch band at 937 cm^{-1}. There are two additional double bands at 1110 and 1085 cm^{-1}, a shoulder (or two small peaks) next to the 937 cm^{-1} band, and two minor bands between the 460 and 627 cm^{-1} bands.

It is necessary to magnify the y axis at its baseline to observe the presence of small quantities of impurities as shown in Figure 3.6b. Sample 1 contains about 0.3% NO_3^- (determined by IC) and shows a small shoulder at 1050 cm^{-1}, which is the characteristic vibrational band for NO_3^-. Sample 2 contains about 1% SO_4^{2-} and shows a vibrational band at ~985 cm^{-1}, which is characteristic of SO_4^{2-}. Sample 3 contains about 1% ReO_4^- and shows two sharp vibrational bands at 970 and 332 cm^{-1}, which are characteristic of $CsReO_4$. If impurities are identified at relatively large concentrations (e.g., >1%), samples may have to be reprocessed by dissolving the salts and repeating the extraction, elution, and recovery as described earlier. In general, the NO_3^- and SO_4^{2-} impurities can be easily removed, but the separation and removal of ReO_4^- can be more difficult because of its similar chemical behavior to ClO_4^-.

Fortunately, rhenium (Re) is among the rarest metals on earth and, unlike NO_3^- and SO_4^{2-}, does not normally occur in significant quantities in soils and groundwater.

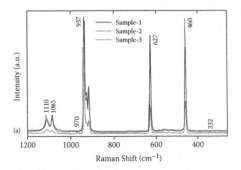 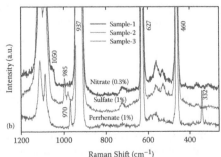

FIGURE 3.6   a. Nondestructive micro Raman spectroscopic analysis of impurities in crystallized $CaClO_4$ samples containing varying amounts of NO_3^-, SO_4^{2-}, and ReO_4. b. Same analysis with magnified y axis (offset for three spectra).

We detected the ReO_4^- only in a few samples from more than a hundred samples processed to date. However, one exception was a soil sample obtained near Death Valley, California, that showed ~75% ReO_4^- in purified salt crystals (Figure 3.7a). Pure $CsReO_4$ shows three strong Raman bands at 970, 332, and 899 cm^{-1} and all were identified in this sample.

ReO_4^-, a large singly charged oxyanion, has similar physicochemical properties to ClO_4^-, and is thus commonly used as a surrogate for ClO_4^-. It is also strongly sorbed by Type I strong base anion exchange resins and desorbed during elution with the $FeCl_3$–HCl solution. The solubility product of $CsReO_4$ [4×10^{-4}] is even lower than that of ClO_4^-, which makes it impossible to physically separate them using the cited extraction–recovery procedures. Additional chemical treatment must be used to separate these two anions. One possibility is to selectively reduce ReO_4^- to lower valent states such as ReO_3^- so it may be separated during extraction and elution. This assumes that ReO_3^- has a lower affinity to be sorbed by the A-530E resin, and the process is still under investigation. Nonetheless, this shows the importance of identifying impurities in ClO_4^- samples intended for stable isotope analysis.

Figure 3.7b illustrates a synthetic sample containing about 20% ReO_4^- and 1% NO_3^-. Again, the presence of ReO_4^- and NO_3^- can be clearly identified by Raman bands at 970 or 332 and 1050 cm^{-1}, respectively. In addition to the aforementioned impurities, singly charged oxyanions such as periodate (IO_4^-), tungstate (WO_4^-), chlorate (ClO_3^-), permanganate (MnO_4^-), and tetrafluoroborate (BF_4^-) may cause problems in the separation and/or purification of recovered ClO_4^-, especially where they are present in soil and groundwater at concentrations higher than or comparable to ClO_4^-. Fortunately, some of these anions (e.g., MnO_4^-, IO_4^-, and ClO_3^-) are not as stable as ClO_4^- and are less likely to exist at substantial concentrations under natural environmental conditions.

Other oxyanions such as arsenate (AsO_4^{3-}), selenate (SeO_4^{2-}), and chromate (CrO_4^{2-}) are not major concerns because they are either not strongly sorbed by the Purolite A-530E resin or do not form precipitates with Cs^+ during the crystallization. Most divalent anions behave like SO_4^{2-} and are readily washed out by 4 M HCl during the initial cleaning process (Figure 3.4). Furthermore, these anions can all be identified

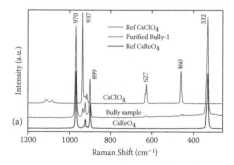 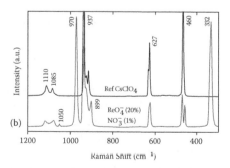

FIGURE 3.7 a. Raman spectroscopic identification of significant amounts (~75%) of perrhenate (ReO_4^-) in a sample from Bully Hill in Death Valley, California. b. Synthetic sample containing about 20% ReO_4^- and 1% NO_3^-. The y axis is offset for clarity of presentation.

by micro Raman analysis. Although they are destructive, techniques such as IC and ICP-MS can also be used to identify these ions or metals (Re and W) if sufficient quantities of recovered ClO_4^- are available. If no impurities are detected or the purity of ClO_4^- approaches 100%, samples are ready for analysis by IRMS.

3.7 ISOTOPIC ANALYSIS AND ENVIRONMENTAL FORENSICS OF PERCHLORATE

The isotopic compositions of Cl and O in the ClO_4^- ion depend on several factors: (1) isotopic compositions of the chemical precursors from which the ClO_4^- is derived; (2) isotopic fractionations associated with the ClO_4^- formation reactions; (3) subsequent isotopic fractionations associated with ClO_4^- transport or degradation in the environment; and (4) isotope effects during sample preparation and analysis. Perchlorate isotopic variations caused by the first three factors can be used to determine ClO_4^- origin and history, provided the effects of the fourth factor can be minimized.

Measurements used for this purpose include the stable isotope ratios $^{37}Cl{:}^{35}Cl$, $^{18}O{:}^{16}O$, and $^{17}O{:}^{16}O$, along with a relative abundance of radioactive ^{36}Cl, thus providing a maximum of four independent parameters. Isotopic data can be used to distinguish synthetic and natural sources of ClO_4^-[4,5,9,14,15,27] and document isotope effects of microbial reduction of ClO_4^- in the environment[14,42]. Isotope effects during sample preparation and analysis may affect the accuracy of measurements, particularly for field samples extracted from low concentration soils and groundwaters, but these effects are minimized by maximizing yields in all chemical procedures.

Systematic tests of simulated samples, in which synthetic perchlorate reference compounds of known isotopic composition are passed through the entire extraction and isotopic analysis in the presence of high concentrations of impurities, have been performed to test the accuracy of isotopic analytical methods. Results indicate the procedures described above and below can yield unbiased isotopic data.

Stable Oxygen Isotope Ratio Analysis—Stable isotope ratios of O in ClO_4^- samples are analyzed by IRMS[4,5,9,27]. The oxygen isotope ratios are expressed as deviations from reference values:

$$\delta^{18}O = [(^{18}O/^{16}O)_{sample}/(^{18}O/^{16}O)_{reference} - 1]$$

$$\delta^{17}O = [(^{17}O/^{16}O)_{sample}/(^{17}O/^{16}O)_{reference} - 1]$$

where the standard reference in both equations is Vienna Standard Mean Ocean Water (VSMOW). Values of $\delta^{18}O$ and $\delta^{17}O$ are reported in parts per thousand (per mil or ‰). By convention, the $\delta^{18}O$ scale is defined by two references, with VSMOW at 0.0 per mil and Standard Light Antarctic Precipitation (SLAP) at a value of −55.5 per mil[44].

Variations in $^{17}O/^{16}O$ and $^{18}O/^{16}O$ caused by most physical–chemical processes, including artificial ClO_4^- synthesis, are related systematically by the relative differences in the isotopic masses: $(1 + \delta^{17}O) = (1 + \delta^{18}O)^\lambda$ with $\lambda \approx 0.51$ to 0.53. However, deviations from mass-dependent variation may be large in some materials, including some types of natural ClO_4^-. Departures from mass-dependent O isotope variation in ClO_4^- can be described by $\Delta^{17}O = [(1 + \delta^{17}O)/(1 + \delta^{18}O)^{0.525}] - 1$. The values of $\Delta^{17}O$ are reported in per mil.

For $\delta^{18}O$ determinations, ClO_4^- salts are reacted with glassy C at 1325°C to produce CO, which is transferred in a He carrier through a molecular sieve gas chromatograph to a Finnigan Delta Plus XP isotope ratio mass spectrometer and analyzed in continuous flow mode by monitoring peaks at m/z 28 and 30 (CO-CFIRMS). Yields of O (as CO) typically are indistinguishable (±2%) for weighed aliquots of H_2O, ClO_4^-, NO_3^-, and SO_4^{2-} reagents and samples, indicating quantitative reaction to CO. The analytical precision of $\delta^{18}O$ values ranges from ±0.1 to 0.3‰ based on replicate analyses of samples and isotopic reference materials.

For $\delta^{18}O$ and $\bar{O}^{17}O$ determinations, ClO_4^- salts are decomposed at 650°C in evacuated quartz glass tubes to produce O_2, and then quenched in air. The O_2 gas is expanded into a liquid N_2 trap and then admitted to a Finnigan Delta Plus XP isotope ratio mass spectrometer and analyzed in dual inlet mode by measurements at m/z 32, 33, and 34 (O2-DIIRMS). Yields of O (as O_2) are typically within ±5% for ClO_4^- reagents, ClO_4^- samples, and measured aliquots of tank O_2. Adjustments are made for minor O isotopic exchange within the glass tubes based on analyses of perchlorate isotopic reference materials with each batch of samples (see below). An alternative decomposition method involves heating ClO_4^- salts in a Pt coil while trapping evolved O_2 in a cooled molecular sieve[4] that reduces the potential for O exchange.

Direct calibration of O isotope measurements against the primary H_2O reference materials defining the O isotope scale (VSMOW, SLAP) may be difficult for other compounds by these methods[45]. Therefore, routine calibrations of ClO_4^- O isotope analyses are done by analyzing a pair of $KClO_4$ isotopic reference materials (USGS37 and USGS38) with contrasting isotopic compositions, a process called normalization:

$$\delta^{18}O_{i/VSMOW} = \delta^{18}O_{37/VSMOW} +$$
$$[\delta^{18}O_{i/rg} - \delta^{18}O_{37/rg}]_{meas.} \cdot [\delta^{18}O_{38/VSMOW} - \delta^{18}O_{37/VSMOW}] / [\delta^{18}O_{38/rg} - \delta^{18}O_{37/rg}]_{meas.}$$

$$\delta^{17}O_{i/VSMOW} = \delta^{17}O_{37/VSMOW} +$$
$$[\delta^{17}O_{i/rg} - \delta^{17}O_{37/rg}]_{meas.} \cdot [\delta^{17}O_{38/VSMOW} - \delta^{17}O_{37/VSMOW}] / [\delta^{17}O_{38/rg} - \delta^{17}O_{37/rg}]_{meas.}$$

where 37 and 38 refer to the ClO_4^- reference materials and rg is an instrumental monitoring gas (CO or O_2) against which all samples and reference materials are analyzed in the mass spectrometer during a single batch of analyses; the isotopic composition of rg need not be known precisely. The isotopic reference materials consist of reagent grade $KClO_4$ salts prepared specifically for calibration of ClO_4^- isotopic analyses as part of this project. The $\delta^{18}O$ scale is based on CO-CFIRMS analyses of the ClO_4^- isotopic reference materials against international H_2O, NO_3^-, and SO_4^{2-} isotopic reference materials as described by Böhlke et al.[46]. All data are referenced to the conventional VSMOW–SLAP scale[44]. For $\delta^{18}O$, the secondary calibration values used to generate provisional ClO_4^- data are −27.9‰ for USGS34 (KNO_3), +25.6‰ for IAEA–N3

(KNO_3), +57.5‰ for USGS35, and +8.6‰ for NBS-127 ($BaSO_4$)[46]. The $\delta^{17}O$ scale for ClO_4^- provisionally is based on the assumption that the normal reagent $KClO_4$ reference material (USGS37) has $^{18}O/^{17}O/^{16}O$ ratios related to those of VSMOW by normal mass-dependent relations ($\Delta^{17}O = 0.0$ ‰] with exponent $\lambda = 0.525^5$. The reproducibility of the $\Delta^{17}O$ measurements was ±0.1‰ or less after normalization, based on replicate analyses of samples and isotopic reference materials.

Stable Chlorine Isotope Ratio Analysis—Stable isotope ratios of Cl in purified ClO_4^- samples are analyzed using IRMS[14,27,40]. The Cl isotope ratios are expressed as deviations from reference values: $\delta^{37}Cl = (^{37}Cl/^{35}Cl)_{sample} : (^{37}Cl/^{35}Cl)_{reference} - 1$ where the reference is Standard Mean Ocean Chloride (SMOC). Values of $\_^{37}Cl$ are also reported in parts per thousand (per mil or ‰). The most common Cl isotope reference material is seawater Cl^-,[47], defined as having a $\overline{O}^{37}Cl$ value of 0.00‰.

For $\delta^{37}Cl$ determinations, ClO_4^- salts are first decomposed at 600 to 650°C in evacuated glass tubes to produce alkali chloride salts that are then analyzed according to well established methods[14,27,40]. The alkali chloride salts produced by ClO_4^- decomposition are dissolved in warm 18.2 MΩ deionized water, and Cl is precipitated as AgCl by the addition of $AgNO_3$. The resulting AgCl is recovered by centrifugation, washed in dilute HNO_3, dried, and reacted in a sealed glass tube with excess CH_3I at 300°C for 2 h to produce CH_3Cl. The resulting CH_3Cl is purified using gas chromatography, cryoconcentrated, and then admitted to a Finnigan Delta Plus XL isotope ratio mass spectrometer and analyzed in dual inlet mode by measurements at m/z 50 and 52. Samples are analyzed in replicate if sufficient amounts are available.

Routine calibration of Cl isotopic analyses is done by using a pair of ClO_4^- isotopic reference materials (USGS37 and USGS38) with contrasting isotopic compositions:

$$\delta^{37}Cl_{i/VSMOW} = \delta^{37}Cl_{37/VSMOW} + [\delta^{37}Cl_{i/rg} - \delta^{37}Cl_{37/rg}]_{meas.} \cdot$$

$$[\delta^{37}Cl_{38/VSMOW} - \delta^{37}Cl_{37/VSMOW}] / [\delta^{37}Cl_{38/rg} - \delta^{37}Cl_{37/rg}]_{meas.}$$

where 37 and 38 refer to the ClO_4^- USGS37 and USG38 reference materials and rg is an internal laboratory reference gas [CH_3Cl] against which all samples and reference materials are analyzed in the mass spectrometer during a single batch of analyses. The $\delta^{37}Cl$ scale is based on isotopic analyses of the USGS ClO_4^- isotopic reference materials against SMOC. After normalization, the analytical precision of $\delta^{37}Cl$ values ranges from ±0.1 to 0.3‰, based on replicate analyses of samples and isotopic reference materials.

36Cl Analysis—This analysis of 36 Cl in ClO_4^- is performed by accelerator mass spectrometry (AMS)[14,27,40]. Prior to AMS, the ClO_4^- is decomposed and the Cl^- is recovered as AgCl. The AgCl is dissolved and the Cl^- is purified twice by anion chromatography (using a method developed by the PRIME Laboratory at Purdue University) to ensure removal of trace amounts of S that may cause isobaric interference at mass 36. Purified Cl^- is then precipitated as AgCl for AMS measurement. Analysis of seawater Cl^- provides a reference datum of $^{36}Cl:Cl = 0.0 \times 10^{-15}$.

3.7.1 Differentiating Perchlorate Sources by Isotopic Methods

Comprehensive measurements of stable isotope ratios (^{37}Cl:^{35}Cl and ^{18}O:^{17}O:^{16}O) and radioactive ^{36}Cl of ClO_4^- from known synthetic and natural sources revealed systematic differences in isotopic characteristics related to formation mechanisms [4,5,27,40]. In addition, isotopic analyses of ClO_4^- extracted from water samples using Purolite A-530E resin demonstrated the feasibility of identifying ClO_4^- sources in contaminated environments based on this technique[5,9,27,40]. Both natural and synthetic sources of ClO_4^- were identified in groundwater samples from sites across the U.S. by the isotopic method. In addition, the Cl and O isotope effects of microbial degradation were investigated in both laboratory and field studies[14,42], providing baseline data for ClO_4^- natural attenuation monitoring and constraints on possible complications for stable isotope forensics.

Representative stable isotopic compositions of ClO_4^- from various sources, along with fractionation trends during microbial ClO_4^- reduction, are summarized in Figure 3.7. Synthetic ClO_4^- has relatively uniform δ^{37}Cl and more variable δ^{18}O, reflecting its production by electrolysis of NaCl brine. The NaCl feed material typically has a narrow range of δ^{37}Cl and is almost completely oxidized to ClO_4^-, resulting in relatively constant δ^{37}Cl in the ClO_4^- product. In contrast, the H_2O feed material can have a range of δ^{18}O and some of the O is lost during the process, resulting in variable δ^{18}O in the ClO_4^- product. Natural ClO_4^- from the Atacama Desert has distinctive isotopic characteristics, including unusually low δ^{37}Cl and elevated δ^{17}O, presumably reflecting ClO_4^- production by reactions involving O_3 and Cl species in the atmosphere[4,5,27]. With the exception of one sample with δ^{18}O = $-25\permil$ and Δ^{17}O = $+4\permil$[4], reported ClO_4^- stable isotopes are similar in natural Atacama salt samples and processed Atacama nitrate fertilizer samples, indicating little or no modification of the natural isotopic composition during processing of the ore into fertilizer in the past (at least until around 2000, when the processing method was reportedly modified to decrease ClO_4^- in the finished fertilizer).

^{36}Cl data combined with δ^{37}Cl values provided complete discrimination of the three principal sources of ClO_4^- in the southwestern U.S., namely synthetic ClO_4^-, Atacama natural ClO_4^-, and indigenous natural ClO_4^-, as shown in Figure 3.8 [27]. In natural ClO_4^-, high ^{36}Cl:Cl values (indigenous material) may indicate ClO_4^- formed largely in the upper atmosphere, whereas lower values (Atacama) could be consistent with postformation radioactive decay over longer time scales[27]. Low ^{36}Cl:Cl values in the synthetic ClO_4^- were consistent with production from terrestrial sources of Cl^-.

Microbial reduction of ClO_4^- causes substantial fractionation of Cl and O isotopes (Figure 3.8). This provides a useful tool for detecting natural or artificial attenuation of ClO_4^-, but it also has the potential to cause ambiguity in the application of isotopes for source identification. Microbial ClO_4^- reduction is common under reducing conditions, such as in aquifers where dissolved oxygen and NO_3^- have also been reduced. However, experiments to date indicate that relative fractionation factors for the different isotopes are predictable, and the multidimensional approach to isotope forensics (Figures 3.8 and 3.9) is not seriously compromised by the degradation patterns.

Isotopic analysis of ClO_4^- is proving useful for identifying ClO_4^- sources in groundwater and surface water. Synthetic ClO_4^- has been identified isotopically in several

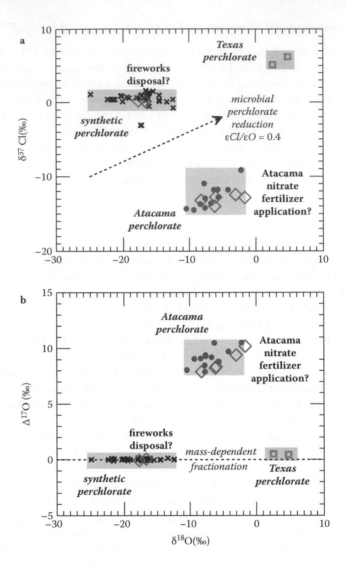

FIGURE 3.8   a. Cl and O stable isotope variations in ClO_4^-. (*Source:* Modified from Böhlke et al. *Environ. Sci. Technol.* 2009, 43, 5619–5625.) Data for ClO_4^- recovered from contaminated groundwater, Long Island, New York (open diamond symbols) compared with major ClO_4^- occurrences described previously. (*Sources:* Böhlke et al. *Anal. Chem.* 2005, 77, 7838–7842. With permission. Sturchio et al. In *Perchlorate Environmental Occurrences, Interactions, and Treatment,* Gu, B. et al., Eds. Springer: New York, 2005, 93–109. With permission. Unpublished data.) b. Synthetic ClO_4^- (crosses) from laboratory reagents, fireworks, road flares, and other products. Atacama ClO_4^- (dots) from samples of natural salt deposits and processed nitrate fertilizer products. Texas ClO_4^- (squares) from groundwater in southern High Plains Aquifer. Microbial ClO_4^- reduction trend is from field and laboratory experiments expected to follow approximately the mass-dependent fractionation trend. (*Sources:* Sturchio, N.C. et al. *Environ. Sci. Technol.* 2007, 41, 2796–2802. Hatzinger, P.B. et al. *Environ. Chem.* 2009, 6, 44–52. With permission.)

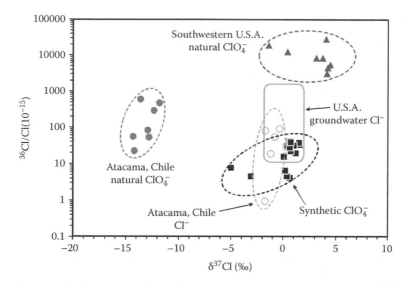

FIGURE 3.9 ^{36}Cl:Cl atom ratio versus $\delta^{37}Cl$ (‰) in representative samples of synthetic ClO_4^- reagents and products, natural ClO_4^-, and Cl^- extracted from soil and groundwater from the Atacama Desert in Chile, and natural ClO_4^- extracted from groundwater and soil from the southwestern U.S. Sizes of symbols exceed analytical errors. Delineated rectangular area shows ranges of ^{36}Cl:Cl ratios and $\delta^{37}Cl$ values for Cl^- in U.S. groundwater. (*Source:* Modified from Sturchio, N.C. et al. *Environ. Sci. Technol.* 2009, 43, 6934–6938.)

locations where such sources were known or suspected, thus confirming the stability of the isotopic compositions during transport under oxic conditions for years to decades. For example, synthetic ClO_4^- was identified isotopically in oxic groundwater contaminated by leaching of ClO_4^- from a fireworks disposal site on Long Island, New York (Figure 3.8). Synthetic ClO_4^- also was identified isotopically in the discharge area of a major contaminant plume and nearby surface water near Henderson, Nevada, supporting other evidence for this location as a major source of ClO_4^- in the Colorado River[5].

In contrast, stable isotope data indicate that Atacama ClO_4^- is present in groundwater affected by agricultural activity at some locations in the U.S., presumably reflecting past application of natural nitrate fertilizer containing ClO_4^- from Chile. This source was identified, for example, in relatively old groundwater (>20 years after recharge) beneath current or former agricultural land on Long Island where the ClO_4^- concentrations commonly exceeded the state drinking water guidance level of 5 µg/L (Figure 3.8).

Similar features may be common elsewhere in groundwater that recharged beneath fields that received Atacama nitrate fertilizer from the late nineteenth through the twentieth centuries[9]. In contrast, groundwater samples from the southern high plains in western Texas contain ClO_4^- of apparently natural origin that reveals isotopic composition distinct from that of the Atacama Desert materials[5,27,40.] Continuing research is designed to determine the mechanisms by which ClO_4^- acquires different isotopic characteristics, identifies additional sources of the natural ion, and tests the stability of ClO_4^- isotopic compositions during transport in the environment.

ACKNOWLEDGMENTS

This work was supported in part by the U.S. Department of Defense's Environmental Security Technology Certification Program and Strategic Environmental Research and Development Program. Helpful reviews of the manuscript were provided by Colleen Rostad of the U.S. Geological Survey and Srinath Rajagopalan of Tamil Nadu, India. The Oak Ridge National Laboratory is managed by UT–Battelle LLC for the U.S. Department of Energy under Contract DE–AC05–00OR22725. Any use of trade, product, or firm names is for descriptive purposes only and does not imply endorsement by the U.S. government.

REFERENCES

1. Gu, B. and Coates, J.D. *Perchlorate Environmental Occurrence, Interactions and Treatment.* Springer: New York, 2006.
2. Gu, B.H., Brown, G.M., and Chiang, C.C. Treatment of perchlorate-contaminated groundwater using highly selective, regenerable ion exchange technologies. *Environ. Sci. Technol.* 2007, 41, 6277–6282.
3. Ericksen, G.E. *Geology and Origin of the Chilean Nitrate Deposits.* Professional Paper 1188, U.S. Geological Survey: Washington, D.C., 1981.
4. Bao, W. and Gu, B. Natural perchlorate has its unique oxygen isotope signature. *Environ. Sci. Technol.* 2004, 38, 5073–5077.
5. Böhlke, J.K., Sturchio, N.C., Gu, B. et al. Perchlorate isotope forensics. *Anal. Chem.* 2005, 77, 7838–7842.
6. DasGupta, P.K., Dyke, J.V., Kirk, A.B. et al. Perchlorate in the United States: Analysis of relative source contributions to the food chain. *Environ. Sci. Technol.* 2006, 40, 6608–6614.
7. Dasgupta, P.K., Martinelango, P.K., Jackson, W.A. et al. Origin of naturally occurring perchlorate: Role of atmospheric processes. *Environ. Sci. Technol.* 2005, 39, 1569–1575.
8. Jackson, W.A., Anandam, S.K., Anderson, T. et al. Perchlorate occurrence in the Texas southern high plains aquifer system. *Ground. Mon. Remed.* 2005, 25, 137–149.
9. Böhlke, J.K., Hatzinger, P.B., Sturchio, N.C. et al. Atacama perchlorate as an agricultural contaminant in groundwater: isotopic and chronologic evidence from Long Island, New York. *Environ. Sci. Technol.* 2009, 43, 5619–5625.
10. Gu, B., Brown, G.M., Alexandratos, S.D. et al. Efficient treatment of perchlorate (ClO_4^-)-contaminated groundwater by bifunctional anion exchange resins. In *Perchlorate in the Environment*, Urbansky, E. T., Ed. Kluwer Plenum: New York, 2000, 165–176.
11. Gu, B., Brown, G.M., and Chiang, C.C. Treatment of perchlorate-contaminated groundwater using highly selective, regenerable ion exchange technologies. *Environ. Sci. Technol.* 2007, 41, 6277–6282.
12. Gu, B., Brown, G.M., Maya, L. et al. Regeneration of perchlorate ClO_4^--loaded anion exchange resins by novel tetrachloroferrate $FeCl_4^-$ displacement technique. *Environ. Sci. Technol.* 2001, 35, 3363–3368.
13. Gu, B., Ku, Y., and Brown, G.M. Treatment of perchlorate-contaminated water using highly selective, regenerable ion exchange technology: A pilot scale demonstration. *Fed. Facil. Environ. J.* 2003, 14, 75–94.
14. Sturchio, N.C., Böhlke, J.K., Beloso, A. D. et al. Oxygen and chlorine isotopic fractionation during perchlorate biodegradation: Laboratory results and implications for forensics and natural attenuation studies. *Environ. Sci. Technol.* 2007, 41, 2796–2802.

15. Sturchio, N.C., Böhlke, J.K., Gu, B. et al. Stable isotopic compositions of chlorine and oxygen in synthetic and natural perchlorates. In *Perchlorate Environmental Occurrences, Interactions, and Treatment*, Gu, B. et al., Eds. Springer: New York, 2005, 93–109.

16. Batista, J.R., McGarvey, F.X., and Vieira, A.R. Removal of perchlorate from waters using ion exchange resins. In *Perchlorate in the Environment*, Urbansky, E.T., Ed. Kluwer Plenum: New York, 2000, 135–145.

17. Tripp, A.R. and Clifford, D.A. Treatability of perchlorate in groundwater using ion exchange technology. In *Perchlorate in the Environment*, Urbansky, E.T., Ed. Kluwer Plenum: New York, 2000, 123–134.

18. Gu, B. and Brown, G.M. Recent advances in ion exchange for perchlorate treatment, recovery and destruction. In *Perchlorate Environmental Occurrences, Interactions, and Treatment*, Gu, B. et al., Eds. Springer: New York, 2006, 209–251.

19. Gu, B., Ku, Y., and Brown, G. Sorption and desorption of perchlorate and U(VI) by strong base anion exchange resins. *Environ. Sci. Technol.* 2005, 39, 901–907.

20. Gu, B., Ku, Y., and Jardine, P.M. Sorption and binary exchange of nitrate, sulfate, and uranium on an anion exchange resin. *Environ Sci. Technol.* 2004, 38, 3184–3188.

21. Urbansky, E.T. Perchlorate chemistry: Implications for analysis and remediation. *Bioremed. J.* 1998, 2, 81–95.

22. Bonnesen, P.V., Brown, G.M., Bavoux, L.B. et al. Development of bifunctional anion exchange resins with improved selectivity and sorptive kinetics for pertechnetate. 1. Batch equilibrium experiments. *Environ. Sci. Technol.* 2000, 34, 3761–3766.

23. Gu, B., Brown, G.M., Bonnesen, P.V. et al. Development of novel bifunctional anion exchange resins with improved selectivity for pertechnetate sorption from contaminated groundwater. *Environ. Sci. Technol.* 2000, 34, 1075–1080.

24. Moyer, B.A. and Bonnesen, P.V. Physical factors in anion separations. In *Supramolecular Chemistry of Anions*, Bianchi, A. et al., Eds. VCH: New York, 1997.

25. Brown, G.M., Bonnesen, P.V., Moyer, B.A. et al. Design of selective resins for the removal of pertechnetate and perchlorate from groundwater. In *Perchlorate in the Environment*, Urbansky, E.T., Ed. Kluwer Plenum: New York, 2000, 155–164.

26. Gu, B. and Brown, G.M. Field demonstration using highly selective, regenerable ion exchange and perchlorate destruction technologies for water treatment. In *Perchlorate Environmental Occurrences, Interactions, and Treatment*, Gu, B. et al., Eds. Springer: New York, 2006, 253–278.

27. Sturchio, N.C., Caffee, M., Beloso, A.D., Jr. et al. Chlorine ⁻ 36 as a tracer of perchlorate origin. *Environ. Sci. Technol.* 2009, 43, 6934–6938.

28. Venkatesh, K.R., Klara, S.M., Jennings, D.L. et al. Removal and destruction of perchlorate and other anions from ground water using the ISEP+™ system. In *Perchlorate in the Environment*, Urbansky, E. T., Ed. Kluwer Plenum: New York, 2000, 147–153.

29. Gu, B., Dong, W., Brown, G.M. et al. Complete degradation of perchlorate in ferric chloride and hydrochloric acid under controlled temperature and pressure. *Environ. Sci. Technol.* 2003, 37, 2291–2295.

30. Rajagopalan, S., Anderson, T.A., Fahlquist, L. et al. Widespread presence of naturally occurring perchlorate in high plains of Texas and New Mexico. *Environ. Sci. Technol.* 2006, 40, 3156–3162.

31. Krynitsky, A.J., Niemann, R.A., and Nortrup, D.A. Determination of perchlorate anion in foods by ion chromatography–tandem mass spectrometry. *Anal. Chem.* 2004, 76, 5518–5522.

32. Sanchez, C.A., Krieger, R.I., Khandaker, N. et al. Accumulation and perchlorate exposure potential of lettuce produced in the Lower Colorado River region. *J. Agric. Food Chem.* 2005, 53, 5479–5486.

33. Kirk, A.B., Martinelango, P.K., Tian, K. et al. Perchlorate and iodide in dairy and breast milk. *Environ. Sci. Technol.* 2005, 39, 2011–2017.
34. Blount, B.C., Pirkle, J.L., Osterloh, J.D. et al. Urinary perchlorate and thyroid hormone levels in adolescent and adult men and women living in the United States. *Environ. Health Persp.* 2006, 114, 1865–1871.
35. Blount, B.C., Valentin-Blasini, L., Osterloh, J.D. et al. Perchlorate exposure of the U.S. population, 2001–2002. *J. Exp. Sci. Environ. Epidemiol.* 2007, 17, 400–407.
36. Dyke, J.V., Ito, K., Obitsu, T. et al. Perchlorate in dairy milk: comparison of Japan versus the United States. *Environ. Sci. Technol.* 2007, 41, 88–92.
37. Kirk, A.B., Dyke, J.V., Martin, C.F. et al. Temporal patterns in perchlorate, thiocyanate, and iodide excretion in human milk. *Environ. Health Persp.* 2007, 115, 182–186.
38. Pearce, E.N., Leung, A.M., Blount, B.C. et al. Breast milk iodine and perchlorate concentrations in lactating Boston area women. *J. Clin. Endocrinol. Metab.* 2007, 92, 1673–1677.
39. Böhlke, J.K., Ericksen, G.E., and Revesz, K. Stable isotope evidence for an atmospheric origin of desert nitrate deposits in northern Chile and southern California, USA. *Chem. Geol.* 1997, 136, 135–152.
40. Sturchio, N.C., Böhlke, J.K., Gu, B. et al. Stable isotopic compositions of chlorine and oxygen in synthetic and natural perchlorates. In *Perchlorate Environmental Occurrences, Interactions, and Treatment*, Gu, B. et al., Eds. Springer: New York, 2006, 93–109.
41. Urbansky, E.T. and Brown, S.K. Perchlorate retention and mobility in soils. *J. Environ. Mon.* 2003, 5, 455–462.
42. Hatzinger, P.B., Böhlke, J.K., Sturchio, N.C., et al. Fractionation of stable isotopes in perchlorate and nitrate during in situ biodegradation in a sandy aquifer. *Environ. Chem.* 2009, 6, 44–52.
43. Hatzinger, P.B., Diebold, J., Yates, C.A. et al. Field demonstration of in situ perchlorate bioremediation in groundwater. In *Perchlorate Environmental Occurrences, Interactions, and Treatment*, Gu, B. et al., Eds. Springer: New York, 2006, 311–341.
44. Gonfiantini, R. Standards for stable isotope measurements in natural compounds. *Nature* 1978, 271, 534–536.
45. Brand, W.A., Coplen, T.B., Aerts-Bijma, A.T. et al. Comprehensive inter-laboratory calibration of reference materials for $\delta^{18}O$ versus VSMOW using various online high temperature conversion techniques. *Rapid Comm. Mass Spectr.* 2009, 23, 999–1019.
46. Böhlke, J.K., Mroczkowski, S.J., and Coplen, T.B. Oxygen isotopes in nitrate: New reference materials for $^{18}O:^{17}O:^{16}O$ measurements and observations on nitrate–water equilibration. *Rapid Comm. Mass Spectr.* 2003, 17, 1835–1846.
47. Godon, A., Jendrzejewski, N., Eggenkamp, H.G. et al. International cross calibration over a large range of chlorine isotope compositions. *Chem. Geol.* 2004, 207, 1–12.

4 Influence of Ion Exchange Resin Particle Size and Size Distribution on Chromatographic Separation in Sweetener Industries

Tuomo Sainio, Ari Kärki, Jarmo Kuisma, and Heikki Mononen

CONTENTS

ABSTRACT

The sweetener industry is a major user of ion exchange resins for chromatographic separation. The glucose–fructose separation in the manufacture of high fructose corn syrup (HFCS) is one of the most successful applications of the simulated moving bed technology. This chapter focuses on the influences of particle size and particle size distribution on separation efficiency in low pressure systems. The effect of particle size on separation efficiency is well known. The role of particle size distribution has been studied to a lesser extent. Experimental data are discussed for glucose–fructose separation and desugarization of molasses. A mathematical model for simulating chromatographic separation in the case of wide particle size distribution is presented. The productivity of a batch column for glucose–fructose separation was found to decrease almost linearly with increasing particle size. The benefit of narrow particle size distribution was also demonstrated experimentally and by simulations. With large particles, the risk of viscous fingering increases with very narrow particle size distribution.

4.1   INTRODUCTION

In industrial chromatography, the homogeneity and symmetry of a system (including the column and flow distributors, velocity and temperature profiles, resin, all external piping, etc.) are essential for efficient separation. In the case of chromatographic separation using ion exchange resins, the effects of particle size and particle size distribution of the stationary phase are especially interesting.

Most applications (apart from water softening and gas adsorption) that employ ion exchange resins use particle diameters between 0.2 and 0.4 mm[1]. Typically, the cross-link density of the resin in large volume applications is 5 to 8 wt-%[2]. The reasons for this are rather obvious. The optimum particle size is a trade-off between pressure drop (investment and operating costs) and mass transfer rate (separation efficiency). A high cross-link density gives a resin a better mechanical stability but requires smaller particles to counterbalance the higher mass transfer resistance. These aspects are well understood and can be incorporated relatively easily into process simulators to find an economical optimum. The question of optimum width of particle size distribution with respect to process performance is somewhat more controversial.

The uniformity coefficient of the stationary phase that describes the width of the particle size distribution varies typically between 1.05 and 1.15 in industrial chromatography. Besides mass transfer homogeneity, particle size distribution also affects packing properties such as bed porosity and the amplitude of swelling–shrinking cycles of the bed during the separation process. These effects are difficult to investigate on a small scale or to incorporate into dynamic simulation models. Unfortunately, information in the open literature about the influence of particle size distribution on industrial scale separations is scarce. Several reports studying these effects for small scale separations [high performance liquid chromatography (HPLC)] are available[3,4,5]. Under such conditions, many cost studies concluded that low-quality packing materials are actually more expensive to use than high-quality ones based on decreased production rates and increased purification costs[6]. Conversely, some studies found no influence of particle size distribution on column performance[7].

The manufacture of industrial grade ion exchange resins falls into two categories based on polymerization technology: the classical suspension polymerization combined with sieving and so-called jetting polymerization. The classical technique produces a relatively wide and nearly Gaussian particle size distribution due to coalescence (and breaking) of droplets during agitation. In addition, the coalescence frequency depends on the position in the reactor. The jetting technique aims to achieve a narrow particle size distribution by feeding the reactor with droplets formed by fractionating a jet of a monomer solution[8].

Both technologies display advantages and disadvantages. Since jetting produces a higher yield with a given mean particle size, it is economical for mass production. However, the use of this technology may have limited the spectrum of available ion exchange resins, especially in chromatography applications. From a manufacturing view, the obvious disadvantage of the Gaussian suspension polymerization technique is the lower yield of product at a given mean particle size. However, the classical method allows more degrees of freedom due to easy adaptability in manufacture of chromatographic resins. The fine tuning of resin performance based on, for example, degree of cross-linkage, porosity, and mean particle size is well established today. In addition, the particle size distribution can be customized for customers' needs.

Experience has shown that theoretical predictions based on small scale experiments and empirical knowledge accumulated by fine tuning industrial scale separation plants appear to be contradictory because the behaviors of packed beds with particles of different sizes in cyclic processes are difficult to predict. As an example, the influence of particle size distribution on mass transfer effects in a separation column is relatively straightforward to investigate by numerical simulations.

The purpose of this chapter is to investigate the influence of particle size and particle size distribution on chromatographic separation in low pressure systems with ion exchange resins as packing materials. The model systems were chosen from those used in the sweetener industry where ion exchange resins are used in very large production volume plants. Only batch operation of fixed bed columns is discussed. Packing procedures and the use of simulated moving and expanded beds are not discussed here.

4.2   MEAN PARTICLE SIZE

Before demonstrating the influence of particle size with experimental and simulated data, the meaning of the mean particle size concept should be clarified. Several definitions are commonly used, but they provide different values, depending on the underlying particle size distribution[9]. A general equation can be written as:

$$\bar{d}_p = \frac{\sum_{j=1}^{N} d_{p,j}^{k+1} n_j}{\sum_{j=1}^{N} d_{p,j}^{k} n_j} \tag{4.1}$$

where N is the number of particle size fractiles and d_p and n represent the particle diameter and number of particles in fractile j. For $k = 0$, Equation (4.1) yields a number average diameter. For $k = 2$ and $k = 3$, the surface average and volume average particle sizes are obtained. The number average and volume average mean values are the most useful and thus the most commonly reported.

A fourth definition of mean particle size is particularly useful when mass transfer effects are of interest. Mass transfer rate in and out of resin particles is proportional to the total surface area of the solid phase in contact with the liquid phase. The molar number of adsorbates in the particles is proportional to the volume of the solid phase. Therefore, a combination of these two properties is needed to characterize the mass transfer properties of a given batch of resin. The Sauter mean diameter is defined as the diameter of a particle that has the same volume-to-surface area ratio as the entire batch of particles:

$$\bar{d}_{p,S} = 6\frac{V_{tot}}{A_{tot}} = \frac{\sum_{j=1}^{N} d_{p,j}^{3} n_j}{\sum_{j=1}^{N} d_{p,j}^{2} n_j} \tag{4.2}$$

where V_p and A_p are the total volume and total surface area of the batch. The Sauter diameter is preferred in simulation of mass transfer controlled physical phenomena and unit operations, especially if incorporating a detailed particle size distribution in a model is computationally expensive. For example, the mass transfer properties of two batches of resins with different size distributions (but otherwise identical physical properties) are very similar if their Sauter diameters are equal.

4.3   OPERATIONAL ASPECTS

Mass transfer resistance and dispersion flatten the concentration gradients in the column, leading to increased dilution and reduced separation efficiency. Such effects are detrimental to productivity and should be minimized. We are interested in the influence of particle size and size distribution on dispersive effects in chromatography columns. Optimization of the material properties should be carried out by considering the process economics: the extent to which the particle size can be reduced before the pressure drop of the bed becomes unacceptably large.

4.3.1   BAND BROADENING

At a macroscopic level, flow through a packed bed in a cylindrical column can be described with a mass balance:

$$\frac{\partial c}{\partial t} = -u^{L} \frac{\partial c}{\partial z} - F \frac{\partial \overline{q}}{\partial t} + D_{L} \frac{\partial^{2} c}{\partial z^{2}} + \frac{1}{r} \frac{\partial}{\partial r} \left(D_{T} r \frac{\partial c}{\partial r} \right) \tag{4.3}$$

In Equation (4.3), c is the liquid phase concentration, $\overline{q}$ is the solid phase concentration (averaged over the radial direction), u^{L} is the superficial velocity, F is the phase ratio, and t is time. The axial and radial coordinates are denoted by z and r, and the dispersion coefficients in the corresponding directions by D_{L} and D_{T}.

4.3.1.1   Particle Size and Size Distribution

The influence of particle size on axial dispersion is well known: the smaller the particles, the smaller the dispersion. In absence of experimental data, D_{L} can be estimated from correlations such as the model proposed by Chung and Wen[10]. Note that particle size per se exerts a negligible effect on the axial dispersion coefficient, provided that the product of the flow rate and the particle size (i.e. the denominator in the definition of the Peclet number) is constant[11]. In contrast, particle size distribution affects the dispersion coefficient. Stronger axial dispersion is obtained when particle size distribution is wider[12].

Particle size distribution has very little effect on the transverse (radial) dispersion coefficient as long as the packing density (porosity) and mean particle size remain constant[12]. However, D_{T} is strongly influenced by the quality of the packing, and very different results can be obtained with similar volume average properties of the bed (i.e., porosity and tortuosity)[13]. Nevertheless, transverse dispersion has a much smaller effect on separation efficiency than axial dispersion in conventional fixed bed and SMB separations mainly because of high velocities in the longitudinal direction.

More severe problems are caused by uneven flow patterns due to imperfect flow distributors. Only in special applications such as continuous rotating annular chromatography[14] is the transverse dispersion clearly more significant than axial dispersion. Delgado[13] recently reviewed the literature on dispersion in packed beds, including the influence of particle size and size distribution. The influence of particle size on mass transfer resistance is well known. Doubling the particle size results in a four-fold increase in the time needed for the transfer of a certain quantity into or out of the resin. As noted above, the effective particle size of a batch of resin is conveniently approximated by the Sauter diameter. Both sum terms on the right hand side of Equation (4.2) are dominated by the large particles in the batch. For a batch of resin with normally distributed particle size, the Sauter diameter is therefore larger than the number average mean particle size, and the overall mass transfer rate is lower than expected.

4.3.1.2 Thermodynamics

It should be noted that thermodynamics (adsorption equilibrium) can lead to broadening of the concentration fronts. The role of thermodynamics is conveniently demonstrated by using the ideal model of chromatography that assumes local phase equilibrium between the liquid and solid phases and neglects axial and radial dispersion. For a single solute (under isothermal, isocratic conditions), the ideal model can be written:

$$\frac{\partial c}{\partial t}\left(1 + F\frac{\partial q(c)}{\partial c}\right) = -u^L\frac{\partial c}{\partial z} \tag{4.4}$$

Equation (4.4) is a first-order hyperbolic partial differential equation, the solution of which is usually obtained by integrating it with respect to the independent variables. This yields the concentrations c and q at a certain location z in the column at a certain time t. However, the ideal model also defines (implicitly) the rate at which the location z of concentration state c changes with time. In other words, it is possible to determine the propagation velocity u_c of concentration state c:

$$u_c = \frac{u^L}{1 + F\frac{\partial q(c)}{\partial c}} \tag{4.5}$$

This equation shows the well known fact that the propagation velocity depends on the slope of the adsorption isotherm. For convex isotherms (e.g., Langmuir), low concentrations propagate slower than high concentrations. This results in sharpening of concentration fronts (shocks) in the course of loading and tailing (simple waves) during elution. For concave isotherms the situation is reversed. Thermodynamics and dispersive effects can thus either be cooperative or counteract each other. In the case of convex isotherms, cooperation results in increased eluent consumption and dilution both in batch and SMB chromatography. This holds also for multicomponent systems with competitive (ad)sorption, but the resulting concentration profiles are more complex since the components affect each other's propagation velocities. For the linear isotherms common in sugar separations, propagation velocity is independent of concentration. Band broadening is thus governed by dispersion and mass transfer resistance rather than thermodynamic effects.

4.3.2 Pressure Drop

To achieve an industrially feasible production rate, the permeability of the packing material should be as high as possible because the linear flow rate in the column is often limited by the maximum allowable pressure drop. The permeability of a column can be characterized by experimentally determining the permeability coefficient. In practice, the pressure drop over the bed is determined as a function of the linear flow velocity, and the permeability coefficient is calculated from:

$$k = \frac{\Delta p}{\mu L u^{L}} \qquad (4.6)$$

where k is the permeability coefficient, Δp is the pressure drop across the bed, m is the viscosity, L is the column length, and u^{L} is the linear flow rate[7]. Obviously, the permeability coefficient depends on the particle size and bed porosity. In order to compare the flow resistances of different materials, the effect of particle size can be eliminated by calculating the resistance factor, ϕ, as shown in Equation (4.7).

$$\phi = \frac{d_{p}^{2}}{k} = \frac{\Delta p d_{p}^{2}}{\mu L u^{L}} \qquad (4.7)$$

For incompressible particles and fluids, the pressure drop due to friction forces can be approximated by using the well known Kozeny–Carman equation[15]. When the particles are perfectly spherical, the equation can be written as:

$$\frac{\Delta p}{L} = \frac{150 \mu u^{L}}{d_{p}^{2}} \frac{(1-\varepsilon)^{2}}{\varepsilon^{3}} \qquad (4.8)$$

With ion exchange resins, however, the resistance factor usually depends on flow rate. Nonlinear behavior of the pressure drop as a function of flow rate is caused by an elastic deformation of the particles under pressure. Compression of the particles decreases the porosity of the bed, which in turn leads to a lower permeability coefficient in the denominator of Equation (4.7).

Soshina and Zaostrovskii[16] found that the porosity of a PS-DVB resin bed decreases exponentially with a pressure drop across the bed. Similar results were reported by Keener et al.[17] for several agarose and polymethacrylate resins used in ion exchange chromatography and hydrophobic interaction chromatography. On the other hand, Danilov et al.[18] found a nearly linear relationship between the porosity and pressure drop for relatively soft size exclusion gels (Sephadex G–25).

Due to the elastic deformation of particles, theoretical prediction of pressure drop in an ion exchange resin bed is not straightforward. Several attempts with varying degrees of rigor have been discussed by Watler et al.[15]. Keener et al.[17] derived an empirical model that extends the Kozeny–Carman equation to the case of an elastic packing material as follows:

$$\frac{\Delta p}{L} = \frac{150 \mu u^{L}}{d_{p}^{2}} \frac{\left(1 - \varepsilon_{0} \exp\left[b\left(1 - CF\right)\right]\right)^{2}}{\left(\varepsilon_{0} \exp\left[b\left(1 - CF\right)\right]\right)^{3}} \qquad (4.9)$$

where ε_{0} is the porosity of the bed at $u^{L} = 0$, and b and CF are empirical parameters. The compression factor, CF, is obtained by determining the bed height as a function of flow rate, and b is obtained from mechanical compression data where the porosity of the bed under compression is recorded.

Hongisto[19,20] noted three decades ago that when elastic ion exchange resins are used in industrial chromatography columns for sugar and polyol production, the value of the permeability coefficient, k, depends also on the dimensions of the bed (i.e., height and width). According to Keener et al., Equation (4.9) can be applied to predict pressure drops for beds of different scales after the parameters b and CF are obtained at one scale. Unfortunately, these parameters are not available for styrenic ion exchange resins used commonly in the sweetener industry. According to Hongisto, the permeability coefficients of such resins are in the order of 10^{-4} to $4 \cdot 10^{-4}$ m^{-2} when the column diameter is larger than 1.0 m, bed height exceeds 3.0 m, and the particle size is 0.25 to 0.50 mm[19].

4.3.2.1 Particle Size Distribution

Considering that particle size distribution affects the packing density of the particles and the size of the flow channels in a bed, it is expected that the pressure drop depends also on the width of the particle size distribution. This aspect has been studied to some extent with HPLC packing materials but less with ion exchange resins for industrial scale chromatography.

Verzele and Dewaele[21] studied the influence of particle size distribution on pressure drop in preparative HPLC. Particles of two sizes, $d_{p1} < d_{p2}$, were mixed in different proportions to pack chromatography columns with varying average particle sizes. The pressure drop over the bed was measured at a fixed flow rate. Figure 4.1 shows the results. The increase in pressure is calculated as shown in Equation (4.10) where y denotes how much higher a pressure drop is observed for the mixture of:

$$y = \frac{\Delta p_{col}^{mix} - \Delta p_{col}^{avg}}{\Delta p_{col}^{avg}} \times 100\% \qquad (4.10)$$

As seen in Figure 4.1, the pressure drop over a mixed bed is always larger than the drop for a bed containing particles of one size only. The figure indicates that the pressure drop increases rapidly when the fraction of small particles is low. In other words, the pressure drop is sensitive to the number of small particles in the bed. The highest pressure drop is recorded when small and large particles are mixed in approximately equal proportions. This means that the pressure drop is largest when the particle size distribution is as wide as possible.

4.4 INFLUENCE OF PARTICLE SIZE ON SEPARATION

To demonstrate the influence of mean particle size on chromatographic separation process performance, two case studies were carried out at laboratory scale. The systems studied were the separation of glucose and fructose and recovery of sucrose from beet molasses. In terms of production volume, the former is probably the most important true binary separation utilizing conventional four-zone SMB units at industrial scale. Sucrose recovery was chosen to include a ternary separation in which the target product (sucrose) is eluted between the other two components (salts and betaine). Such ternary separations are very common. The glucose–fructose separation and desugarization of molasses are discussed below.

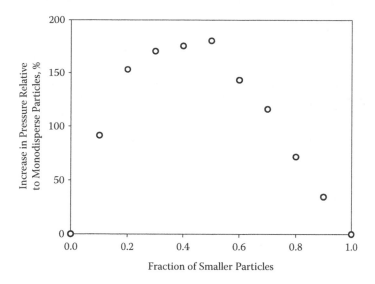

FIGURE 4.1   Increase in pressure drop in a mixed bed composed of particles of two sizes $(d_{p1} < d_{p2})$ relative to a bed of monodisperse particles with $d_p = fd_{p1} + (1 - f)d_{p2}$, where f is the fraction of smaller particles. (*Source:* Verzele, M. and Dewaele, C. *Preparative High Performance Liquid Chromatography: A Practical Guideline.* RSL/Alltech Europe, Eke, 1985. With permission.)

4.4.1   Glucose–Fructose Separation for HFCS

High fructose corn syrup (HFCS) is widely used as a sweetener in beverages. In fact, its consumption in the United States is almost equal to that of sugar. The fructose content of HFCS is typically 55% of dry solids. HFCS is manufactured by hydrolyzing corn starch to glucose and other sugars, highly purifying the glucose, and inverting it to fructose by using immobilized glucose isomerase. The fructose content of the isomerized liquor is 42% on a dry solids basis. It is chromatographically enriched to 55% by strong acid cation exchange resins[22]. To reduce the volumes processed in the separation unit, the chromatographic glucose–fructose separation is usually carried out to 90% fructose, which is then blended with 42% fructose syrup to produce the desired 55% HFSC.

4.4.1.1   Experimental Methods and Resin Properties

In this case study, the influence of the particle size of the ion exchange resin on the glucose–fructose separation efficiency was investigated. The feed solution consisted of 232 g/L of glucose and 168 g/L of fructose in water, which corresponds to the composition of inverted sugar. The ion exchange resin CS11GC (Finex Oy, Finland) in Ca^{2+} form was chosen as the stationary phase. It is a strong acid cation exchange resin (PS-DVB) and has a nominal cross-linking density of 5.5 wt-%.

Several batches of the resin with different mean particle sizes were supplied by the manufacturer. The properties of these batches are listed in Table 4.1. The ion exchange capacity (in H^+ form) and water content were determined by standard

TABLE 4.1

Properties of Batches of Finex CS11GC Resins

Na⁺ Mean Particle Size (μm)	Result within ±20 %	Cross-Link Density (wt-%)	H⁺ Volume Capacity.	Water Content (Na⁺, wt-%)	Bed Porosity
110	91.7	5.5	1.54	54.7	0.29
250	96.1	5.5	1.50	53.9	0.30
270	94.6	5.5	1.51	54.5	0.31
328	93.8	5.5	1.50	54.3	0.31
354	98.3	5.5	1.56	52.8	0.30
393	98.4	5.5	1.55	53.1	0.31
518	95.6	5.5	1.49	53.5	0.31

Note: Mean particle size and size distribution are volume-based properties.

techniques. Bed porosities were determined by pulse injection of blue dextrane (with M_w = 2000 000 g/mol). The resin was converted to Ca^{2+} form via standard techniques.

The mean particle size and particle size distributions of the batches of CS11GC resin (in Na⁺ form) were determined in a water suspension using a Pamas SVSS 16 particle size analyzer. The values reported are volume average particle size (Equation 4.1) and volume-based size distribution. As seen in the table and in Figure 4.2, the resins are monobead type (particle size distributions are very narrow for industrial grade resins).

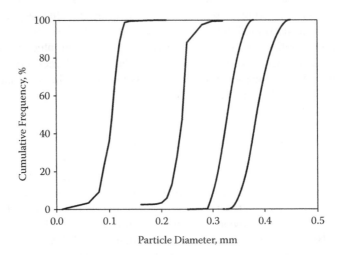

FIGURE 4.2 Cumulative particle size distributions corresponding to selected batches of CS11GC resin (see Table 4.1).

4.4.1.2 Results

Typical chromatograms appear in Figure 4.3. Fructose is more retained than glucose by the resin because fructose forms a slightly more stable complex with the resin in the Ca^{2+} form[23,24]. As expected, the smaller particle size yields better separation.

The influence of particle size on separation was quantified using the method of moments[25] to calculate bed porosity, apparent separation factor, HETP, and resolution. The resolution and apparent separation are shown as functions of mean particle size in Figure 4.4.

High resolution means that the peaks are narrow and overlap slightly; a high separation factor means that the mass centers of the peaks are widely separated. Nearly linear trends were observed for both quantities. The resolution decreased rapidly with increasing particle size due to increasing mass transfer resistance, as expected. The separation factor, however, should be independent of particle size. The apparent change in the separation factor originates from the asymmetry of the peaks: the mass center of the pulse shifts to the right as tailing increases due to mass transfer effects. Since fructose profiles exhibit stronger tailing than glucose profiles, the separation factor appears to improve with increasing particle size, although the separation actually becomes worse.

The influence of particle size on the yield of fructose is shown in Figure 4.5 for two flow rates and purity constraints of the product fraction. Decreasing the particle size strongly influenced the yield. Again, the trend was nearly linear, except when the yield was already high (above 90%).

When the flow rate was increased from 0.6 to 1.6 BV/h, the yield remained constant only if the purity requirement of fructose was lowered. The filled circles and open

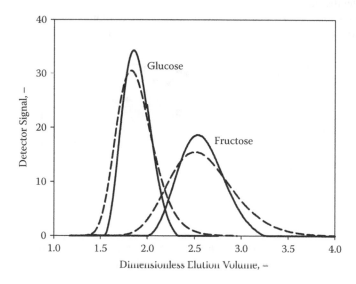

FIGURE 4.3 Influence of mean particle size on separation of glucose and fructose with CS11GC resin (Finex Oy, Finland) in Ca^{2+} form. Experimental data. Solid lines: $\bar{d}_p = 250$ μm; dashed lines: $\bar{d}_p = 328$ μm.

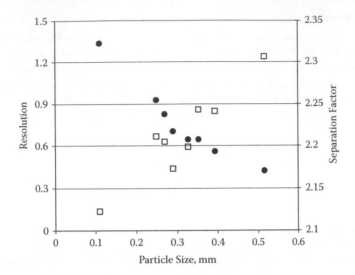

FIGURE 4.4 Influence of particle size on resolution (filled circles) and apparent separation factor (open squares) for separation of glucose and fructose in a batch column. Experimental data. The stationary phase utilized Finex CS11GC ion exchange resin in Ca_{2+} form. Feed concentrations were 232 g/L of glucose and 168 g/L of fructose. Flow rate was 0.6 BV/h.

squares in Figure 4.5 nearly coincide, indicating that the purity constraint should be decreased from 90 to 70% to maintain the desired yield of fructose. Since the product specifications cannot usually be changed, separation throughput can be increased only by reducing the particle size.

The productivity of the batch separation was calculated by assuming that the interval between consecutive injections was such that the collection of the glucose

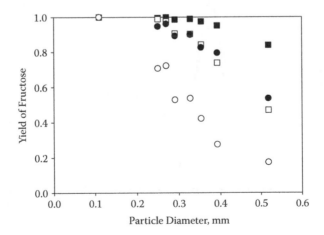

FIGURE 4.5 Influence of particle size on yield of fructose in a batch column (see Figure 4.4 caption). Flow rates: squares = 0.6 BV/h, circles = 1.6 BV/h. Purity constraints for fructose fraction: filled symbols = 70% fructose, open symbols = 90% fructose.

fraction of injection $N + 1$ would start when the collection of the fructose fraction of injection N ended. In other words, part of the tail of the fructose peak is collected in the fraction recycled during the process. This minimizes cycle time and thus maximizes the productivity, PR, which was calculated as:

$$PR_{fructose} = \frac{m_{fructose}^{feed} Y_{fructose}}{V_{col} t_{cycle}} \tag{4.11}$$

where Y is the yield per cycle, m is the amount of feed, and t_{cycle} is the cycle time. Since the amount of feed (loading) and column size were not varied in these experiments, the productivity was normalized such that PR was equal to unity for the highest flow rate and smallest particle size (1.6 BV/h, 110 µm).

The normalized productivity is presented as a function of particle size in Figure 4.6. As expected, small particles resulted in better productivity than large particles. Note that at low flow rates (squares in the figure), the productivity is little affected by the purity constraint of the fructose fraction. In addition, the purity constraint has no effect on the productivity at $d_p = 100$ µm because the cycle time does not depend strongly on the choice of the purity constraint when the target compound (fructose) is the more retained one. Moreover, the yield of fructose does not decrease rapidly with particle size at low flow rates. When the mass transfer resistance is more pronounced, however, the productivity is strongly influenced by the purity constraints.

The particle size clearly influenced productivity: the smaller the particles, the higher productivity is achieved. This is mainly due to the improved yield, as seen in Figure 4.5. The trend is again nearly linear, at least above 250 µm. The particle size was found to have a greater effect on the productivity when the flow rate was high

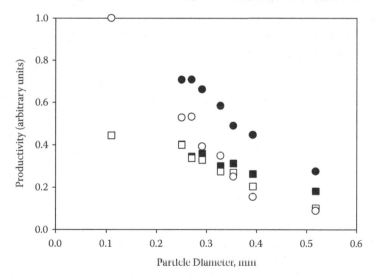

FIGURE 4.6 Influence of particle size on productivity of batch separation (see Figure 4.4 caption). Flow rates: squares = 0.6 BV/h, circles = 1.6 BV/h. Purity constraints for fructose fraction: filled symbols = 70% fructose, open symbols = 90% fructose.

(circles in Figure 4.6). These results demonstrate that the first option for increasing productivity is to increase the flow rate. However, the particle size should be as small as possible (limited by the pressure drop) to further improve productivity.

4.4.2 Molasses Desugarization

The viscous syrup left from crystallization in sugar cane and sugar beet processing is called molasses. The composition varies, depending on the raw material and number of crystallization steps. Typically, molasses consists mostly of sucrose and inorganic salts but contains also organic acids, amino acids, and betaine (trimethylglycine)—a derivative of the glycine amino acid that has commercial value for example as an additive in fodder[26].

According to Paananen[27], the world's first industrial scale molasses desugarization plant was built in Kotka, Finland, by Finnish Sugar Co. in 1965. The development and installation of batch processes continued until the mid 1980s, when Mitsubishi built the first simulated moving bed molasses plant for Hokobu Sugar Co. in Okinawa, Japan. This was soon followed by a plant delivered by Amalgamated Sugar for another Japanese sugar company in 1985. Amalgamated's process, which was employed in several installations in the U.S. during the 1990s, is a classical four-zone SMB with a continuous flow of all streams (feed, eluent, raffinate, extract, and internal recycle). This process splits the feed stream into two fractions. The Finnsugar sequential SMB introduced in 1989 is capable of separating molasses into several product fractions (sucrose, betaine, and salts). Similar concepts were later developed by Mitsubishi and Japan Organo.

4.4.2.1   Experimental Methods

Chromatographic separation experiments were conducted in a batch column to study the influence of ion exchange resin particle size on the recovery of sucrose from molasses. Synthetic solutions containing sulfate salts, sucrose, and betaine were separated at 358 K. The bed volume was 100 mL ($L = 26$ cm, $d_{col} = 2.2$ cm) and flow rate was 1.9 BV/h. The volumetric loading of the column was 5 vol-%, and the feed consisted of 120 g/L sucrose, 10 g/L betaine, and 70 g/L salts (30% Na_2SO_4 and 70% K_2SO_4).

The total amount of dry solids, 200 g/L, was somewhat lower than is typical for industrial molasses desugarization processes (approximately 600 g/L[28]). The columns were packed (manually) with Finex CS11GC resin in Na^+ form via a slurry packing technique. The mean particle size of the resin varied from 110 to 518μm. Table 4.1 lists the properties of the resins.

4.4.2.2   Results

Figure 4.7 illustrates the influence of particle size on the elution profiles. Salts are eluted first due to the electrolyte exclusion phenomenon. The concentration profiles in the figure are typical for anti-Langmuir isotherms. The particle size has only a small effect on the salt profile because the salts are effectively excluded from the resin and mass transfer resistance does not play a significant role. The differences in the salt elution profiles mainly reflect the influence of particle size on the axial

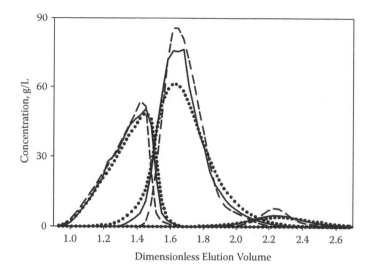

FIGURE 4.7   Desugarization of molasses using Finex CS11GC resin. Dashed line: $\overline{d}_p$ = 110 μm. Solid line: $\overline{d}_p$ = 250 μm. Dotted line: $\overline{d}_p$ = 328 μm. See text for details on column dimensions and feed composition.

mixing in the column. The concentration profile of sucrose (eluted in the middle) is strongly influenced by particle size. As expected, the (apparent) distribution coefficient is constant but the profiles become wider and increasingly disperse when particle size increases. The same is true for betaine—the most strongly retained component in the synthetic molasses feed.

Typically, desugarization of molasses with a simulated moving bed process provides a 90% recovery yield with a purity of 90% for the sucrose fraction [27]. Although sucrose is the main target component, recovery of betaine (along with amino acids, inositol, and other products) can significantly improve the economy of the process[28]. Therefore, when calculating the yield of sucrose based on experimental batch chromatography data, the recovery yield of betaine was chosen as one constraint. The other constraint was the salt content of the sucrose fraction (< 3 wt-% of dry solids).

Figure 4.8 displays the influence of particle size on the yield of sucrose. The yield increases very rapidly when the particle size decreases from 400 to 300 μm. Below this limit, the increase of the yield with decreasing particle size is more gradual. Such a strong influence of particle size arises from the recovery of sucrose as a "center cut" in the elution profile. The purity of the product fraction is affected by the band broadening (due to mass transfer resistance) from both sides.

If maximum allowable pressure drop limits the productivity of the process, the first thought is often to increase the particle size. In principle, it would be possible to maintain the yield at an acceptable level by simultaneously increasing the bed height or decreasing the loading. However, the stability of the flow becomes an issue with large particles and viscous feeds because of the range of flow rates, outside which viscous fingering easily occurs.

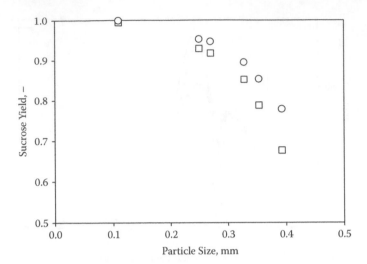

FIGURE 4.8   Influence of mean particle size on sucrose yield in desugarization of molasses. Circles: yield of betaine = 70%. Squares: yield of betaine = 90%.

Viscous fingering is flow instability in a chromatography column (or any packed bed) due to large differences in the density and viscosity of the eluent and the feed. It leads to excessive dispersion—obviously detrimental for a separation process. These effects are minimized when the linear velocity is 0.5 to 2.0 times a critical velocity defined by the system parameters[19]. The critical velocity, u_{crit}^{L} , can be expressed as:

$$u_{crit}^{L} = \frac{g(\rho_2 - \rho_1)}{k(\mu_2 - \mu_1)}$$
(4.12)

where g is the gravity constant (9.81 m s^{-2}) and ρ is density. The subscripts 1 and 2 refer to the solvent and feed pulse. As we will show, the particle size of the ion exchange resin affects the critical velocity through the permeability coefficient. As seen in Equation (4.12), the limits of the safe flow rate range depend on the permeability of the bed. In the molasses desugarization experiments with a mean particle size of 518 µm, the permeability of the bed had become too large, and flow instabilities destroyed the separation. This is shown in Figure 4.9. Similar effects were observed, for example, when separating sulfuric acid from monosaccharides by using ion exchange resins.

The resins used in this study were monobead types with very narrow particle size distributions (Table 4.1) necessary for properly investigating the influence of particle size on separation. It is expected that a wider particle size distribution would have improved flow stability under these conditions, but would also lead to a larger pressure drop. The influence of particle size distribution on separation is discussed below.

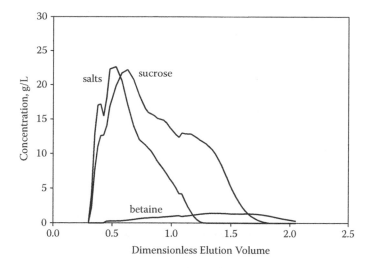

FIGURE 4.9 Poor separation due to viscous fingering in a chromatography column packed with large particles with narrow particle size distribution (d_p = 518 μm).

4.5 INFLUENCE OF PARTICLE SIZE DISTRIBUTION ON SEPARATION

The influence of particle size distribution on the efficiency of glucose–fructose separation was studied by using Finex CS13GC resin—a commercial strong acid cation exchanger that has a nominal cross-link density of 6.5 wt-% DVB. This is often considered optimal for large scale separation of monosaccharides in the sweetener industry.

4.5.1 EXPERIMENTAL RESULTS

Two batches with equal volumetric mean particle size, $\overline{d}_p$ = 120 μm, but different particle size distributions were tested. In the first batch, 70% of the total volume of the particles differed less than 20% from the volume average particle size (which was within the range of 111 to 128 μm). In the second batch, 90% of the particle volume was within these limits.

The resin was converted to the Ca^{2+} form via standard techniques and packed manually into 100 mL glass columns (L = 50 cm, d_{col} = 1.6 cm) by using slurry packing. Pulse experiments were carried out at room temperature using a chromatography station (ÄKTA Purifier, Amersham Pharmacia Biotech). The feed contained 232 g/L glucose and 168 g/L fructose. Samples of 1.0 mL were collected at the column outlet, and the concentrations of glucose and fructose were analyzed with HPLC (Agilent 1100, Bio-Rad Aminex HPX-87N column). Two flow rates were used in the pulse experiments: 0.6 and 1.6 BV/h.

Chromatograms corresponding to small injections (0.5% of bed volume) are displayed in Figure 4.10. Chromatographic performance parameters are listed in

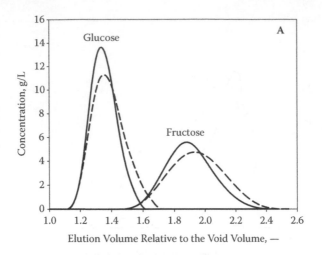

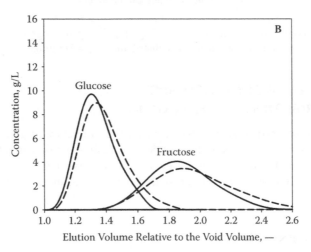

FIGURE 4.10   Influence of width of particle size distribution on chromatographic separation of glucose and fructose on Finex CS16GC (Ca^{2+}) resin. Flow rate: 0.6 BV/h (A); 1.6 BV/h (B). Dashed line: 70% within ±20% of mean particle size/. Solid line: 90% within ±20% of mean particle size. Feed concentration: 232 g/L of glucose and 168 g/L of fructose.

Table 4.2. Room temperature was chosen for these experiments to create conditions under which the mass transfer effects limit the separation efficiency despite the relatively small mean particle sizes of the resins.

Comparison of the solid lines (narrow particle size distribution) and dashed lines (broader distribution) in Figure 4.10 clearly demonstrates that optimizing the particle size distribution has a positive effect on separation efficiency. The product fractions are less dilute with the narrow size distribution resin, and the cycle time is slightly shorter. Moreover, the resolution is much better with the narrow size distribution, as seen in Table 4.2 where the number of theoretical plates (NTP) quantifies the band

TABLE 4.2

Influence of Particle Size Distribution on Glucose–
Fructose Separation (Bed Height = 0.5 m)

		Flow Rate			
		0.6 BV/h		1.6 BV/h	
Size distribution within ±20%		70%	90%	70%	90%
Number of Theoretical Plates					
	Glucose	175	263	98	122
	Fructose	131	161	60	83
c_{max}, g/L					
	Glucose	11.3	13.6	9	9.7
	Fructose	4.7	5.6	3.4	4
Resolution		1.03	1.18	0.73	0.83

broadening due to mass transfer and axial dispersion. Noted earlier, the axial dispersion coefficient increases with the width of the particle size distribution. Since both glucose and fructose have nearly linear isotherms, thermodynamic effects do not affect the band broadening.

Nevertheless, these results should not be interpreted such that the particle size distribution should be as narrow as possible in all cases because monodisperse particles tend to result in larger volume changes (swelling and shrinking) during process cycles. Large variations in bed height shorten the lifetime of the resin due to stronger mechanical attrition. In addition, shrinking of the bed often leads to excessive backmixing in the liquid layer above the column inlet.

4.5.2 Simulation Results

The experimental results above were obtained with a small loading of the column and a small mean particle size. Numerical simulations were performed to better analyze how particle size distribution affects the optimum loading of a batch column.

4.5.2.1 Theory and Calculations

The size heterogeneity of the ion exchange resin was modeled by assuming that the particle size was normally distributed around the mean and dividing the particles into 10 discrete size fractiles representing this distribution. In absence of a correlation for the axial dispersion coefficient in a packed bed with a wide particle size distribution, axial dispersion was generated numerically by using a first-order accurate method to calculate the spatial derivatives. The total material balance for component i in a volume element in the column was therefore written as:

$$\frac{\partial c_i}{\partial t} = -u^L \frac{\partial c_i}{\partial z} - F \frac{\partial \overline{q}_i}{\partial t} \tag{4.13}$$

where F is the phase ratio and $\overline{q}$ denotes the average solid phase concentration in the volume element (i.e., averaged over all the fractiles). Mass transfer between the liquid and the solid phases was described with the linear driving force approximation:

$$\frac{\partial \hat{q}_{i,k}}{\partial t} = \frac{60D_i^S}{d_{p,k}^2}\left[q_{i,k}^*\left(c_i\right) - \hat{q}_{i,k}\right] \tag{4.14}$$

where the asterisk denotes solid phase concentration at phase equilibrium and the hat denotes volume average concentration in fractile k. D_i^S is the diffusion coefficient in the ion exchange resin. The temporal derivative of $\overline{q}$ in Equation (4.13) was obtained by summation over all fractiles:

$$\frac{\partial \overline{q}_i}{\partial t} = \sum_{k=1}^{N} \phi_k \frac{\partial \hat{q}_{i,k}}{\partial t} \tag{4.15}$$

where N is the number of particle size fractiles and ϕ_k is the volume fraction of the particles in fractile k. Dirichlet boundary conditions were used at the column inlet and outlet (first-order PDE). Equations (4.13) to (4.15) were solved by the method of lines. The density of the calculation grid points was 3000/m. The column length and diameter were 0.5 m and 0.05 m, respectively. The flow rate was 0.73 BV/h and bed porosity was 0.40. The diffusion coefficients of glucose and fructose in CS13GC were estimated from the experimental column data obtained with small injections. Values of 1.3×10^{-11} m^2 s^{-1} and 9.5×10^{-12} m^2 s^{-1} were used for glucose and fructose, respectively.

Two normally distributed particle size distributions with a mean particle size of 300 μm were used. In the first batch, 70% of the volume was within ±20% of the mean. In the second batch, 90% of volume was within ±20% of the mean. In addition, simulations were carried out as references with monodisperse batches of particles.

4.5.2.2   Results

A purity constraint of 90% was chosen for the fructose fraction, corresponding to the typical outlet composition in HFCS manufacture. This value should be understood as a yield per cycle because, in a real industrial batch process, the glucose-rich fraction would be recycled to the isomerization reactor. The simulation results are summarized in Figures 4.11 and 4.12, which show the influence of column loading on the yield of fructose and the productivity of the batch separation process.

Although the results obtained for a batch process are not directly comparable to those from a continuous simulated moving bed process, the influence of particle size distribution on the process performance is similar. Before discussing the results, it should be noted that the yield of fructose in Figure 4.11 does not reach 100% when the column loading approaches zero because it is not economical to collect the long tail of fructose in the product fraction because of the need for excessive dilution and long cycle time. The stop criterion employed in these simulations was $c_{\text{fructose}}^{\text{outlet}} < 0.005 c_{\text{fructose}}^{\text{feed}}$.

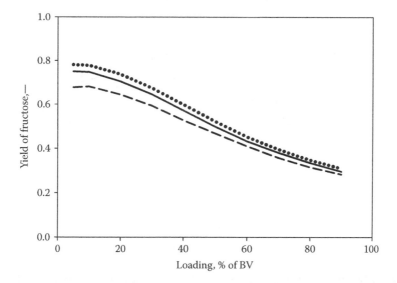

FIGURE 4.11 Influence of particle size distribution on process performance: loss of yield with increasing loading. Simulated results for glucose–fructose separation in batch column. Design constraint: 90% purity for second product fraction. Dashed line: 70% within ±20% of mean. Solid line: 90% within ±20% of mean. Dotted line: monodisperse particles.

As observed in Figure 4.11, polydisperse particles lead to a reduced yield of fructose. At a column loading of 20%, for example, the resin with broader particle size distribution (dashed line) provides a yield per cycle of 64%, whereas the more narrow size distribution (solid line) provides a 71% yield. The theoretical maximum yield (i.e., monodisperse particles, dotted line) is 74% under these conditions. Also in Figure 4.11 note that the loading that results in a certain yield is more strongly influenced by the width of the particle size distribution than the yield obtained with a certain loading. For example, if a fructose yield of 64% is chosen as the design constraint, using the narrow size distribution resin allows the operator to increase the volumetric loading from 20 to 30% of the bed volume. In other words, the loading can be increased by 50% by tailoring the particle size distribution of the separation material.

The productivity of the batch process at constant loading is also strongly affected by the particle size distribution. This is observed in Figure 4.12 showing productivity of the separation process as a function of the column loading. The optimum loading (approximately 50 vol-% of bed volume) is practically independent of the width of the size distribution because the adsorption isotherms are linear and the propagation velocities of the concentration fronts are thus independent of concentration. Consequently, dilution due to wide particle size distribution does not affect the propagation velocities. At optimum loading, the narrow size distribution resin results in a 12% higher productivity than the other resin. On the average, the narrow particle size distribution resin provides some 5% lower productivity than the (hypothetical) monodisperse resin. The resin with the broader particle size distribution gives approximately 15% lower productivity than the reference.

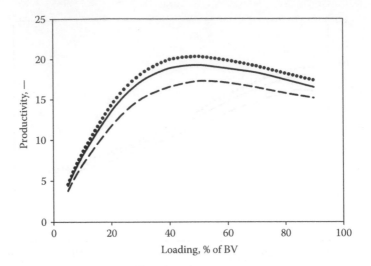

FIGURE 4.12 Influence of particle size distribution on process performance: dependence of productivity on loading. Simulated results for glucose–fructose separation in a batch column. Design constraint: 90% purity for second product fraction. Dashed line: 70% within ±20% of mean. Solid line: 90% within ±20% of mean. Dotted line: monodisperse particles.

It is interesting to see how much of the fructose yield must be sacrificed to improve the productivity of the separation process. The loss of yield per cycle corresponds to increasing volumes of recycle streams and reservoirs, which increases both investment and operation costs. This is illustrated in Figure 4.13, in which productivity is plotted as a function of yield. According to these results, maximum productivity is associated with a higher yield when the particle size distribution is made narrower. Therefore, a narrow particle size distribution provides the option to either reduce volumes of the recycle streams and pumping costs (at a constant flow rate) or increase the throughput (at a constant yield per cycle).

Based on these results, employing a narrow particle size distribution has several positive effects compared to using a wide particle size distribution. First, a higher loading can be employed to achieve a given yield per cycle. Second, the productivity at a given loading is higher. Together these two effects indicate that the productivity at a given yield depends strongly on the width of the particle size distribution. In addition, a bed packed with polydisperse particles has a higher pressure drop than a bed of particles of the same size (Figure 4.1). Therefore, a higher flow rate can be employed when size distribution is narrow. Finally, smaller volumes of recycle streams are required to achieve maximum productivity with a high quality resin.

Although batch chromatography is not the method of choice for industrial scale separation of glucose and fructose, these results demonstrate the benefits of tailoring the particle size distribution of an ion exchange resin to the target application.

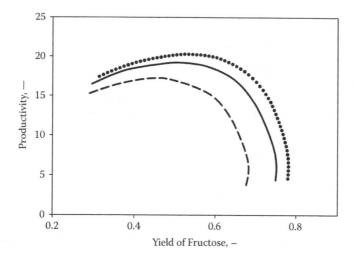

FIGURE 4.13 Influence of particle size distribution on process performance: dependence of productivity on yield of fructose. Simulated results for glucose–fructose separation in batch column. Design constraint: 90% purity for second product fraction. Dashed line: 70% within ±20% of mean. Solid line: 90% within ±20% of mean. Dotted line: monodisperse particles.

4.6 CONCLUSIONS

The influence of ion exchange resin particle size and size distribution on chromatographic separation in the sweetener industry was investigated. Glucose–fructose separation was chosen as a model system because of its utility in the manufacture of HFCS. Desugarization of molasses (to recover sucrose and betaine) was chosen to represent a process in which one product (sucrose) is collected as a center cut between two other fractions (salt and betaine).

The results demonstrate that small scale experiments and numerical simulations advocate narrow particle size distribution because large particles in a batch of polydisperse resin exert a strong influence on the overall mass transfer rate. Small particles in the batch increase the pressure drop, which may limit the production rate. The economics of chromatographic separation processes can be improved by tailoring the particle size and size distribution to each application.

Note, however, that the methods used here underestimate or neglect packing density variations during processing. Particles of uniform size produce larger volume changes, for example, than particles with some size distribution. Therefore, comparing ion exchange resins based on uniformity coefficient alone is not sufficient, and what is inside the bead becomes more important than size distribution.

ACKNOWLEDGMENTS

We thank Dr. Jari Tiihonen, M.Sc. Katriina Liimatainen, and M.Sc. Eeva-Liisa Peuha for conducting the experiments and analyzing part of the data.

REFERENCES

1. Paillat, D., Cotillon, M., and Théoleyre, M.A. Technology of chromatographic separation in glucose syrup processes, AVH Association Seventh Symposium, Reims, 2000, 62–67.
2. Schoenrock, K. Performance limits in industrial chromatographic separation processes, *Zuckerindustrie* 122 (1997), 22–27.
3. Billena, J., Guillarme, D., Rudaz, S. et al. Relation between the particle size distribution and the kinetic performance of packed columns: Application to a commercial sub-2 μm particle material, *J. Chromatogr. A* 1161 (2007), 224–233.
4. Halasz, I. and Naefe, M. Influence of column parameters on peak broadening in high-pressure liquid chromatography. *Anal. Chem.* 44 (1971), 76–84.
5. Neue, U.D. *HPLC Columns: Theory, Technology and Practice*, Wiley-VCH, New York, 1997.
6. Colin, H. Large scale high-performance PLC. In *Preparative and Production Scale Chromatography*.
7. Dewaele, C. and Verzele, M. Influence of the particle size distribution of the packing material in reversed-phase high performance liquid chromatography, *J. Chromatogr.* 260 (1983), 13–21.
8. Timm, E.E. Process and apparatus for preparing uniform size polymer beads. U.S. Patent 4,444,961.
9. Jung, S., Ehlert, S., Mora, J.A. et al. Packing density, permeability, and separation efficiency of packed microchips at different particle-aspect ratios. *J. Chromatogr. A* 1216 (2009), 264–273.
10. Chung, S.F. and Wen, C.Y. Longitudinal dispersion of liquid flowing through fixed and fluidized beds. *AIChE J.* 14 (1968), 857.
11. Guedes de Carvalho, J.R.F. and Delgado, J.M.P. Effect of fluid properties on dispersion in flow through packed beds. *AIChE J.* 49 (2003), 1980–1985.
12. Han, N.W., Vhakta, J., and Carbonell, R.G. Longitudinal and lateral dispersion in packed beds: Effect of column length and particle size distribution. *AIChE J.* 31 (1985), 277–288.
13. Delgado, J.M.P. A critical review of dispersion in packed beds. *Heat Mass Trans.* 42 (2006), 279–310.
14. Sarmidi, M.R. and Barker, P.E. Simultaneous biochemical reaction and separation in a rotating annular chromatograph, *Chem. Eng. Sci.* 48 (1993), 2615–2623.
15. Watler, P., Yamamoto, S., Kaltenbrunner, O. et al. In *Scale-Up and Optimization in Preparative Chromatography: Principles and Biopharmaceutical Applications*, Marcel Dekker, New York, 2002.
16. Soshina, L.P. and Zaostrovskii, F.P. Hydraulic drag in a stationary bed of ion exchange resin, *Inzhen. Fizisch. Zh.* 25 (1973), 1013–1018.
17. Keener, R.N., Fernandez, E.J., Maneval, J.E. et al. Advancement in the modeling of pressure flow for the guidance of development and scale-up of commercial-scale biopharmaceutical chromatography. *J. Chromatogr. A* 1190 (2008), 127–140.
18. Danilov, A.V., Vagenina, I.V., Mustaeva, L.G. et al. Liquid chromatography on soft packing material, under axial compression: size-exclusion chromatography of polypeptides. *J. Chromatogr. A* 773 (1997), 103–114.
19. Hongisto, H.J. Chromatographic separation of sugar solutions: The Finnsugar molasses desugarization process, Part I. *Int. Sug. J.* 79 (1977), 100–104.
20. Hongisto, H.J. Chromatographic separation of sugar solutions: The Finnsugar molasses desugarization process, Part II. *Int. Sug. J.* 79 (1977), 131–134.
21. Verzele, M. and Dewaele, C. *Preparative High Performance Liquid Chromatography: A Practical Guideline*. RSL/Alltech Europe, Eke, 1985.

22. Godshall, M. A. Sugar and other sweeteners. In *Riegel's Handbook of Industrial Chemistry*, 10th ed. Kluwer Academic, New York, 2003.
23. Dorfner, K. *Ion Exchangers*. de Gruyter, New York, 1991.
24. Goulding, R.W. Liquid chromatography of sugars and related polyhydric alcohols on cation exchangers: Effect of cation variation. *J. Chromatogr.* 103 (1975), 229–239.
25. Ruthven, D.M. *Principles of Adsorption and Adsorption Processes*, Wiley, New York, 1984.
26. Eklund, M., Bauer, E., Wamatu, J. et al. Potential nutritional and physiological functions of betaine in livestock. *Nutr. Res. Rev.* 18 (2005), 31–48.
27. Paananen, H. A. Trends in the chromatographic separation of molasses. *Zuckerindustrie* 122 (1997), 23–33.
28. Paananen, H. and Kuisma, J. Chromatographic separation of molasses components. AVH Association Seventh Symposium, Reims, 2000, 17–21.

5 Dual Temperature Methods of Separation and Concentration of Elements in Ion Exchange Columns

Ruslan Kh. Khamizov, Vladimir A. Ivanov, and Nikolay A. Tikhonov

CONTENTS

ABSTRACT

Ion exchange plays an important role in many technological areas because of its advantages, particularly selectivity of separation. Traditional ion exchange processes impose restrictions related to the use of large quantities of reagents in cyclic sorption and desorption operations, leading to secondary liquid wastes. To eliminate the problems of sorbent regeneration, we sought other ways, including parametric methods that do not require reagents and are based on regulating selectivity through physical parameters (temperature, pressure, and electric and magnetic fields). This chapter reviews a special class of reactant-less ion exchange processes regulated by temperature changes. Analysis of published results and new data; thermodynamic characteristics of various ion exchange reactions; and theoretical descriptions of equilibrium, kinetics, and dynamics of nonisothermal conditions are covered. The authors present experimental data and attempt to classify temperature-driven methods based on fundamental physical and chemical influences of temperature on ion exchange in columns.

5.1   INTRODUCTION

Ion exchange processes using either polymeric exchangers or inorganic ion exchangers such as natural and synthetic zeolites play important roles in water treatment, hydrometallurgy, chemical analysis, and many other scientific and technological applications. Unfortunately, the handling of large amounts of auxiliary reagents and generation of secondary wastes offer disadvantages from the use of ion exchange processes. Development of advanced reagent-free technologies like membrane separation is therefore gradually replacing traditional ion exchange processes in many applications.

Traditional ion exchange has an inherent limitation: the more selective the sorbent made for extraction of certain target components from feed solutions, the more difficult the desorption of components and regeneration of the sorbent ion exchanger. To overcome the problems of sorbent regeneration, other ways of ion exchange organizing are under investigation; dual parametric methods are of particular interest. Dual parametric ion exchange and molecular sorption methods of separation of electrolytes are based on controlling selectivity through physical parameters such as temperature, pressure, and electric and magnetic fields. They do not require auxiliary chemical reagents for desorption and regeneration of the sorbent.

The most advanced parametric methods known to date are dual temperature separations based on temperature controlled interphase mass exchange equilibrium.[1-7] Concentrating deuterium and separating it from protium in the process of heavy water extraction from natural water became the first and best known example of the large scale application of dual temperature methods to chemical isotope exchanges in liquid–gas systems.[8,9] The dual temperature method is the most economically feasible way to produce heavy water from a water–hydrogen sulfide system at the industrial production level.

However, the dual temperature method is not yet popular for ion exchange separation of liquid solutions due to its requirement for a considerable amount of thermal energy.[3] Nevertheless, research in these areas has continued during the past four decades[10–32] since the pioneering works of Andreev et al.[10,11] and the later works of Wilhelm et al. on parametric pumping in sorption systems[12], and B. Bolto et al. on Sirotherm processes[13].

In the wake of more stringent regulations to meet ecological requirements, it is expected that methods using far fewer or even no reagents will gain priority in industrial technologies. Significant future roles for dual temperature ion exchange and sorption methods are expected in extracting valuable components from the oceans and other waters in environmentally benign ways.[17–25] Other possible applications include the high grade purification of substances (because auxiliary reagents in traditional ion exchange can be additional sources of impurities)[4–6] and concentration of trace quantities of isotopes of various elements.[8,9] However, purification of natural and industrial waste waters[26–31] is considered the most promising field of application of temperature-driven processes.

Many references on dual temperature sorption processes are available in reviews in serial monographs and journals.[1–7,33] The perfect and detailed descriptions of theory and practice of different dual temperature methods written by Wancat, Grevillot, and Tondeur[1–3] are often used and cited by researchers. Except for a few reviews and articles published by Gorshkov and Ivanov and their co-authors,[4–6] the works of many Russian scientists in this field remain unpublished in English.

At present, despite a great number of publications about dual temperature ion exchange methods, we have no fundamental classification based on objective principles covering the effects of temperature on separation of elements in columns. Overviews of methods are usually compiled historically,[1–3,6] using existing vocabulary that is not capable of defining the new methods. Such empirical approaches do not allow the analysis of various methods from unified positions—a vital part of extracting basic and secondary information; predicting ways to develop existing approaches; and devising new dual temperature and other reagent-free ion exchange methods for concentration and separation.

In line with the classification proposed earlier,[5,6] dual temperature ion exchange falls into two broad process categories: (1) single dual temperature separation methods that use the separating effects of a single-step change of selectivity of an ion exchanger by temperature; and (2) processes involving multiplication of single effects of separation, for example, cascading set-ups of single dual temperature separations, parametric pumping, dual parametric (dual temperature) separations using counter flow columns for different temperature sections, and the simulation of dual temperature separation in a set of columns with stationary beds of ion exchangers.

The focus of this paper is to demonstrate new experimental results of dual temperature separation and create more appropriate classifications of dual temperature methods based on the fundamentals of physical and chemical influences of temperature on ion exchange—mainly, on the equilibrium and dynamics of separation in columns.

5.2 TEMPERATURE EFFECT ON EQUILIBRIUM AND DYNAMICS OF ION EXCHANGE

To describe temperature effects on the equilibrium, kinetics, and dynamics of ion exchange, we will rely on well known fundamental theories and note only important correlations required for clarity of the concepts relating to classification.

5.2.1 EQUILIBRIUM OF ION EXCHANGE

Following the latest recommendations on unification of nomenclature, formulations, and experimentation for ion exchange,[34,35] the exchange reaction in which equivalents of ions A of charges z_A and B of charges z_B participate is given by:

$$\frac{1}{z_B} B^{z_B} + \frac{1}{z_A} \bar{A}^{z_A} = \frac{1}{z_B} \bar{B}^{z_B} + \frac{1}{z_A} A^{z_A} \tag{5.1}$$

The important equilibrium parameters of ion exchange are:
Separation factor:

$$\alpha = \frac{\bar{x}_B \cdot x_A}{\bar{x}_A \cdot x_B} \tag{5.2}$$

Selectivity coefficient:

$$k = \frac{\bar{x}_B^{1/z_B} \cdot x_A^{1/z_A}}{\bar{x}_A^{1/z_A} \cdot x_B^{1/z_B}} \tag{5.3}$$

Equilibrium coefficient:

$$\tilde{\tilde{K}} = \frac{\bar{c}_B^{1/z_B} \cdot c_A^{1/z_A}}{\bar{c}_A^{1/z_A} \cdot c_B^{1/z_B}} \tag{5.4}$$

Corrected equilibrium coefficient:

$$\tilde{K} = \frac{\bar{c}_B^{1/z_B} \cdot a_A^{1/z_A}}{\bar{c}_A^{1/z_A} \cdot a_B^{1/z_B}} = \tilde{\tilde{K}} \frac{\bar{f}_A^{1/z_A}}{\bar{f}_B^{1/z_B}} \tag{5.5}$$

Thermodynamic equilibrium constant:

$$K = \frac{\bar{a}_B^{1/z_B} \cdot a_A^{1/z_A}}{\bar{a}_A^{1/z_A} \cdot a_B^{1/z_B}} = \tilde{K} \frac{\bar{f}_B^{1/z_B}}{\bar{f}_A^{1/z_A}} = \tilde{\tilde{K}} \frac{\bar{f}_B^{1/z_B} \cdot f_A^{1/z_A}}{\bar{f}_A^{1/z_A} \cdot f_B^{1/z_B}} \tag{5.6}$$

Here $\bar{a}_i$ and a_i are the thermodynamic activities of ion i (A or B); $\bar{c}_i$ and c_i are the concentrations in molarity, molality, or mole fractions (the water content of the ion exchanger and sorbed electrolyte should be considered); $\bar{f}_i$ and f_i are the activity coefficients; and $\bar{x}_i$ and x_i are the equivalent fractions in ion exchanger and solution, respectively. The over bar represents the exchanger (solid phase). For calculations according to Equations (5.5) and (5.6), the mean ionic activity coefficients of electrolytes in solutions are used in the following proportion:

$$\frac{\gamma_A^{1/z_A}}{\gamma_B^{1/z_B}} = \frac{\gamma_A^{1/z_A} \cdot \gamma_X^{1/z_X}}{\gamma_B^{1/z_B} \cdot \gamma_X^{1/z_X}} = \frac{\left(\gamma_{\pm AX}\right)^{(1/z_A + 1/z_X)}}{\left(\gamma_{\pm BX}\right)^{(1/z_B + 1/z_X)}} \tag{5.7}$$

where: X is a co-ion with the charge z_X, of the same sign as the sign of the fixed group of the ion exchanger; electrolytes $A_{z_X} X z_A$ and $B_{z_X} X z_B$ are designated by subscripts AX and BX. The thermodynamic constant K relates to standard Gibbs energy ΔG^o and standard enthalpy ΔH^o according to the expressions:

$$\Delta G^o = \overline{\mu}_B^0 - \overline{\mu}_A^0 + \mu_A^0 - \mu_B^0 = -RT \ln K \tag{5.8}$$

$$\left(\frac{\partial \ln K}{\partial T}\right)_p = \frac{\Delta H^0}{RT^2} \tag{5.9}$$

$\overline{\mu}_i^0$ and μ_i^0 are standard chemical potentials of ions in phases. ΔG^0 and ΔH^0 values characterize the substitution of one equivalent of ion B contained in the ion exchanger with one equivalent of ion A from the solution, while ions A and B are in standard states in both phases.[36–38]

Mono-ionic ion exchangers are usually chosen as standard states of exchanging ions in the ion exchanger phase.[36] Values ΔG^0 and ΔH^0 are considered as integral, i.e., they correspond to transformation of the ion exchanger from the mono-ionic form of component A to the mono-ionic form of component B. The integration of Equation (5.9) results in correlation of equilibrium constants with two different temperatures. Again, ΔH^0 is related to temperature according to the following equation:[39]

$$\frac{d(\Delta H^0)}{dT} = \Delta c_p^0 \tag{5.10}$$

Combining Equations (5.9) and (5.10) and simplifying at constant Δc_p^o:

$$\ln \frac{K_{T_2}}{K_{T_1}} = \frac{T_2 - T_1}{RT_1 T_2} \cdot \Delta H_{T_1}^0 + \frac{\Delta c_p^o}{R}\left(\ln \frac{T_2}{T_1} - \frac{T_1(T_2 - T_1)}{T_2 - T_1}\right) \tag{5.11}$$

However, in practice, the temperature dependence of enthalpy is mostly ignored and the next approximation is used:

$$\ln \frac{K_{T_2}}{K_{T_1}} = \frac{T_2 - T_1}{RT_1 T_2} \cdot \Delta H^0 \tag{5.12}$$

where ΔH^0 is the mean value in the temperature interval T_1 to T_2.

The activities of ions A and B in the ion exchanger (or the activities of resins) and the thermodynamic constant can be defined by the method of Gaines and Thomas[40] using experimental values of the corrected equilibrium coefficient measured along a whole range of exchanger ionic composition: $0 < \bar{x}_B < 1$. This method was considered in detail in a number of publications.[36–38,41] Accordingly, by disregarding the change of resin swelling during ion exchange, the correlations of the thermodynamic constant and standard Gibbs energy with the corrected equilibrium coefficient are written as:

$$ lnK = \int_{\bar{x}_B=0}^{\bar{x}_B=1} ln\tilde{K}_A^B \cdot d\bar{x}_B + \int_{\bar{c}_A=\bar{c}_A^o,\bar{c}_B=0,}^{\bar{c}_A=0,\bar{c}_B=\bar{c}_B^o} \frac{d(\bar{c}_A+\bar{c}_B)}{z_A\bar{c}_A+z_B\bar{c}_B} + ln\frac{\bar{c}_A^o}{\bar{c}_B^o} \tag{5.13} $$

$$ \Delta G^o = -RT\,lnK = -RT \int_{\bar{x}_B=0}^{\bar{x}_B=1} ln\tilde{K}_A^B \cdot d\bar{x}_B - RT \int_{\bar{c}_A=\bar{c}_A^o,\bar{c}_B=0,}^{\bar{c}_A=0,\bar{c}_B=\bar{c}_B^o} \frac{d(\bar{c}_A+\bar{c}_B)}{z_A\bar{c}_A+z_B\bar{c}_B} - RT\,ln\frac{\bar{c}_A^o}{\bar{c}_B^o} \tag{5.14} $$

The mono-ionic forms of the ion exchanger in equilibrium with the solution are chosen as the standard states ($\bar{f}_i = 1/\bar{c}_i^o$ where $\bar{c}_i^o$ is the concentration of ion i in such a mono-ionic form). Writing the last two terms in the right parts of Equations (5.13) and (5.14), we take into account that in a general case of heterovalent ion exchange, molar concentrations in solutions can change. For the exchange of equally charged ions, we have the more usual formulas of Gaines and Thomas[40] (without the second and third terms of the cited equations).

As noted earlier,[37,42] standard thermodynamic functions do not yield information about the degree of exchange because they relate to a complete substitution of ion A by ion B. More detailed information about regularities of ion exchange can be obtained from differential thermodynamic functions. Two main approaches can be used for the definition of a differential Gibbs energy. In line with the first one,[42] writing the chemical potentials of ions in equilibrium phases, as: $\bar{\mu}_i = \bar{\mu}_i^o + RT\,ln\,\bar{a}_i$ and $\mu_i = \mu_i^o + RT\,ln\,a_i$ and taking

$$ \Delta G_{\bar{n}} = \frac{1}{z_B}\bar{\mu}_B - \frac{1}{z_A}\bar{\mu}_A + \frac{1}{z_A}\mu_A - \frac{1}{z_B}\mu_B = 0 $$

for an equilibrium system:

$$ RT\,ln\tilde{K} = -\left(\frac{1}{z_B}\bar{\mu}_B - \frac{1}{z_A}\bar{\mu}_A + \frac{1}{z_A}\mu_A^0 - \frac{1}{z_B}\mu_B^0\right) + RT\,ln\frac{\bar{c}_B^{1/z_B}}{\bar{c}_A^{1/z_A}} = -\Delta G_{\bar{n}} + RT\,ln\frac{\bar{c}_B^{1/z_B}}{\bar{c}_A^{1/z_A}} \tag{5.15} $$

where:

$$ \Delta G_{\bar{n}} = \frac{1}{z_B}\bar{\mu}_B - \frac{1}{z_A}\bar{\mu}_A + \frac{1}{z_A}\mu_A^0 - \frac{1}{z_B}\mu_B^0 = RT\,ln\,a_B^{1/z_B}/a_A^{1/z_A} $$

is the differential Gibbs energy of the exchange reaction, which refers to an ion exchanger of fixed composition (with fixed numbers of moles of exchangeable ions, co-ions, and water and ion exchange centers in the exchanger). This value can be interpreted as the Gibbs energy of the reaction in which one equivalent of ion B^+ in solution is substituted by one equivalent of ion A^+ from an infinite amount of ion exchanger of a specific composition when ions A^+ and B^+ in solution are at their standard states. Comparing Equation (5.13) with Equation (5.15) reveals the relation between standard and differential Gibbs energies:

$$\Delta G^o = \int_{\overline{x}_B=0}^{\overline{x}_B=1} \Delta G_n \cdot d\overline{x}_B - RT \int_{\overline{x}_B=0}^{\overline{x}_B=1} ln\frac{\overline{c}_B^{1/z_B}}{\overline{c}_A^{1/z_A}} \cdot d\overline{x}_B - RT \int_{\overline{c}_A=\overline{c}_A^o,\overline{c}_B=0,}^{\overline{c}_A=0,\overline{c}_B=\overline{c}_B^o} \frac{d(\overline{c}_A + \overline{c}_B)}{z_A\overline{c}_A + z_B\overline{c}_B} - RT\ ln\frac{\overline{c}_A^o}{\overline{c}_B^o}$$

(5.16)

Considering exchange of equally charged ions when the molar concentration of ions in the exchanger phase does not change, we have a more compact formula:

$$\Delta G^o = \int_{\overline{x}_B=0}^{\overline{x}_B=1} \Delta G_n \cdot d\overline{x}_B - RT \int_{\overline{x}_B=0}^{\overline{x}_B=1} ln\frac{\overline{c}_B^{1/z_B}}{\overline{c}_A^{1/z_A}} \cdot d\overline{x}_B$$

(5.17)

Soldatov[37] gives another definition for differential Gibbs energy: he designates the whole right part of Equation (5.13), as

$$\int_{\overline{x}_B=0}^{\overline{x}_B=1} \Delta G_n \cdot d\overline{x}_B,$$

leading to the more simple

$$\Delta G^o = \int_{\overline{x}_B=0}^{\overline{x}_B=1} \Delta G_n \cdot d\overline{x}_B .$$

In this case, the physical meaning of ΔG_n is not clear enough to be determined from chemical potentials. Differentiating Equation (5.15) with respect to temperature at constant pressure and constant composition, and using the well-known expression:[43]

$$\left[\frac{\partial(\overline{\mu}_i/T)}{\partial T}\right]_{p,\overline{n}_j} = -\overline{h}_i \cdot \frac{1}{T^2}$$

(5.18)

for the exchanger phase and analogous expression for the solution ($\overline{h}_i$ is the partial molar enthalpy; subscripts p and $\overline{n}$ designate the constancy of pressure and phase composition):

$$\left(\frac{\partial \ln \tilde{K}}{\partial T}\right)_{p,\bar{n}_j} = \frac{\bar{h}_B - \bar{h}_A + h_A^0 - h_B^0}{RT^2} + \left[\frac{\partial \ln\left(\bar{c}_A^{1/z_A}/\bar{c}_B^{1/z_B}\right)}{\partial T}\right]_{p,\bar{n}_i} = \frac{\Delta H_{\bar{n}}}{RT^2} + \left[\frac{\partial \ln\left(\bar{c}_A^{1/z_A}/\bar{c}_B^{1/z_B}\right)}{\partial T}\right]_{p,\bar{n}_i}$$

$$(5.19)$$

The value $\Delta H_{\bar{n}} = \bar{h}_B - \bar{h}_A + h_A^0 - h_B^0$ is the *differential enthalpy* of the exchange reaction (1). This function refers to an ion exchanger of fixed composition and the corresponding chemical reaction is interpreted similarly as for $\Delta G_{\bar{n}}$. Using the last term in the right part of Equation (5.19), we take into account the possible change of ion concentrations due to the influence of temperature on the volume of the ion exchanger containing fixed amounts of moles of all components. Usually, the value of this term is negligibly small as compared to the first one. As a result, it is possible to estimate the differential enthalpies with the use of a reduced formula which is similar to van`t-Hoff`s Equation (5.9) at constant pressure.

$$\left(\frac{\partial ln\tilde{K}}{\partial T}\right)_{p,\bar{x}_j} = \frac{\Delta H_{\bar{n}}}{RT^2}$$

$$(5.20)$$

The subscript $\bar{x}_j$ is the constancy of equivalent fractions of exchanging ions in an exchanger phase. The integral change of enthalpy of ion exchange reaction can be estimated by:

$$\Delta H^0 = \int_{\bar{x}_B=0}^{\bar{x}_B=1} \Delta H_{\bar{n}} d\bar{x}_B$$

$$(5.21)$$

The influence of temperature on ion exchange equilibria of some inorganic materials like natural and synthetic zeolites, strong acid cationic resins like Dowex-50 and KU-2, and strong base resins, like Dowex-1 (AV-17), and Dowex-2 of polystyrene type was studied starting almost 50 years ago and much data appears in the open literature.[44–49] One of the most complete thermodynamic analyses of the pattern of ion exchange on strong acid resins was given by Soldatov.[48] For the sake of brevity, we summarize only the significant findings here. For most studied ionic systems, the equilibria of sulfonic cation exchange resins and tetraalkyl ammonium strong basic anionic resins were found to submit to principles described below.

1. Temperature only slightly influences the selectivity of ion exchange on polymeric resins, excepting exchanges of metal ions to hydrogen ions. More significant influence of temperature on exchange of metal ions on zeolites is found.
2. Increased degrees of cross linking of ion exchanger resins [containing more divinylbenzene (DVB) in polymer matrices] enhance the influence of temperature on ion exchange equilibrium.

3. During exchange of ions with equal charges, elevation of temperature brings down selectivity of sorbents and reduces equilibrium coefficients of ion exchange.
4. For exchange of ions with different charges, the elevation of temperature enhances selectivity and also increases equilibrium coefficients.
5. At the same temperature, the selectivity and equilibrium coefficients of ion exchange diminish with the increase in proportion of more strongly sorbed ions in the exchanger.

Real interest advanced to practical uses of dual temperature methods about 20 years ago.[50] Many subsequent works[51–60] confirmed that the influence of temperature on ion exchange equilibrium in weak acid cationic resins and polyampholytes is more pronounced than for sulfonic resins. This corresponds to Helfferich's observation that strong heat effects in ion exchange may arise from reactions of complexing or formation of weakly dissociating compounds.[44]

Many experimental data concerning equilibrium on weak acid cation exchanging resins and polyampholytes, inorganic ion exchangers, and anionic resins appear in the literature.[4,7,17,23,25,28,32,50–85] Tables 5.1 through 5.4 present significant portions of data that illustrate the basic principles of the influence of temperature on ion exchange equilibrium. They show that the fundamentals of influence of temperature on exchange equilibria of strong acid resins apply also to other ion exchangers. Note that a strong effect of temperature for ion exchange equilibria of differently charged metal ions on weak acid cation exchangers (Table 5.1) was found. In many cases, the selectivity coefficient changed two to four times with temperature in the interval from 343 K to 353 K. Data in Tables 5.2 and 5.3 also indicate that resins impregnated by extra agents, inorganic ion exchangers, and strong base anion exchangers are also very promising for dual temperature separation.

To analyze the details of temperature influence ion exchange, we include a brief overview of heat effects on ion exchange reactions in Tables 5.5 and 5.6. Along with Tables 5.1 through 5.3, these tables show that exothermic effects ($\Delta H < 0$) occur during the exchange of evenly charged ions at $\bar{K}_A^B > 1$. In contrast, the opposite effect may seen for the exchange of ions of uneven charges, e.g., exchange reactions of selective sorption of divalent ions and desorption of monovalent ones are characterized by the endothermic effect ($\Delta H > 0$).

For exothermic reactions of exchange, partial dehydration of ions in the sorbent phase rather than in an external liquid solution can explain the temperature dependences of selectivity.[81] The entropy change is not significant and serves as the main driving force ($\Delta G \sim \Delta H$). Elevation of temperature reduces the hydration of ions in external solution that may lead to reduction of the exothermic effect of the ion exchange reactions. In essence, it would decrease both selectivity and the thermal coefficient of selectivity. Based on calorimetric measurements of ion exchange reactions of unevenly charged ions on polystyrene sulfonic resin KU-2 (analogous to Dowex-50) with different contents of DVB, it was concluded that the integral heat correlates with the values of differences between radii of corresponding hydrated ions.[82]

TABLE 5.1

Coefficients of Equilibrium of Ion Exchange on Weak Acid Cationic Resins and Polyampholytes at Different Temperatures[*]

Sorbent	Exchanging ions $B^{zB} - A^{zA}$	T (°C)	Solution composition		$\tilde{K}_A^B$	References
			C_Σ (eqv/l)	C_A/C_Σ (equiv. part)		
KB-4	Ca^{2+}–Na$^+$	13	0.51	$2 \cdot 10^{-2}$	4.5	50
	Ca^{2+}–Na$^+$	80	0.51	$2 \cdot 10^{-2}$	10.9	50
	Ca^{2+}–Na$^+$	10	2.58	$3 \cdot 10^{-2}$	5.0	51–54
	Ca^{2+}–Na$^+$	45	2.58	$3 \cdot 10^{-2}$	7.6	51–54
	Ca^{2+}–Na$^+$	90	2.58	$3 \cdot 10^{-2}$	11.5	51–54
	Zn^{2+}–Na$^+$	20	2.51	$4 \cdot 10^{-3}$	14.2	4,55–57
	Zn^{2+}–Na$^+$	90	2.51	$4 \cdot 10^{-3}$	19.2	4,55–57
	Ni^{2+}–Na$^+$	20	2.51	$4 \cdot 10^{-3}$	9.5	4,55–57
	Ni^{2+}–Na$^+$	90	2.51	$4 \cdot 10^{-3}$	20.9	4,55–57
	Co^{2+}–Na$^+$	20	2.51	$4 \cdot 10^{-3}$	9.0	4,55–57
	Co^{2+}–Na$^+$	90	2.51	$4 \cdot 10^{-3}$	23.8	4,55–57
	Ca^{2+}–Mg^{2+}	13–16	0.1	$3.3 \cdot 10^{-1}$	1.4	51,54,58
	Ca^{2+}–Mg^{2+}	77–80	0.1	$3.3 \cdot 10^{-1}$	1.9	51,54,58
KB-4P2	Ca^{2+}–Na$^+$	4–6	2.58	$3 \cdot 10^{-2}$	3.5	51–54
	Ca^{2+}–Na$^+$	90	2.58	$3 \cdot 10^{-2}$	10.3	51–54
KB-4 × 16	Ca^{2+}–Na$^+$	20	2.58	$3 \cdot 10^{-2}$	7.2	51–54
	Ca^{2+}–Na$^+$	90	2.58	$3 \cdot 10^{-2}$	11.5	51–54
KB-2 × 10	Ca^{2+}–Na$^+$	4–6	2.58	$3 \cdot 10^{-2}$	4.2	51–54
	Ca^{2+}–Na$^+$	80	2.58	$3 \cdot 10^{-2}$	16.3	51–54
SG-1	Ca^{2+}–Na$^+$	20	2.58	$3 \cdot 10^{-2}$	2.8	51–54
	Ca^{2+}–Na$^+$	90	2.58	$3 \cdot 10^{-2}$	7.5	51–54
KMD	Ca^{2+}–Na$^+$	20	2.58	$3 \cdot 10^{-2}$	4.8	51–54
	Ca^{2+}–Na$^+$	90	2.58	$3 \cdot 10^{-2}$	6.5	51–54
ANKB-50	Ca^{2+}–Na$^+$	6–8	2.58	$3 \cdot 10^{-2}$	14.5	51–54
	Ca^{2+}–Na$^+$	76–78	2.58	$3 \cdot 10^{-2}$	16.5	51–54
	Zn^{2+}–Na$^+$	20	2.51	$4 \cdot 10^{-3}$	30.0	4,55–57
	Zn^{2+}–Na$^+$	90	2.51	$4 \cdot 10^{-3}$	26.5	4,55–57
	Ni^{2+}–Na$^+$	20	2.51	$4 \cdot 10^{-3}$	27.8	4,55–57
	Ni^{2+}–Na$^+$	90	2.51	$4 \cdot 10^{-3}$	23.3	4,55–57
	Co^{2+}–Na$^+$	20	2.51	$4 \cdot 10^{-3}$	27.2	4,55–77
	Co^{2+}–Na$^+$	90	2.51	$4 \cdot 10^{-3}$	26.2	4,55–57
	Ca^{2+}–Mg^{2+}	18–19	0.1	$3.3 \cdot 10^{-1}$	2.2	51,54,58
	Ca^{2+}–Mg^{2+}	76–79	0.1	$3.3 \cdot 10^{-1}$	1.5	51,54,58
VPK	Ca^{2+}–Na$^+$	20	2.58	$3 \cdot 10^{-2}$	7.7	51–54
	Ca^{2+}–Na$^+$	90	2.58	$3 \cdot 10^{-2}$	9.3	51–54
FFC-1.4/0.7	Cs$^+$–Rb$^+$	25	0.1	$5 \cdot 10^{-1}$	2.3	58,59
	Cs$^+$–Rb$^+$	65	0.1	$5 \cdot 10^{-1}$	2.0	58,59

TABLE 5.1 (Continued)

Coefficients of Equilibrium of Ion Exchange on Weak Acid Cationic Resins and Polyampholytes at Different Temperatures*

Sorbent	Exchanging ions $B^{zB} - A^{zA}$	T (°C)	Solution composition C_Σ (eqv/l)	C_A/C_Σ (equiv. part)	$\tilde{K}_A^B$	References
Lewatit R249-K	Ca²⁺–Na⁺	20	0.93	9·10⁻²	7.1	4,26,64
	Ca²⁺–Na⁺	40	0.93	9·10⁻²	10.4	4,26,64
	Ca²⁺–Na⁺	80	0.93	9·10⁻²	13.8	4,26,64
	Mg²⁺–Na⁺	20	0.93	5·10⁻¹	3.9	4,26,64
	Mg²⁺–Na⁺	40	0.93	5·10⁻¹	5.8	4,26,64
	Mg²⁺–Na⁺	80	0.93	5·10⁻¹	7.5	4,26,64
Lewatit R250-K	Ca²⁺–Na⁺	20	0.93	9·10⁻²	3.4	4,26,64
	Ca²⁺–Na⁺	40	0.93	9·10⁻²	5.5	4,26,64
	Ca²⁺–Na⁺	80	0.93	9·10⁻²	7.8	4,26,64
	Mg²⁺–Na⁺	20	0.93	5·10⁻¹	1.9	4,26,64
	Mg²⁺–Na⁺	40	0.93	5·10⁻¹	3.1	4,26,64
	Mg²⁺–Na⁺	80	0.93	5·10⁻¹	4.4	4,26,64
Lewatit R252-K	Cu²⁺–Zn²⁺	10	0.3	1.2·10⁻¹	1.3	28
	Cu²⁺–Zn²⁺	80	0.3	1.2·10⁻¹	1.2	28
Lewatit TP207	Cu²⁺–Zn²⁺	20	0.3	1.2·10⁻¹	9.2	28
	Cu²⁺–Zn²⁺	40	0.3	1.2·10⁻¹	8.6	28
	Cu²⁺–Zn²⁺	80	0.3	1.2·10⁻¹	7.5	28

* KB-4P2, KB-4, and KB-4 × 16 are gel type polymethacrylic cationic resins corresponding with 2%, 6.5%, and 16% of divinylbenzene (DVB) linking agent. KB-2 × 10 is a gel type polyacrylic resin. ANKB-50 is a polyampholyte of a polymethacrylic gel with amino and carboxylic functional groups. SG-1 is a gel type polymethacrylic resin linked by triethylene glycol-methacrylate. KMD is a gel type styrene divinylbenzene with iminodiacetic groups. VPK is a gel type polyampholyte containing 14% DVB with functional groups of α-picolinic acid. FFC-1.4/0.7 is a macroporous phenol–formaldehyde cationic resin with phenol functional groups, synthesized at a mole proportion of phenol to formaldehyde of 1.4:1. Lewatit macroporous ion exchangers are produced by Bayer AG, a German company. Lewatit R249-K is a polyacrylic cationic resin. Lewatit R252-K is a polyampholyte aminomethyl phosphonic resin. Lewatit TP207 is a polyampholyte resin from hepatoiminodiacetate. (*Source*: Soldatov V.S. *Simple Ion Exchange Equilibria*, Nauka, Minsk, 1972. With permission.).

** Equilibrium solution is a mixture of hydroxides of elements.

*** Equilibrium solution is a salt mixture with cations (eqv/l): 0.45 Mg²⁺ + 0.4 Na⁺ + 0.08 Ca²⁺ and anions 0.73Cl⁻ + 0.20 SO₄²⁻.

**** Equilibrium solution is a sulfate mixture in sulfuric acid at pH = 1.9.

TABLE 5.2

Coefficients of Equilibrium of Ion Exchange on Composite Materials: Cationic Resins Impregnated by Organic Extracting Agents and Immobilized Biopolymers

Sorbent	Exchanging ions $B^{zB} - A^{zA}$	T (°C)	Solution composition		$\tilde{\tilde{K}}_A^B$	References
			C_Σ (eqv/l)	C_A/C_Σ (equiv. part)		
Amberlite-XAD-2-DEHPA	$Zn^{2+}–Cu^{2+}$	20	0.31	1.10–1	1.3	7,68,69
	$Zn^{2+}–Cu^{2+}$	50	0.31	1.10–1	1.6	7,68,69
Amberlite-XAD-2-DEHDTPA	$Cu^{2+}–Zn^{2+}$	20	0.31	1.10–1	1.6	7,68,69
	$Cu^{2+}–Zn^{2+}$	50	0.31	1.10–1	2.1	7,68,69
DNA-PAAG	$Cu^{2+}–Ca^{2+}$	20	2.10–3	5.10–1	1.4	70,71
	$Cu^{2+}–Ca^{2+}$	40	2.10–3	5.10–1	1.6	70,71
	$Cu^{2+}–Ca^{2+}$	80	2.10–3	5.10–1	2.8	70,71

[*] Amberlite-XAD-2-DEHPA is a polysterene–divinylbenzene cationic resin. Amberlite-XAD-2 is impregnated with a 50% solution of di-2-ethylhexyl phosphorus acid (DEHPA)[52] in ethanol. Amberlite-XAD-2-DEHDTPA is a polysterene–divinylbenzene cation. Amberlite-XAD-2, impregnated with a 12.5% solution of di-2-ethylhexyl phosphorus acid[52] in ethanol. DNA-PAAG is an ion exchanger with a capacity 0.1 meqv/l obtained by immobilization of sodium salt of DNA (with functional phosphoric acid groups and nitrogen-containing base) in polyacrylamide gel.

[**] Equilibrium solution is a mixture of sulfates of exchanging elements in 0.025 N sulfuric acid.

TABLE 5.3

Coefficients of Equilibrium of Ion Exchange on Synthetic and Natural Zeolites at Different Temperatures

Sorbent	Exchanging ions $B^{zB} - A^{zA}$	T (°C)	Solution composition		$\tilde{\tilde{K}}_A^B$	References
			C_Σ (eqv/l)	C_A/C_Σ (equiv. part)		
Zeolite	$Ca^{2+}–Na^+$	20	1.0	$3 \cdot 10^{-1}$	1.2	72,74
	$Ca^{2+}–Na^+$	70	1,0	$3 \cdot 10^{-1}$	1.2	72,74
Clinoptilolite	$K^+– Na^+$	13	0.55*	$2 \cdot 10^{-2}$	26.4	72,75,76
	$K^+– Na^+$	70	0.55*	$2 \cdot 10^{-2}$	12.5	72,75,76
	$Ca^{2+}–Na^+$	13	0.55*	$4 \cdot 10^{-2}$	1.1	72,75,76
	$Ca^{2+}–Na^+$	70	0.55*	$4 \cdot 10^{-2}$	1.7	72,75
	$Mg^{2+}–Na^+$	13	0.55*	$2 \cdot 10^{-1}$	0.6	72,75,76
	$Mg^{2+}–Na^+$	70	0.55*	$2 \cdot 10^{-1}$	0.9	72,75,76
Mordenite	$Cs^+– Na^+$	25	0.1	$5 \cdot 10^{-1}$	29.2	77
	$Cs^+– Na^+$	70	0.1	$5 \cdot 10^{-1}$	18.3	77

[*] Model solution of seawater.

TABLE 5.4
Coefficients of Equilibrium of Ion Exchange on Strong Base Anionic Resins[*] at Different Temperatures

Sorbent	Exchanging ions $B^{zB} - A^{zA}$	T (°C)	Solution composition		$\tilde{K}_A^B$	References
			C_Σ (eqv/l)	C_A/C_Σ (equiv. part)		
Dowex-1 × 1	Br⁻–Cl⁻	9	0.5	$1.6 \cdot 10^{-3}$	2.5	19,78,79
	Br⁻–Cl⁻	90	0.5	$1.6 \cdot 10^{-3}$	1.9	19,78,79
Dowex-1 × 4	Br⁻–Cl⁻	9	0.5	$1.6 \cdot 10^{-3}$	3.8	19,78,79
	Br⁻–Cl⁻	90	0.5	$1.6 \cdot 10^{-3}$	2.5	19,78,79
AV-17 × 8, AM, Dowex-1 × 8	Br⁻–Cl⁻	9	0.5	$1.6 \cdot 10^{-3}$	4.9	19,78,79
	Br⁻–Cl⁻	9	0.5	$1.6 \cdot 10^{-1}$	4.8	19,78,79
	Br⁻–Cl⁻	25	0.5	$1.6 \cdot 10^{-3}$	4.2	19,78,79
	Br⁻–Cl⁻	90	0.5	$1.6 \cdot 10^{-3}$	2.8	19,78,79
	SO_4^{2-}–Cl⁻	9	0.53[**]	$5 \cdot 10^{-2}$	0.09	19,78,79
	SO_4^{2-}–Cl⁻	90	0.53[**]	$5 \cdot 10^{-2}$	0.16	19,78,79
	I⁻–Cl⁻	2	1.00	$2.4 \cdot 10^{-4}$	66.4	80
	I⁻–Cl⁻	80	1.00	$2.4 \cdot 10^{-4}$	12.4	80
	I⁻–Cl⁻	12	1.00	$2.4 \cdot 10^{-2}$	45.0	80
	I⁻–Cl⁻	**80**	1.00	$2.4 \cdot 10^{-2}$	15.9	80
Dowex-1 × 16	Br⁻–Cl⁻	9	0.5	$1.6 \cdot 10^{-3}$	6.9	19,78,79
	Br⁻–Cl⁻	90	0.5	$1.6 \cdot 10^{-3}$	3.5	19,78,79

[*] Resins bases of copolymer of styrene and divinylbenzene with benzyl trimethyl ammonium functional groups containing 1, 4, 8, and 16% DVB.

[**] Model solution of seawater.

[***] Model solutions of underground brines.

For mono- and divalent ion exchanges, the entropy change is more significant. At the same time, dehydration of ions moving from external solution to the exchanger phase is also very important and predetermines the endothermic effects of the processes.[60] In other words, endothermic effect increases with increase of temperature,[54] and observation of the increase of differential enthalpy validates this premise[86] for exchange of di- and monovalent ions on polystyrene sulfonic resins.

More significant temperature effects on the enthalpies of endothermic reactions were confirmed through experiments with the exchanges of Ca^{2+}–Na^+ and Ni^{2+} Na^+ on polymethacrylic and polyacrylic resins.[87,88] In the Ca^{2+}–Na^+ exchange on polymethacrylic resins with 2.5 to 6% DVB, differential enthalpy increased linearly with temperature from 7 to 8 kJ/equiv at 20°C up to 25 to 28 kJ/equiv at 130 to 140°C. In the Ni^{2+}–Na^+ exchange, differential enthalpy also increased linearly with temperature from 5 to 7 kJ/equiv at 20 °C up to 22 kJ/equiv at 95°C.

TABLE 5.5

Standard and Integral Enthalpies of Ion Exchange Reactions on Different Exchangers

Sorbent	Exchanging Ions $B^{zB} - A^{zA}$	$\Delta H°$ (298 K) kJ/eqv	References
KY-2 × 2	Li^+–Na^+	−0.65	32
KY-2 × 8	Li^+–Na^+	−1.46	32
	Mg^{2+}– K^+	2.05	32
	Ba^{2+}–Mg^{2+}	−0.75	32
KY-2 × 24	Li^+–Na^+	−1.83	32
Dowex-1 × 1	Br^-–Cl^-	−3.05	19,78,79
Dowex-1 × 4	Br^-–Cl^-	−4.68	19,78,79
Dowex-1 × 8	Br^-–Cl^-	−5.70	19,78,79
Dowex-1 × 8	I^-–Cl^-	−15.9	80
Dowex-1 × 16	Br^-–Cl^-	−7.04	19,78,79
Lewatit TP207	Co^{2+}–Cu^{2+}	−4.9	7,28,68
	Co^{2+}–Ni^{2+}	−6.1	7,28,68
	Co^{2+}–Zn^{2+}	−0.1	7,28,68
	Ni^{2+}–Cd^{2+}	−7.5	7,28,68
	Ni^{2+}–Cu^{2+}	−1.2	7,28,68

TABLE 5.6

Differential Enthalpies of Ion Exchange Reactions on Different Exchangers

Sorbent	Exchanging Ions $B^{zB} - A^{zA}$	ΔH_n (kJ/eqv)	References
KB-4	Ca^{2+}–Na^+	7–9*	50–52
	Sr^{2+}–Na^+	6*	50–52
	Mg^{2+}–Na^+	5*	50–52
	Ca^{2+}–Mg^{2+}	3.3**	50–52
	Ca^{2+}–Mg^{2+}	3.6**	50–52
ANKB-50	Ca^{2+}–Mg^{2+}	−6.3**	50–52
	Ca^{2+}–Mg^{2+}	−8.3**	50–52

* $X_A = $ Cons t. $c_A = 2.5$ eqv/l.

** $Y_A = $ Cons t. $Y_A = 0.4$ (1); $Y_A = 0.75$ (2).

Note that data on thermodynamic constants of equilibrium and standard and integral enthalpies of ion exchange reactions available in the open literature reveal variations because authors use different scales of standards and reference states of components, i.e., sorbents and solutions. Such variations were also found in previously mentioned studies in which the pure mono-ion form of ionite[40,47] was chosen as the measuring state. These variations are also possible with the other approach in which thermodynamic constants of ion exchange equilibrium are equal or close to one (neglecting osmosis factors), and selectivity of ionite to different ions is defined only by the proportion of coefficients of activity of components of the ion exchange reaction.[89–91] This approach is less suitable for analyzing the influence of temperature on the equilibrium of ion exchange. In brief, the main factor of temperature influence on ion exchange is the changing of equilibrium distribution of ions between the solution and sorbent phases.

5.2.2 DYNAMICS OF ION EXCHANGE

The dynamics of ion exchange, as fundamentals of the formation and movement of concentration profiles of components along an ion exchange column, were studied extensively. The underlying principles of dual temperature processes are already known[92–100]. For the ion exchange column liquid phase of an electrolyte solution moving through a stationary phase of a sorbent, the general mass balance equations can be written:[97]

$$\varepsilon \frac{\partial c_i}{\partial t} = -\frac{\partial j_i}{\partial x} + \sigma_{s,i} \qquad \text{and} \qquad (1-\varepsilon)\frac{\partial \overline{c}_i}{\partial t} = -\sigma_{s,i} \qquad (5.22)$$

where j_i is the solution flow density of substance along the column, $\overline{c}_i(x,t)$ and $c_i(x,t)$ are the concentrations of the i component in the sorbent and solution phases, ε is the porosity of the exchanger bed in the column, σ_s reflects the speed of interphase mass exchange (i.e., kinetics of ion exchange), t is the time, and x is the distance along the column from the solution inlet point.

To understand the dynamic temperature effects, we must briefly analyze the equilibrium approximation of the dynamics of ion exchange. We assume that the mass exchange between phases is infinitely quick (an equilibrium approach) and neglect the axial dispersion in the column. The evolution of concentration profiles in the column is determined by boundary conditions at the inlet and by the speed of solution transmission through the column. When the speed of interphase ion exchange is infinitely high, the equilibrium of phases takes place at any cross-section of the column and the systems of Equation (5.15) transforms to the equation of the mass balance of substances in the column:

$$v\frac{\partial c_i}{\partial x} + \varepsilon \frac{\partial c_i}{\partial t} + \frac{\partial \overline{c}_i}{\partial t} = 0 , \quad i = 1,2,3 \ldots \qquad (5.23)$$

and the equilibrium equation.

$$f\left(c_1, \overline{c}_1 \ldots c_i, \overline{c}_i \ldots\right) = 0$$

where v is the linear speed of liquid motion in the column. Equation (5.22), after simple transformations, can be expressed as:

$$-\frac{\partial c_i}{\partial t} = \left(\frac{v}{\varepsilon + \partial \overline{c}_i / \partial c_i} \right) \frac{\partial c_i}{\partial x}$$

(5.24)

Further use of the chain equation $\left(\frac{\partial c_i}{\partial t} \right)_x \cdot \left(\frac{\partial t}{\partial x} \right)_{c_i} \cdot \left(\frac{\partial x}{\partial c_i} \right)_t = -1$ transforms Equation (5.24):

$$u_{c_i} = \left(\frac{\partial z}{\partial t} \right)_{c_i} = \frac{v}{\varepsilon + \partial \overline{c}_i / \partial c_i}$$

(5.25)

The expression in parentheses in Equation (5.24) is the speed of progress of the points of fixed concentration ($c_i(x,t) = const$ or $\overline{c}_i(x,t) = const$) along the column. Equation (5.24) illustrates very simply the role of temperature in the dynamics of sorption through its influence on ion exchange equilibrium. The derivative $\partial \overline{c}_i / \partial c_i$ in the denominator of Equation (5.25) is the inclination of equilibrium dependence of $\overline{c}_i$ on c_i at a certain point ($\overline{c}_i, c_i$), i.e., it depends on the equilibrium constant of the ion exchange reaction (influenced by temperature when the enthalpy of the exchange is not equal to zero). As follows from Equation (5.25), temperature influences the speed of progress of exchange frontier points along the column. For isotherms of Langmuir, Equation (5.4) can be reduced at the requirement of equal valences of exchanging ions i and $j(z = z_i = z_j)$, and we have:

$$\frac{\partial \overline{c}_i}{\partial c_i} = \frac{\left(\tilde{K}_j^i \right)^z \overline{c}_\Sigma / c_\Sigma}{\left\{ 1 + \left[\left(\tilde{K}_j^i \right)^z - 1 \right] c_i / c_\Sigma \right\}^2}$$

(5.26)

where $\overline{c}_\Sigma$ and c_Σ are the total concentrations of ions in the ion exchanger and solution phases, respectively. When the component with index i is a microcomponent ($c_i \rightarrow 0$), we have:

$$\frac{\partial \overline{c}_i}{\partial c_i} = \left(\tilde{K}_j^i \right)^z \frac{\overline{c}_\Sigma}{c_\Sigma} = \Gamma_i(T)$$

(5.27)

where $\Gamma_i(T) = \overline{c}_i / c_i$ is the equilibrium distribution coefficient of component i (Henry's law constant). As follows from Equation (5.25) in Henry's law approximation (5.27), all points of the sorption frontier move with the same speed:

$$u_{c_i} = \left(\frac{\partial z}{\partial t} \right)_{c_i} = u(T) = \frac{v}{\varepsilon + \Gamma_i(T)}$$

(5.28)

Accordingly, the length and the shape of the sorption frontier should remain constant as it propagates along the column. The speed u_{c_i} depends on temperature due to the influence of temperature on selectivity of the sorbent: u_{c_i} decreases with the selectivity increase. In the area of "saturation" of isotherm $(c_i/c_\Sigma \rightarrow 1)$, at the condition $(\tilde{K}_j^i)^z \gg 1$:

$$\frac{\partial \bar{c}_i}{\partial c_i} = \frac{1}{\tilde{\tilde{K}}_j^i} \approx 0 \tag{5.29}$$

In this case, all points of the sorption frontier move also with the same constant speed

$$u_{c_i} = \left(\frac{\partial z}{\partial t}\right)_{c_i} = u = \frac{v}{\varepsilon}$$

and this does not depend on equilibrium proprieties or temperature. Comparison of these two cases $c_i \rightarrow 0$ and $(c_i/c_\Sigma \rightarrow 1)$ shows that the most significant influence of temperature on the dynamic sorption process occurs when we are considering a microcomponent in a mixed solution. In multicomponent systems, we can see more complex behavior of concentration profiles (waves) transferred in the column due to interdependence of the components.

The principles discussed above remain valid for nonequilibrium dynamics, in which the rate of interphase mass transfer is not a quick process, and the equations that describe it [(5.22) and (5.23)] are complicated by corresponding kinetic interrelations. Particulars of nonequilibrium dynamics will be discussed later. The main difference of nonequilibrium and equilibrium dynamics is the "kinetic erosion" of concentration profiles (frontiers). However, from the view of separation processes, the kinetic effects are secondary as compared with the equilibrium effects.[98–100] Elevation of temperature makes diffusion processes faster in most cases and weakens the degradation of concentration profiles due to deviation of the system from equilibrium behavior. Accordingly, the length of the sorption frontier usually decreases with the temperature increase. This rule usually works in most ion exchange systems except those[52] connected to intensive shrinking of weak acid cation exchanging resins of polyacrylic and polymethacrylic types with temperature hikes of solutions containing salts of divalent metals. Experimental results for sorption of divalent ions on these resins taken in initial sodium form indicated that the lengths of frontiers did not diminish with temperature.

The analysis presented earlier shows that temperature affects the equilibrium of ion exchange dynamics in two ways: (1) the component distribution, i.e., proportions of concentrations in solution and sorbent phases, changes with temperature, and (2) the speed of transmission of concentration profiles (waves) along the columns.

Another important influence of temperature on dynamics of sorption and ion exchange in connection with dual temperature separation[17,72] has been investigated. Let us imagine that as a result of some processes, a concentration band of a component was formed in some layer of a sorbent bed where the concentration differed from concentrations in other layers. This band can be a concentration peak in elution chromatography or may be generated when some component is desorbed from the ion exchanging bed. Such a concentration profile contains two waves—the first and last frontiers commonly called peaks or concentration peaks. Assume that the sorbent bed has different temperature zones (at least two), as shown in Figure 5.1. The distinguishing effect is the compression of the peak (band) as it transfers from the temperature zone with lesser sorbing ability (selectivity) of the ion exchanger to a zone with larger selectivity.

It is easy to demonstrate this effect by considering the linear isotherm (Henry's law approximation). Based on Equation (5.28), when $\Gamma(T_2) < \Gamma(T_1)$, $u_x(T_2) > u_x(T_1)$. When the concentration peak moves along the column (Figure 5.1), its back frontier goes along the T_2 part of the column longer than the upper frontier. Conversely, the upper frontier spends more time along the T_1 part than the back frontier. As a result of difference of speeds at two temperatures, the concentration peak (band) compresses, the peak width becomes narrow, and height increases. Based on this overview, the following three rules regulating dual temperature sorption and ion exchange processes have been devised:

1. As temperature changes, the distribution of components between the sorbent (ion exchanger) and solution phase also changes (distribution rule).
2. As temperature changes, the speed of motion of the concentration profile in the sorption bed also changes; speed and selectivity change in an opposite manner (rule of change of the speed of transfer of concentration wave with temperature).
3. As the concentration peak of a component is transferred in a sorption bed consisting of two or more temperature zones, it compresses when transferring from a zone with less selectivity to the component to a zone of higher selectivity (rule of change of width of concentration peak).

All known dual temperature ion exchange methods follow these rules. Rigorous analysis and classification of separation processes on the basis of these rules was not performed earlier. It is appropriate to note that Zhukhovitsky and coauthors[101–103] who developed "chromathermography" initially for gas and later for liquid systems

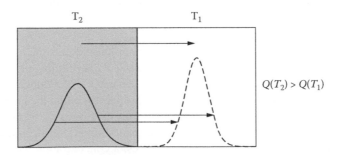

FIGURE 5.1   Compression of concentration peak at its transition through temperature zones of sorbent bed.

indirectly validated two of the fundamental rules. They combined the temperature effects on component distribution in phases and the speed of progress of concentration profiles in columns and described large effects of separation in nonisothermal chromatography in terms of the influence of the movement of thermal waves along the column on chromatographic peaks.

The classification and analysis of the known dual temperature sorption methods presented here are based on the described rules. This approach permits the study of dual temperature methods from a unified view and also helps define the fundamental and derived factors. This classification also permits distinction of sibling connections between methods of dual temperature separation and predicted new methods.

5.3   SEPARATION METHODS REGULATED BY COMPONENT DISTRIBUTION IN PHASES

5.3.1   PROCESSES BASED ON SINGLE EFFECT OF SEPARATION

In ion exchange, the single dual temperature effect of separation can be defined as the redistribution of component concentrations in phases when the system sorbent-solution is transformed from one equilibrium state to another due to change of temperature from T_1 to T_2. The single effect can be explained using the isotherms of dual component (binary) exchange shown in Figure 5.1, which presents dependencies:

$$Y_i = \frac{\alpha X_i}{(\alpha - 1) X_i + 1}$$

(5.30)

which can be derived from Equation (5.2). $Y_i = \bar{c}_i / \bar{c}_\Sigma$ (equivalent fraction in sorbent phase), $X_i = c_i / c_\Sigma$ (equivalent fraction in solution phase), and separation factor, $\alpha = \alpha (T)$. $\alpha (T_1)$ and $\alpha (T_2)$ are separation factors at temperatures T_1 and T_2. Any point $\{X_i, Y_i\}$ on the isotherm corresponds to equilibrium state, i.e., concentrations of component i in solution and sorbent at a given temperature. All the possible ways of system transition from one equilibrium state to another can be imagined as lines between points belonging to different isotherms. For example, one way of transforming the system of original state A taking place at heating from T_1 to T_2 can be imagined by the transition from starting point $\{X_i^0, Y_i^0\}$ to ending point $\{X_i', Y_i'\}$ of redistribution of concentrations in equilibrium phases. Such a transformation can take place if the conditions for material balance are satisfied:

$$\bar{c}_\Sigma \cdot W \cdot Y_i^0 + V \cdot c_\Sigma \cdot X_i^0 = \bar{c}_\Sigma \cdot W \cdot Y_i' + V \cdot c_\Sigma \cdot X_i'$$

(5.31)

where V and W are the volumes of solution and sorbent, respectively. Depending on the chosen volumes of phases, the inclination of the line that matches the isotherm between T_1 and T_2 is defined by the proportion:

$$tg\phi = \delta = -\frac{Vc_\Sigma}{Wc_\Sigma} \tag{5.32}$$

When the capacity of sorbent is high and the concentration of component is low, heating or cooling the system strongly changes the composition of the solution (e.g., the trajectory

$$A\left\{X_i^0, Y_i^0\right\} - B\left\{X_i^", Y_i^0\right\}$$

in Figure 5.2). Another extreme variant arises when the sorbent is in equilibrium in a large amount of solution. The solution composition remains almost the same, but the sorbent composition changes. In this case, the equilibrium condition reaches the point $C\{X_i^o, Y_i^"\}$ (Figure 5.2). These procedures can be performed under dynamic conditions in columns containing equilibrated phases of sorbent bed and initial solution in interspace volume. After a change of temperature from T_1 to T_2 and transformation of the starting equilibrium point following the trajectory, for example,

$$A\left\{X_i^0, Y_i^0\right\} - B\left\{X_i^", Y_i^0\right\},$$

the initial solution at T_2 can pass through the bed, displacing some volume of the solution of a new composition and transforming the sorbent contents to equilibrium at temperature T_2. For example, with the passage of a large amount of solution of mole fraction X_i^0 of i component at temperature T_2 through an ion

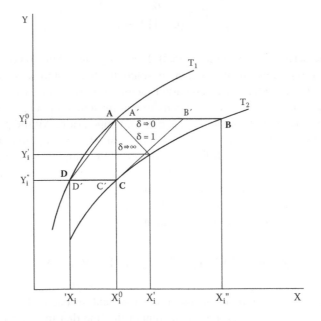

FIGURE 5.2   Ion exchange isotherms in dual temperature process.

exchanger bed previously equilibrated with the same solution at temperature T_1, the maximal value of single concentration that corresponds to the maximal point of elution curve can be evaluated by the proportions in Equation (5.30) for $\alpha(T_1)$ and $\alpha(T_2)$:

$$\frac{X_i^"}{X_i^0} = \frac{\alpha(T_1) + Y_i^0 \left[\alpha(T_1) - 1\right]}{\alpha(T_2) + Y_i^0 \left[\alpha(T_2) - 1\right]} \tag{5.33}$$

In case of exchange of microcomponents ($X_i \rightarrow 0$ or $Y_i \rightarrow 0$), a practical characteristic of maximal single separation effect follows from Equations (5.30) and (5.33):

$$K_{T_2}^{T_1} = \frac{\alpha(T_1)}{\alpha(T_2)} \approx \frac{X_i^"}{X_i^0} \approx \frac{Y_i^0}{Y_i^"} \tag{5.34}$$

In equilibrium dynamics, the system passes through corresponding points of isotherms. In an actual dual temperature dynamic process, the separation effect is defined by equilibrium lines and by the working line $Y = f(X)$ in the area between two isotherms. We now focus on processes of practical interest that use the single separation effect.

5.3.1.1 Simple Cyclic Process

Figure 5.3 illustrates a simple cyclic dual temperature process. This type of separation can be classified as the temperature swing mode. The flow of feed solution containing components i and j is continuously passed through the sorbent bed at a temperature that changes from time to time at the expense of external heating or cooling of the column, or by heating and cooling of the feed solution.

If the selectivity of the sorbent to component i decreases with the increase of temperature ($k_{T_2}^{T_1} > 1$), its concentration in the solution decreases at T_1 (cooling), while enrichment is observed at T_2 (heating). The trajectory of transformation of system composition through points A through D in Figure 5.2 is based on a simple cyclic process. The bed of sorbent, previously equilibrated with original solution at T_1 (point A) is heated to T_2 and the composition of the system transforms to a new state (point B). The original feed solution passes through the sorbent bed to equilibrium at T_2 (point C), then the bed cools to T_1 (D) and equilibrates again with the original feed solution at T_1 (A). During the passing of the original feed solution (X_i^0) at T_2, we obtain the enriched solution with variable concentration from X_i^0 to $X_i^"$, and at T_1, we get a lean solution with variable concentration from X_i to X_i^0. The pattern $A'B'C'D'$ shows

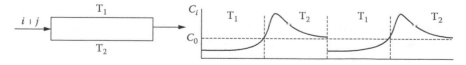

FIGURE 5.3 Simple cyclic dual temperature process.

a simple cyclic process between two tie lines $A'D'$ and $B'C'$ for a process that is not passing through equilibrium points.

Where the selectivity of the target component increases with temperature, a phenomenon opposite of the one in Figure 5.3 occurs. Figures 5.4a and 5.4b show examples of experimental elution in simple cyclic processes. The original feed solution is the concentrated brine of sodium chloride that contains a minor impurity of calcium chloride. As the experimental elution curves show, sorption processes (or purification of solution from calcium) are observed at elevated temperatures, and desorption of excessive calcium from the sorbent (or enrichment of the solution with calcium) takes place at low temperatures.

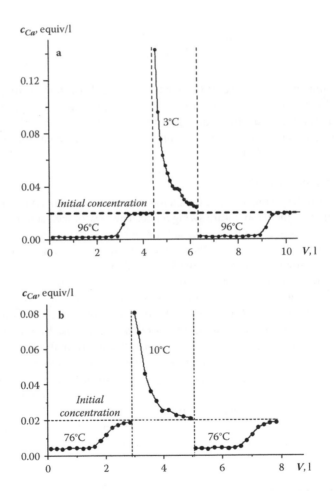

FIGURE 5.4 Concentration histories of Ca^{2+} for ion exchange dual temperature purification of concentrated solutions of sodium chloride from the impurities of calcium chloride: a. original mixture: 2.5 N NaCl and 0.02 N CaCl$_2$; cation exchanger: KB-4; sorbent bed characteristics: = 70 cm; 0.785 cm^2; b. 2.5 N NaCl and 0.02 N CaCl$_2$; Amberlite IRC 50; = 60 cm (Na form); 0.62 cm^2.

Such a process was developed[4,104,105] for reagent-free purification of a concentrated solution of sodium salt from calcium salt on carboxylic cation exchanging resins that demonstrated that selectivity for bivalent metal cations significantly increased with a rise of temperature. In a simple cyclic process, the high temperature phase at 70 to 95°C reduces five to ten times the calcium concentration (purification of original feed solution from calcium); the low temperature phase at ~20°C corresponds to obtaining a solution enriched five to eight times. In real processes, the sorption and regeneration are achieved via equilibrium between sorbent and original solution in each phase. They depend on the technical specifications required for the specific process. The experimental process described earlier is useful for preparing a solution of sodium chloride for membrane electrolysis and for processing brines after seawater desalination to produce salt components of commercial grade.

This simple cyclic process can also be applied successfully for reagent-free concentration of iodides from underground brines, especially geothermal waters, owing to their heat potential.[80] This possibility arises from the fact that the selectivity of strong base anion exchangers to iodide (compared to chloride) decreases extremely with temperature. Figure 5.5 shows concentration profiles obtained at sorption of iodide on a polystyrene type strong base anion exchanging resin AV-17 × 8 in chloride form for a model solution of geothermal water containing 60 g/L of NaCl and 30 mg/L of NaI. The dynamic experiments were performed at different temperatures; all other conditions remained identical. The elevation of temperature in the sorption system led to two important observations: (1) a strong decrease of the capacity of resin toward iodide (up to 6.5 times); and (2) increased steepness of dynamic curves due to diminution of the influence of the kinetic "erosion" effect.

Figure 5.6a shows dual temperature breakthrough curves of three successive cycles for the model feed solution of geothermal water (60 g/L of NaCl and 30 mg/L of NaI). The feed solution was continuously passed through the column with a strong base anion exchange resin AV-17. The temperature of feed solution was periodically changed from

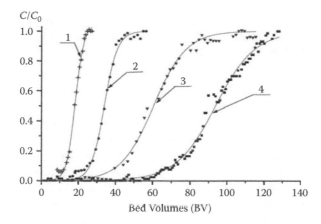

FIGURE 5.5 Concentration histories of I⁻ at different temperatures on strong base anionic resin AV-17 × 8. Model solution of hydrothermal water: 60 g/L NaCl + 30 mg/L NaI. Flow rate = 5 BV/h.

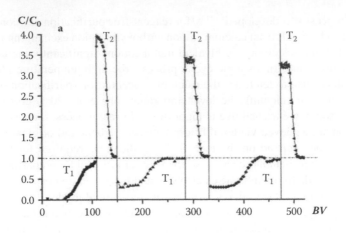

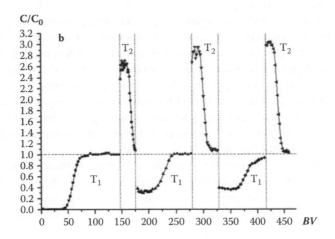

FIGURE 5.6   Cyclic process of I^- concentration on anionic resin AV-17 × 8 at $T_1 = 15°C$, $T_2 = 75°C$. Flow rate = 5 BV/h. Initial composition: a. 60 g/L NaCl + 30 mg/L NaI; b 60 g/L NaCl + 300 mg/L NaI.

$T_1 = 11°C$ to $T_2 = 75°C$ and vice versa. A solution enriched with iodide more than 3.4 times (at maximal point) was produced in each hot half-cycle of the simple cyclic process. Cold half-cycles yielded an exhausted (lean) solution that is discarded.

It is evident that the use of two columns operating in counterphases can provide continuous production of enriched concentrate and processed brine is discharged. The important technological advantage of this (and any other dual temperature ion exchange process) is that the total equivalent concentration of the enriched solution is not raised and remains equal to the original. The process can be used for the utilization of poor hydromineral sources presently recognized as not feasible in view of available standard technologies for recovering iodine from underground brines. The process can also be applied to concentrate iodide by using dual temperature cascade schemes that will be described later in this chapter. See Figure 5.6b. A model feed solution of one of the intermediate concentrates of brine processing containing 60 g/L of NaCl and 300 mg/L of NaI was used in the experiment. The maximal concentration of iodide in the repetitive cycles was ~3.

Another example demonstrating the possibilities of the dual temperature method in reagent-free processes consisting of simple repetitive cycles is purification of drinking water by removing boron.[106,107] To produce fresh water through desalination of sea water (around 3 km^3/year worldwide by reverse osmosis [RO]), permeation of boron through RO membranes is a perpetual problem and requires additional purification from boron. Also, groundwater of different regions is contaminated with boron. Figure 5.7 shows concentration histories for three successive cycles of the dual temperature process. Desalinated water (total salinity of 0.9 g/L) containing 1.5 mg/L of boron was continuously passed through a column containing a weak base anion exchanging resin SB-1 with glucoamine functional groups. The temperature of the original solution was periodically changed from $T_1 = 15°C$ to $T_2 = 72°C$. During sorption of the first cycle, the sorbent exhibits a large capacity for boron due to the extra-equivalent molecular sorption which is irreversible in dual temperature processes. Around 150 relative bed volumes of original water can be purified from boron at cold half-cycles, whereas desorption can be carried out with 25 relative bed volumes of hot water.

One very promising application of simple cyclic processes is the extraction of potassium from seawater. Investigations were carried out in U.S., China, and Japan for the development of technologies to recover potassium and produce potassium fertilizers from seawater by using artificial and natural sorbents.[108–110] Natural zeolites appeared to be the most promising sorbents for ion exchange to concentrate potassium from seawater because of their selectivity properties and low cost.[111,112] These sorption materials are easily available. For example, deposits of clinoptilolite (a natural zeolite) are widespread and estimated in tens of billions of tons. Today, clinoptilolites are introduced into the international market as commercial products because they have several useful properties. Additionally, they are good sorbents for application in water and wastewater purification.

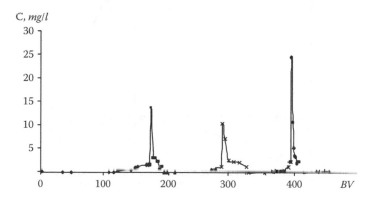

FIGURE 5.7 Cyclic process of boron removal from permeate of seawater desalination using weak base anionic resin SB-1 with N-methyl glucamine groups. $T_1 = 12°C$, $T_2 = 70°C$.

Clinoptilolites are also used in agriculture to improve soils, for example, by increasing their ion exchange capacities, improving their structures and drainage characteristics, maintaining their humidity, and reducing their acidity. The distribution coefficients for potassium from seawater on clinoptilolites are ~50 to 75—not adequate for a one-step purification technology. The main problem is to find an efficient chemical agent for potassium desorption and sorbent regeneration. Another issue is that clinoptilolite also sorbs other elements including sodium, calcium, and strontium from seawater and they must be separated from the potassium. However, both the problems can be resolved by using the specific properties of clinoptilolites to significantly change their selectivity with changes of temperature.[75]

Figure 5.8 shows the concentration histories for macrocomponents of seawater in a dual temperature process that passes seawater through a bed of natural zeolite, first, at $T_1 = 20°C$ and then, at $T_2 = 75°C$. Heating leads to a triple increase of the concentration of potassium in the effluent. Furthermore, the oppositely directed thermoselectivity effect takes place for divalent cations such as calcium and strontium, with the selectivity increasing with temperature. These effects made it possible to develop environmentally friendly dual temperature technology that requires no technology to concentrate potassium from seawater and produce concentrates purified from calcium and strontium.[76]

Figure 5.9 demonstrates a two-stage process passing seawater through a column with a clinoptilolite bed: during the first stage, cold seawater is passed and potassium is concentrated in a sorbent; in the second stage, hot seawater is passed and desorption of potassium occurs concurrently with the removal of divalent ionic impurities from the obtained liquid concentrate. Based on this concept, two technologies were proposed. The first involves the production of natural zeolites enriched with potassium that can be applied as prolonged action fertilizers and artificial soils.[113,114] The second is the production of chlorine-free mixed NK or NKP fertilizers of standard forms.

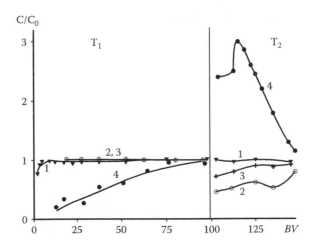

FIGURE 5.8 Concentration histories of seawater cationic components in a dual temperature process on natural clinoptilolite. 1. Na⁺; 2. Ca²⁺; 3. Mg²⁺; 4. K⁺.

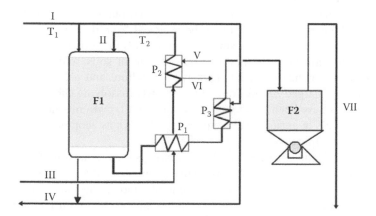

FIGURE 5.9   Flow chart of pilot unit for production of clinoptilolite-based potassium fertilizers from sea water. I and III = cold seawater. II = hot seawater. IV = exhausted water. V and VI = steam from and to power plant. F_1 and F_2 = sorbent beds. P_1 and P_3 = heat exchangers.

Figure 5.9 demonstrates the proposed two-step concentration process. Cold seawater below 25°C is passed for 15 h through thermally isolated ion exchange filters loaded with 20 tons of clinoptilolite. Potassium is sorbed on the clinoptilolite at a flow rate of 100 m³/h (total volume passed = 1500 m³/h). The exhausted seawater (effluent from column) is discarded. After this stage, the process continues; hot seawater at a flow rate of 33 m³/h is passed for 15 hours through the sorbent bed. In this stage, cold seawater is preliminarily passed through the assembly of heat exchangers: first, through the recuperator P_1 and second, through the heater P_2 that operate with separately prepared steam. The temperature of water processed changes in the following manner: it is elevated to 50°C after the recuperator P_1 and reaches 75°C after the heater P_2. During treatment of the clinoptilolite partially enriched with potassium, desorption of potassium and additional sorption of calcium and strontium takes place. As a result, the filtrate obtained represents a relatively pure liquid concentrate of potassium.

This filtrate is cooled by successively passing through the recuperator P_1 to reduce the temperature to 50°C, and then through the cooler P_3 that runs with original cold seawater. Next, cold concentrate is passed through the second (changeable) bed of fresh clinoptilolite (6.5 tons) equipped with a special open filtering apparatus complete with a drainage system and systems for loading and unloading the sorbent. The clinoptilolite, treated in the second filter and enriched with potassium up to 6 to 8% of its mass, along with other microelements from seawater, as the nutrients, is the ready product that represents a chlorine-free fertilizer for prolonged use. The ready product is changed with a fresh portion of clinoptilolite, and the process is repeated.

For the technological process to be continuous, two similar modules operating in opposite phases are used. By using a system of heat exchangers, the heat consumption is limited to a rise of only 20°C for 33 m³/h flow (continuous for both modules). Additional benefit accrues if the modules are installed near power or desalination plants because that allows partially heated seawater (after cooling the aggregates) to be used. The advantages of material obtained via this technology include: (1) the active nutrients cannot be removed from the clinoptilolite by soil watering and are completely consumed by plants; (2) the fertilization process can be concurrent with the recreation or improvement of soils; and, (3) the product can serve as an artificial soil for intensive agriculture and greenhouse systems. Today, nutrient-charged zeolites mixed with slowly dissolving substances containing phosphorus have been developed as commercial growth media.[115] Application of these materials in agricultural technology is known as zeoponics, zeoponix, or zeoponica; see http://www.zeoponix.com.

Another application of simple cyclic processes is the production of chlorine-free fertilizers of standard types. The distinctive property of this technology is that a stationary (not exchangeable) sorption loading of clinoptilolite serves as the second step. The solution of ammonium nitrate (or ammonium sulfate) is used for potassium desorption and regeneration of the multiuse bed. For ammonium nitrate operations, a liquid concentrate composed of 50 g/L of KNO_3 and 50 g/L of NH_4NO_3 can be produced.

Potassium recovery processes are now undergoing pilot testing in a collaboration of the Vernadsky Institute (Moscow, Russia) and the King Abdul University (Jeddah, Kingdom of Saudi Arabia). Dual temperature methods are promising for solving agrochemical problems in several shore-based countries like Georgia that lack potassium fertilizers.[116–118]

Another prospective application of temperature swing techniques is an ion exchange process for recovering lithium from mixed brines containing lithium, sodium, and potassium by using zeolite-X.[119] One distinctive feature of the zeolite is the strong influence of temperature on its selectivity to sodium and potassium ions as compared to lithium. Experiments were carried out with brine at a 1 M total concentration and composition of 87% lithium and 13% sodium on an equivalent basis. Swinging modes of operation between 20 and 97°C provided 97% of lithium recovery at 99.8% purity.[119]

5.3.1.2 Sirotherm Process

A typical example of a practical dual temperature cyclic process in which separation effects are realized for both anions and cations is the Sirotherm process (CSP) developed for partial desalination of natural brackish waters in Australia as a result of research by the Australian CSIRO operation. The development of CSP was favored by fundamental works on ion exchange and polymer chemistry performed in a laboratory under the supervision of Weiss.[13] The development of CSP technology and its introduction were achieved by Bolto.[120–122] The technology includes passing cold brackish water (20 to 30°C) through a bed of special resin containing both weak acid cation and weak base anion exchangers where desalination takes place, according to the reaction:

$$K_{eff}$$
$$R_3N + R^1CO_2H + Na^+ + Cl^- \Leftrightarrow R_3NH^+Cl^- + R_1CO_2^-Na^+ \qquad (5.35)$$

with an effective constant of equilibrium K_{eff}. Regeneration of sorbent after processing is performed with the same brackish water, but at higher temperatures (70 to 85°C). Sorption and elution curves are analogous to those shown in Figures 5.3 and 5.4.

For the first time, a complete theoretical model for the description of CSP separation was developed by Hamann[123] whose calculations included the equilibrium conditions and titration behaviors of mixtures of weak acid and weak base resins with aqueous solutions of 1:1 and 2:2 electrolytes, assuming the systems obeyed Donnan's equilibrium relationship. New approaches[124] based on the complex LeVan-Vermoulen model[125] for bicomponent sorption of two solutes with Langmuir-like or Freundlich-like isotherms are available in the open literature.

It is also possible to derive the main relationships for dual temperature effects in CSP from the simple Butler-Ockrent model[126] for competitive sorption. The reaction (5.35) is coupled with dissociation of acidic and basic functional groups and exchange of ions such as Na$^+$–H$^+$ and Cl–OH$^-$, characterized by relevant constants $\tilde{K}_{Dc}$, $\tilde{K}_{Da}$ $\tilde{\tilde{K}}_H^{Na}$ and $\tilde{\tilde{K}}_{OH}^{Cl}$. In neutral medium and at small concentrations of free ions H$^+$ и OH$^-$ in sorbent phase (characteristic for weak acid and weak base exchangers), $Y_H \ll Y_{Na}$ and $Y_{OH} \ll Y_{Cl}$:

$$K_{eff} \approx \frac{\tilde{\tilde{K}}_H^{Na}(T) \cdot \tilde{K}_{Dc}(T) \cdot \tilde{\tilde{K}}_{OH}^{Cl}(T) \cdot \tilde{K}_{Da}(T)}{K_W(T) \cdot a_{\Sigma,c} \cdot a_{\Sigma,a}} \cdot \frac{1}{Y_{Na} \cdot Y_{Cl}} \tag{5.36}$$

where

$$\frac{Y_{Na}^2}{1 - Y_{Na}} \approx \frac{c_{NaCl}}{a_{\Sigma,c}} \cdot \frac{\tilde{\tilde{K}}_H^{Na} \cdot \tilde{K}_{Dc}}{\sqrt{K_W}} \tag{5.37}$$

and

$$\frac{Y_{Cl}^2}{1 - Y_{Cl}} \approx \frac{c_{NaCl}}{a_{\Sigma,a}} \cdot \frac{\tilde{\tilde{K}}_{OH}^{Cl} \cdot \tilde{K}_{Da}}{\sqrt{K_W}} \tag{5.38}$$

According to Equation (5.36), a change of K_{eff} in the described dual temperature process on a mixed bed can be connected to a change of the corresponding dissociation and exchange constants and also an increase in water dissociation with an increase of temperature. Obviously, temperature is the main driving force in CSP. At a temperature change from 20 to 80°C, water dissociation increases ~30 times, favors a shift to the left in the reaction (5.35), and increases the degree of regeneration. However, knowledge of the behavior of equilibrium is not enough to reveal the real CSP process in a column. For explanation of the dynamic aspects of a CSP column, the article on dynamic studies of periodic operation with thermally regenerable ion exchange resins by Matsuda and coauthors[127] is recommended.

CSP is an example of high level ion exchange technology based on a unique combination of many factors. It is very difficult to reproduce CSP even in a laboratory using the usual weak acid and weak base ion exchangers. The ion exchanger should have buffer properties in the area of average values of pH. The efficiency of chosen combinations of weak base and weak acid functional groups depends strongly on pH values. A short range of operational pH is defined by the chemical natures and proportions of resins. The physical method of mixing the acid and basic groups also carries great importance. In principle, it is possible to realize CSP on a mixed bed of cationic and anionic exchanger grains of sizes usual for water treatment (with diameters of 0.2 to 1.2 mm). However, such a process is too slow and the capacity too small when compared to other efficient demineralization methods including membrane desalination. A faster process using polyampholyte functional groups distributed on a molecular level[128,129] is possible. Before the emergence of special polyampholytes like Amberlite HD-2 and HD-5 in the 1970s (Rohm & Haas Co., Philadelphia),[130] and polyampholytes like Bio-RAD AG 11[131,132] with carboxylic and secondary amino groups and highly porous structures, it was impossible to achieve significant sorption capacities in Sirotherm processes.

Satisfactory capacities (up to 0.3 meqv/ml) and kinetic properties were achieved by Bolto and colleagues, who created a composite sorbent they called "pudding with plums." [122] In the sorbent, the powder-like cationic and anionic materials were incorporated in required proportions in granules of highly porous inert material of usual size. This sorbent was suitable for the use in columns with stationary beds. The other sorbents used for CSP were composite microgranules (around 50 μ) that contained magneto-active materials. These sorbents were more convenient for use in continuous countercurrent ion exchange processes. All these composite materials were produced by ICI Australia Operation Pty Ltd.[122]

The exploration of new types of efficient and accessible sorption materials for Sirotherm processes still serves as a focus of interest of many researchers.[124, 133,134] CSP was introduced in industrial application in Australia and some information is available about two plants. In Adelaide in South Australia, a plant with stationary bed columns is used to obtain 640 m³ per day of feed water for boilers from saline underground water.[135] In Reff City in Western Australia, an industrial plant uses countercurrent ion exchange columns like Higgins reactors to produce 1000 m³ per day[136] of desalinated water. The research objective was the rational use of CSP in cases in which the original water could be concurrently softened and desalinated. The process of separate removal of divalent and monovalent ions was realized on beds of Sirotherm sorbent installed in series or on the same bed with separated countercurrent regeneration and desorption of salts of corresponding components.[137,138]

In practice, it is more convenient to achieve prior softening by the traditional method of regeneration using concentrated brine. To some degree, it reduces the number of components to be processed by CSP. Slightly saline waters with TDS of 0.5 to 5 g/L containing excess concentrations of sodium ions compared to calcium are optimal for processing via Sirotherm. Another issue with CSP is the need for high level pretreatment of original water for removal of organics and oxygen (to prevent thermo-oxidizing destruction of a weak base anionic resin). For the processing of relatively organic-free and anoxic underground water, CSP efficiency is comparable to reverse osmosis and electrodialysis.[122] In some books about separation technologies and desalination, Sirotherm is considered a classic process.[139,140] Today, CSP is included in encyclopedic dictionaries of named chemical technology processes.[141]

5.3.1.3   Simple Cyclic Processes in Multicomponent Systems

In multispecies systems, the dual temperature separation of two components A and B can be realized even when the separation factor α_B^A does not depend on temperature. It is important that their separation factors toward the third component C, i.e., α_C^A and α_C^B, change with temperature in a similar manner. In this case, temperature influences the capacities of sorbent for ions A and B. If $\alpha_B^A > 1$, and if one of these exchanging ions, for example, B, is a microcomponent, it is possible to get significant separation through a simple cyclic process. Such a process for a system for a sulfate of copper (microcomponent), sulfate of zinc, and sulfuric acid, using the weak acid cation exchanger Amberlite IRC-718 was described.[23,142–144]

At $\alpha_{Cu}^{Zn} \sim 1.20$, elevation of temperature from 15°C to 75°C changed the total sorbent capacities for Cu and Zn from 1.2 to 2.10 mg/g (dry resin of H^+ form). The maximal coefficient of enrichment of copper (c_{max}/c_0) in each cold half-cycle of the process was 1.9, and the degree of purification (c_0/c_{min}) in the hot cycle was 5.

5.3.2   Methods with Multiplication of Single Separation Effects

As to multiplication of single separation effects at the expense of combination of simple cyclic processes (cascades), the temperature swing mode of dual temperature separation yields relatively low effects in a single unit. A series of successive simple cyclic processes is the basis of any method involving multiplication of single separation effects ("methods with multiplication"). Two approaches in this direction can be considered. In the first, each separation stage of a single separating effect can be installed using a separate set-up (for example, column). Multiplication of separation effects is achieved in a cascade of single dual temperature set-ups. The other approach consists of a number of nonrecurring cyclic processes (stages) using a single column. This method will be described in the next two sections.

Cascade processes are consecutive simple cyclic processes realized according to multistep schemes. Any single dual temperature cyclic stage of cascade yields two products: partially enriched and partially exhausted solutions as shown in the previous section. Depending on the arrangement of flows in a cascade scheme, good purification of solution from some component and multihold concentration of the component can be achieved. In real processes, cascade schemes are used only for steady enrichment or purification. In a cascade system, the enriched solution of the first stage is fed to the second; the enriched solution of second stage is directed to the third, and so on. A fraction of the exhausted (lean) solution of each stage is added to the feed solution of the previous stage. Accordingly, the enriched solution of the last stage is the final product and the depleted solution of the first stage is discarded. Direct cascade achieved in single dual temperature stages would improve schemes for "multiplying" separation degrees in geometric progression.

In order to harness the maximum separating effect of each single stage, one proposal is to combine the enriched solution of the previous stage and the depleted solution of the next one (with identical compositions) as demonstrated in an ideal cascade.[145] For decades, the ideal cascade theory has played an important role in isotope separation technologies with the use of large numbers of single separation stages (up to thousands).[8,9] It allows economic optimization of technologies. Dual temperature ion exchange separations do not usually require such large numbers of stages and the optimization problem is not so important.

Figure 5.10 shows an example of a three-step cascade scheme of dual temperature separation. Enriched solutions are obtained at a higher temperature, T_2. Some volume of the initial solution mixed with the exhausted (lean) solution of the second stage is passed through the heat exchanger for cooling, and then fed to the first stage where the target component is sorbed by the ion exchanger at low temperature T_1. The exhausted solution of this cold half stage is discharged. Then, some other volume of the same initial solution is heated and the excessively sorbed component is eluted from the ion exchanger of the first stage with this hot solution. The second and third stages operate in the same manner with the difference that the enriched solution of the last stage is the product of the whole process. Using a sorbent with $\alpha(T_2) > \alpha(T_1)$ in a cascade scheme allows purification of the initial solution.

To obtain two target-products at the same time, solution cleaned from a given component and solution enriched with this component yield a second cascade scheme in line with that shown in Figure 5.10. The solution discharged on the first scheme becomes the original one (feed) for the second cascade. Necessary temperature regimes in columns can be provided at the expense of heating and cooling of incoming solutions (Figure 5.10). Sorption columns work adiabatically when concentration and temperature profiles (frontiers) arise in columns. It is also possible to maintain temperature regimes in columns at the expense of their heating and cooling by external heat carriers and columns that work isothermally without formation of temperature fronts. In the cascade scheme explained earlier, a single stage can be considered an individual ion exchange column with a stationary bed of sorbent.

The theory of ideal cascade which makes it possible to choose optimal conditions for separation, including the amounts of sorbents to be loaded at different stages, as well as the proportion of flow rates at different temperatures at any stage and at different stages, is very complex[146]. Some important parameters can be easily estimated from general principals of separation in simple cyclic processes mentioned earlier. In particular, for a linear isotherm and for the simplest cascade schemes (without recycle of solutions), the optimal proportion of sorbent loading at stages

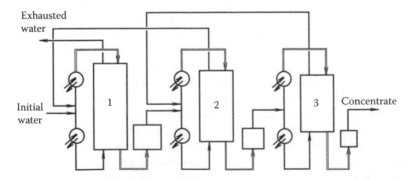

FIGURE 5.10   Dual temperature cascade scheme. 1–3 = columns with sorbent beds ω_1– ω_3.

is: $W_i / W_{i+1} \approx \alpha(T_1) / \alpha(T_2)$. If the cascade scheme is working in a regime of full recycles of solutions, $\omega_i \approx \omega_{i+1}$.

Figure 5.11 shows the separation of a binary system in a multistage cascade process carried out between two isotherms of exchange at temperatures T_1 and T_2. Simple cyclic processes are arranged in cascade mode, the pattern for each following cycle designated as $A_{i+1}B_{i+1}C_{i+1}D_{i+1}$ and the previous cycle as $A_iB_iC_iD_i$. The value of such intersection of cycles is defined by the difference between the values of the extreme concentration of the extracting component in solution X^i_{min} (X^i_{max}) and the average value of $\bar{X}^i$ of the target-product in the total volume of solution produced after I stage (as noted earlier, to simplify, we ignore the volume of liquid phase between grains of sorbent).

Figure 5.11 demonstrates the results of concentration of bromide through three-step cascade processing of seawater on a strong base polystyrene type anion exchanger AV-17 with a ratio of coefficients of equilibrium

$$k_{T_2}^{T_1} = \tilde{K}_{Cl}^{Br}\left(10^\circ C\right) \Big/ \tilde{K}_{Cl}^{Br}\left(75^\circ C\right) = 1.8.$$

The initial concentration of bromine in the Sakhalin coastal water of the Sea of Okhotsk is 56 mg/L. The pilot scale test showed a concentration of 250 mg/L in the third stage enriched solution. This dual temperature cascade process looks promising for the environmentally friendly technology of bromine production from seawater. The existing technologies, such as the Dow-process based on the chlorination and acidification of seawater, are ecologically harmful. The only problem of dual temperature ion exchange is finding cheap sources of heat. Hot seawater wastes (discharges) from power plants on sea shores can be viable alternative sources. However, additional energy consumption for the dual temperature process is offset by the advantages of a water product that has less tendency to form scales and requires no further treatment by chlorination, air stripping, or isolation of bromine.

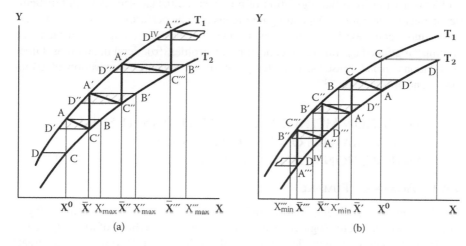

FIGURE 5.11 Separation diagrams for multistep cascade. a) Enrichment with target component. b) Purification of solution from this component.

Wankat et al.[147] described dual temperature schemes as "two-dimension cascades." In the simplest variant, the two-dimension cascade scheme is a system of mass exchanging devices (columns), that form a fixed number of rows, for example, N, and each consists of a fixed number of stages, for example, M. The solution to be treated is fed to each stage of the first row $1_1, ..1_j, ..1_M$ from the bottom to the top until the stages $N_1, ..N_j, ..N_M$ of row N are reached, after which the product requiring different degrees of cleaning and enrichment is obtained.

The carrier (extracting agent) moves from the left to the right of the system: from the first stages of each row $1_1, ..1_i, ..1_N$ to the last stage $M_1, ..M_j, ..M_N$. The solution at the inlet forming part of the first stage, for example $1_1, 1_2$, is fed periodically at temperatures T_1 and T_2 to create temperature waves. While a hot wave moves at T_2 along with the movement of other flows, the "diagonal stripe" of heated zones forms in a two-dimensional cascade. Temperature change can be provided by external heating and cooling of columns in a special order for stages and rows. A two-dimensional cascade can be used in extraction and sorption.

A one-column variant of a two-dimensional cascade can also be realized for dual temperature separation with the use of a rotating column. Martin initially proposed rotating sorption columns to increase separation effects in gas chromatography.[148] Reviews of two-dimensional separation systems, including the use of a rotating mass exchange apparatus for sorption and extraction, are presented in previous studies.[149,150] Dual temperature and two-dimensional chromatography can be effectively used both for analysis and in technological applications. An example of such a process is the separation of acetic acid and water on a rotating sorption column with the use of activated charcoal.[147]

Despite many advantages of cascade schemes including stability of processes to parameter variations, they present a number of limitations for most technical applications (they are used in expensive technologies like separation of isotope mixtures).[146] The most significant drawbacks relate to high equipment (capital) costs, operational problems arising from the regulation of complex multistage schemes, and difficulties in heat recuperation. Depending on the tasks to be performed, cascade schemes can be more complex; for example, each stage can be a countercurrent ion exchange column[52] regulated by transfer of concentration profiles (frontiers) in columns. Other applications include cascade dual temperature ion exchange separation of metal ions[151] and amino acids.[152]

5.4 SEPARATION METHODS REGULATED BY INFLUENCE OF TEMPERATURE ON CONCENTRATION WAVE MOTION ALONG COLUMN

5.4.1 Parametric Pumping

The description of dual temperature methods requires explanation of the main physical, chemical, and technological concepts of noncascade methods of multiplying the single effect. The first such idea was formulated by Pigford[153] while he was creating the cyclic zone absorption method (described later) by analysis of the existing method of parametric pumping (PP). He characterized the main idea as creation of a gradient of concentration of extracted components in ion exchange beds or ion exchange systems.

The most basic principle of PP was first proposed by Wilhelm and Sweed[154,155] (Figure 5.12). X_0 is a molar part (mole fraction) of a component from which the solution should be cleaned. Initially, the ion exchanger in a column is equilibrated with the original solution at temperature T_1. One of the tanks shown in Figure 5.12 is filled with the original solution at volume V. Purification of the solution through extraction of a component is realized at the expense of multiple pumping of volume V through the column in one direction at temperature T_1 and in the reverse direction at T_2. Depending on the proportion of volumes of the solution and sorbent bed (V/W), the concentration profile of the component (exchange frontier) in the column passes through different distances at different temperatures. Consider linear isotherms of component sorption on a sorbent with $\Gamma(T_1) > \Gamma(T_2)$. From the equilibrium of dynamic regulations, it follows Equation (5.28) for phase speed of motion of the concentration profile.

For a column of length L, the concentration profile travels the whole column length over time t:

$$t = \frac{L}{u_{c_i}} = \frac{L\left[\varepsilon + \Gamma(T)\right]}{v} \tag{5.39}$$

Considering that $LS = W$ and $vSt = V$, where S is the area of bed cross-section and V is the volume of solution, it is possible to express:

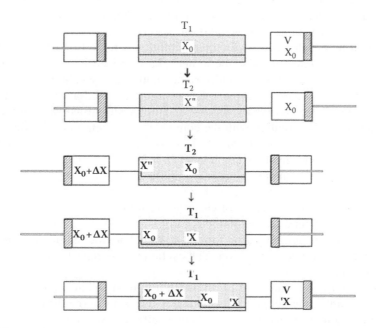

FIGURE 5.12   Stages of parametric pumping.

$$\frac{V}{W} = \varepsilon + \Gamma(n) \tag{5.40}$$

Let us choose any value of V within limits, defined by inequalities:

$$\varepsilon + \Gamma(T_2) \le \frac{V}{W} < \varepsilon + \Gamma(T_1) \tag{5.41, 5.42}$$

This means that by passing a total volume V of solution through the column at temperature T_2 from right to left, the frontier of exchange will travel a distance equal to or longer than the column length. Passing the solution through the column at T_1 (from left to right), the frontier would not achieve the right border of the column. Considering all these facts, let us consider one complete cycle of separation, namely cleaning solution from a given component of the initial molar fraction X_0, according to the PP scheme shown in Figure 5.12.

The solution volume is chosen according to inequalities [Equations (5.41) and (5.42)], for example, equal to the minimal value. At the beginning of the cycle, the temperature in the column changes from T_1 to T_2, and the corresponding molar fraction of the component in solution (at equilibrium) increases to X'' according to the exchange isotherm (Figure 5.2). At temperature T_2, the solution passes from the right tank to the left one and the sorbent equilibrates with the initial solution containing the molar fraction X_0 of the component; the exchange frontier achieves the left border of the column, as explained in Figure 5.12.

An additional quantity of component goes into the left tank and the average concentration in this tank $(n_0 + \Delta n) \le X''$ can be estimated by knowing the volume of the solution and the porosity of the sorbent bed in the column. The half cycle of operation of parametric pumping finishes at this stage.

The next half cycle begins with a temperature decrease to T_1. The component transfers from the liquid phase to the sorbent phase and the distribution of concentration of the component in the equilibrium solution in the column can be evaluated according to isotherms (Figure 5.2). The molar fraction of the component in solution inside the column reduces to $'X$. The fourth or final stage of the cycle includes pumping of solution with the component concentration X'' from the left tank backward to the right one. At the end of this operation at temperature T_1, the exchange frontier profile of the component does not achieve the output of the column and the distribution of concentrations in the equilibrium solution is expected to follow the last (bottom) picture in Figure 5.12. After carrying out one cycle of parametric pumping, the degree of cleaning from the component corresponds to the single dual temperature separation effect (i.e., from X_0 to $'X$). Then, the second cycle of parametric pumping is performed starting with the solution of concentration $'X$ in the right tank, and so on. The cyclic process repeats until the desired level of separation is reached.

This variant (direct mode) of PP in which a limited volume of solution is processed by maintaining necessary temperatures through external heating and cooling of the column is easier to analyze theoretically. To explain the principle of PP, we consider the general theoretical description of the process made by Pigford and collaborators[153] for equilibrium dynamics and linear isotherms of sorption (Henry's law approximation). When Henry's law constant is a function of temperature, we can write the following equation in partial derivatives on time:

$$\frac{\partial \overline{c}}{\partial t} = \Gamma \frac{\partial c}{\partial t} + c \frac{\partial \Gamma}{\partial t} \tag{5.43}$$

Considering Equation (5.43), Equation (5.22) can be transformed into:

$$(\varepsilon + c \cdot \Gamma) \frac{\partial c}{\partial t} + v \frac{\partial c}{\partial x} = -\frac{d\Gamma}{dT} \cdot \frac{\partial T}{\partial t} \cdot - \tag{5.44}$$

Characteristics of Equation (5.44) are the solution of the following standard differential equations:

$$\frac{dt}{\varepsilon + \Gamma} = \frac{dx}{v} \tag{5.45}$$

$$\frac{dt}{\varepsilon + \Gamma} = \frac{-dc}{\left(\dfrac{d\Gamma}{dT}\right) \cdot \left(\dfrac{dT}{dt}\right) c} \tag{5.46}$$

The first equation defines coordinates $x(t)$, for which values c remain constant during the half period. The second represents concentration change with temperature. According to the flow rate of the feeding solution and the column temperature, the value of Γ changes periodically. The following expressions for current parameters v and Γ have been used:[153]

$$v = v^* sq(\overline{\omega} t) \tag{5.47}$$

$$\Gamma = \frac{\Gamma(T_1) + \Gamma(T_2)}{2} - a \, sq(\overline{\omega} t) \tag{5.48}$$

$sq(\overline{\omega} t)$ represents periodic step functions valued at ± 1, $a = [\Gamma(T_1) - \Gamma(T_1)] / 2$, and $\overline{\omega}$ is some angular frequency used to describe easy-to-grasp geometric images of parametric pumping. In the course of the first full cycle of PP, the product $\overline{\omega} t$ "runs" over the values from 0 to 2π; for the second cycle, it progresses from 2π to 4π; and so on. Substituting Equations (5.47) and (5.48) in (5.46) gives:

$$\frac{dx}{dt} = \frac{v^* sq(\overline{\omega} t)}{\varepsilon + \Gamma_0 - a \, sq(\overline{\omega} t)} = \frac{u_{ci} \, sq(\overline{\omega} t)}{1 - b \, sq(\overline{\omega} t)} \tag{5.49}$$

$$\text{where } u_{c_i} = \frac{v^*}{\varepsilon + \Gamma_0}, \ b = \frac{a}{\varepsilon + \Gamma_0},$$

Γ_0 is Γ at $t = 0$. Thus, $dx/dt = f(\overline{\omega}t)$, and this function is defined by the right part of Equation (5.49) and takes the values $+u/(1-b)$ and $-u/(1+b)$. From Equation (5.46), we have:

$$\frac{d\ln c}{dt} = -\frac{d\ln\left[1 - b\,sq(\overline{\omega}t)\right]}{dt} \tag{5.50}$$

Hence:

$$c(x,t) = \varphi\left[x - f(\overline{\omega}t)t, T\right] \tag{5.51}$$

where φ defines the initial conditions in the column. Projection of characteristics on the surface (x, t) constitutes straight lines with the inclination $+u/(1-b)$ in the hot column and $-u/(1+b)$ in the cold column. These lines shown in Figure 5.13 can be considered graphic images of "ways" for ions or molecules of the component to be removed from a solution during PP. This image is helpful for illustrating the principle of separation during PP.

Imagine a separating column shown in Figure 5.12 as a segment [0, L] of a vertical straight line. Any point on this segment is represented as ions located at certain distances of boundary points 0 and L (or as some set of ions of one type at the same boundary distances). In Figure 5.13, rectangles with sides [0, L] and $[n\pi, (n+1)\pi]$, where $n = 0, 1, 2, ..$ are pictured to "scan" the column map with respect to time for different half cycles of the PP process that essentially represent the ion activities during the process. Viewing the three lines inside the rectangles, we can see the histories of three ions that had different starting points at the beginning of the process (at $\overline{\omega}t = 0$): in the middle of the column, at the column inlet (for example, on the right border shown in Figure 5.12), and outside the column beyond the lower border (for example, in the right tank in Figure 5.12). Because the inclination of the right lines in each half cycle at T_2 $[+u/(1-b)]$ is higher than $+u/(1-b)$ at T_1, after a certain number of cycles, the overpass by the ions of the component looks

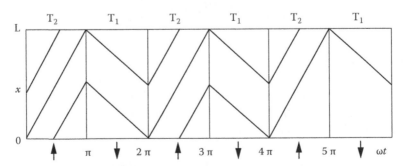

FIGURE 5.13 Ways which are covered in column by ions to be removed by parametric pumping.

like an unbalanced broken curve. The component to be extracted exits the column (beyond the upper border) and never returns. In practice, with the increase of the number of cycles (pumping the solution from right tank to the left and vice versa), purification occurs in the right tank and enrichment with the extracted component takes place in the left one.

We already explained that separation in PP concerns a simple variant: the bath (or closed) mode for which the temperature regimes are provided with the help of external heating or cooling of the column. The initial work of Wilhelm and co-authors[12] was dedicated to more complex variants of PP, one of which was later called the open mode, and another variant without external heating of columns was designated the recuperative mode. In the open mode of PP, continuous feeding with the original solution from the external source occurs at the column inlet along with intake (collection) of hot or cold product at the column outlet.[14,156,157] In recuperative mode, the temperature regimes are created at the expense of periodic heating or cooling of the solution to be processed. The temperature change depending on time is more complex and each half cycle is under adiabatic conditions and not isothermal.[1,158,159]

Correct theoretical descriptions of earlier mentioned variants and separation processes at nonlinear equilibrium and nonequilibrium dynamics require special analysis, but generally they agree with the analysis we have presented. To maintain simplicity, we have not discussed or elaborated on complex variants of PP analyzed in other works.[160-166] PP remains one of the most elegant reagent-less sorption methods for separation of solutes.

Different applications of PP have achieved practical tasks of purification of solutions and substances from organic and inorganic components.[1,158-174] The method is still undergoing development[4,50-60,165,166,175-179] and testing for many species including separation of amino acids,[174] purification of drugs,[178] and treatment of wastewater by removal of phenols[176,177] and heavy metals.[179] The technology has been included in handbooks and monographs[33,180-182] that indicate its potential for applications in chemical engineering. Figure 5.14 demonstrates experimental data on purification of brine (sodium chloride solution) from traces of calcium chloride on a polymethacrylic cation exchange resin with a laboratory PP setup of closed type.[54] As shown in Figure 5.14b, five cycles of operation allow a purification degree around 500.

The prospective application of open (recuperative) PP is demonstrated by Figure 5.15. The flow chart for the unit is elaborated for processing of a concentrated mixed waste solution of NaCl and $CaCl_2$. Such solutions are typically produced in large amounts as wastes by many power plants and other enterprises after the regeneration of water softening columns. Two PP columns operating in successive mode are used. Initial brine containing 18 g/L of $CaCl_2$ and 45 g/L of NaCl is introduced at a flow rate of 8.7 m³/h. Within the unit, the solution is circulated at the increased flow rate to be processed in four successive PP cycles (two cycles for each column). As a result, two separated solutions can be produced at the same flow rates. One solution contains 27.5 g/L of $CaCl_2$, and 35 g/L of NaCl enriched with calcium chloride is discarded. The second solution with a diminished concentration of calcium chloride and rich in sodium chloride (0.6 g/L of $CaCl_2$ and 69 g/L of NaCl) can be reused to regenerate first-step water softening columns. PP units of this type are expected to

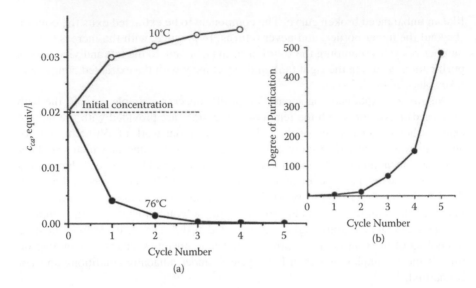

FIGURE 5.14   Purification of sodium chloride brine from calcium chloride by dual temperature parametric pumping. Original mixture: 2.5 N NaCl and 0.02 N CaCl$_2$; cation exchanger: KB-4; sorbent bed characteristics: = 70 cm; 0.79 cm^2.

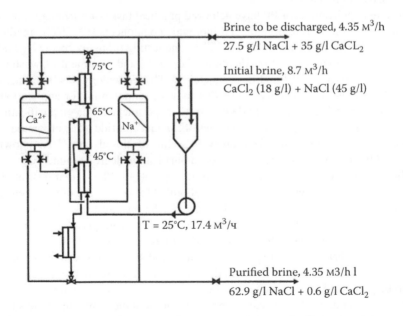

FIGURE 5.15   Pilot plant for recuperation of waste brines produced after regeneration of softening (Na cationite) columns.

reduce the consumption of NaCl two times and achieve an identical reduction of the quantities of brines to be discharged.

Experimental results[175] found PP on zeolites to be efficient for separating a specific alkali metal from mixes with the production of high grade pure salts of one metal. Ordinarily, it is hard to achieve such a separation due to closely spaced chemical and physical properties of these ions. In experiments using relatively small beds (20 to 40 ml) of zeolites of A and Y types, four to six cycles of parametric pumping were sufficient to reduce 100 to 700 times the content of admixtures in representative solutions.[175] Most significant results were obtained for the purification of lithium salts from admixtures of sodium and potassium. These results show that dual temperature can be applied with highly selective zeolites for reagent-less ion exchange separations of alkali metal salts in mixed concentrated solutions.

PP is most suitable for purification of solutions, but it has limitations in concentrating multiple microcomponents from natural waters and other multispecies solutions. These limitations arise because the product of PP separation (cleaned or enriched) is the total volume of processed solution. The method does not allow collection of all the mass of the target component contained in the work solution from small sections of the separating system. The effort needed to create a system to handle that in any part of solution flow or in the sorbent bed would result in narrow zones with high concentrations of microcomponents.

5.4.2 Cycling Zone Adsorption Methods

The cycling zone adsorption (CZA) method was proposed in 1969 by Pigford et al.[153] A number of identical sorbent beds in columns arranged in series are used, and the original solution passes through these successively arranged columns uninterrupted. Cold columns alternate with hot columns. Thus, at any time during the process, a cold column with temperature T_1 is followed by a hot column at temperature T_2. Following a fixed time interval, after passing a definite quantity of solution through the system, the temperatures in columns are changed so that the cold columns become hot and vice versa.

The complete cycle of the process consists of two half cycles (Figure 5.16). $\alpha(T_1) > \alpha(T_2)$, and in each half cycle, the purified solution exits the system if the last column is in cold mode. If the last column is in hot mode, the solution enriched with the given component exits. According to the scheme in Figure 5.16, the regeneration of the first column is performed by the original solution at concentration c_0, the second column is regenerated with partially cleaned solution from the first one, and so on. That is why desorption in each following column is more efficient than the previous one. Similarly, sorption in the first column is performed from the solution at concentration c_0, in the second one from partially enriched solution withdrawn from the first column, etc. With appropriate selection of flow rate of solution and duration of each half cycle depending on equilibrium coefficients $\tilde{K}(T_1)$ and $\tilde{K}(T_2)$, the multiplication of single effect of separation is achieved.

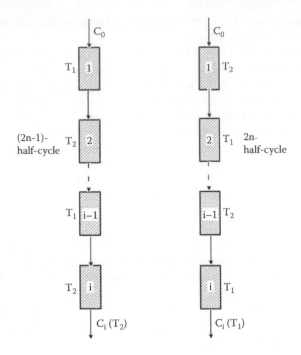

FIGURE 5.16   Stages of cyclic zone adsorption.

The two variants of the method differ in how the temperature regime is maintained. The variant with the external heating or cooling of columns is called CZA, and the variant with the standing temperature wave is designated standing wave.[153,183] The variant in which the change of temperature regime in columns is realized at the expense of periodic change of the temperature of the feeding solution is known as the traveling wave.[153,183]

Theoretical analysis of the principle of separation in CZA does not differ from analysis via PP and can be easily done for the simplest cases of linear isotherm and equilibrium dynamics. CZA is sensitive enough to parameter variations, and even if one parameter (e.g., linear flow rate v or volume of the passed solution) is chosen incorrectly, it would not affect the multiplication of the flash effect of separation, but it would deteriorate separation effect through an increased number of stages. The correct analysis of the method based on the mathematical model for optional isotherms and considering kinetic effects[16,183] is very complex. The overview in the frames of the equilibrium dynamic model is sufficient for defining criteria and requirements of separation since kinetics is a secondary factor. Consequently, many equilibrium models, in particular, for linear isotherms[184] and for Freundlich and Langmuir-type isotherms,[185,186] were developed.

Gupta and co-authors[184] calculated a large number of CZA cycles using a mathematical model for two columns (zones) arranged in series and varying the process parameters and sorbent characteristics. Based on similar calculations, our results performed for four moments of the process are given in Figure 5.17. Figure 5.17a shows the initial concentration profiles in liquid phase; the first zone is cold, the

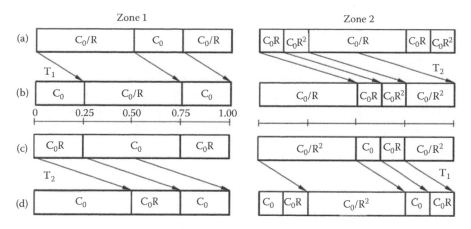

FIGURE 5.17 Concentration profiles of component to be removed at four different stages of the separation cycle of CZA.

second is hot, and the cycle of separation has not yet begun. Figure 5.17b shows the concentration profiles at the moment when the first half cycle is finished. It corresponds to the moment when the concentration profile in the first zone overpasses the distance $\delta/[\varepsilon + \Gamma(T_1)]$, and the profile in the second zone overpasses the distance $\delta/[\varepsilon + \Gamma(T_2)]$, where $\delta = vt/L$ is a dimensionless term.

These movements take place at the expense of feeding the first zone with the fixed portion of original solution of concentration c_0, and feeding the second zone with the same quantity of solution that goes from the first zone (starting from the initial moment in Figure 5.17a). Figure 5.17c represents concentration profiles (bands) after temperature changes in zones. As the temperature changes from T_1 to T_2 in the first zone, desorption of the component from the sorbent to liquid phase takes place and concentrations in all bands increase R times, where $R = [\varepsilon + \Gamma(T_1)]/[\varepsilon + \Gamma(T_2)]$ is the value of the averaged flash effect of separation along the whole band. In the second zone, the temperature changes from T_2 to T_1, and concentrations of all bands in liquid phase decrease R times. Figure 5.17d shows concentration profiles at the end of the second half cycle, after traveling the distance $\delta/[\varepsilon + \Gamma(T_1)]$ by concentration bands in the first zone, and the distance $\delta/[\varepsilon + \Gamma(T_2)]$ in the second one. The four diagrams in the figure were calculated for the following parameters: $\Gamma(T_1)\cdot\varepsilon = 3$, $\Gamma(T_2)\cdot\varepsilon = 1$, $\delta = 1$, and $R = 2$. Note that the first half cycle finishes when the portion of purified solution with the concentration c_0/R passes from the first zone to the second and the enriched portion (several concentration bands with different degrees of enrichment) comes out of the system (second zone). The second half-cycle finishes when the portion of a more purified solution with the concentration c_0/R^2 comes out of the system.

For a steady state process carried out with fixed linear flow rate of feed solution v, the numbers of concentration bands that can be "located" inside one zone and the concentrations in the bands that come out of the system after the first and the second half cycles depend on the value of v. For example, with $\Gamma(T_1)\cdot\varepsilon = 5$, $\Gamma(T_2)\cdot\varepsilon = 2$, and $\delta = 1$ we have a situation in which four different concentration bands are located in one zone (two of them had equal lengths of $\delta/[\varepsilon + \Gamma(T_1)]$

and the other two lengths were $\delta / [\varepsilon + \Gamma(T_2)]$); in each half-cycle, the portion of solution with the concentration c_0 came out of the system. However, with these parameters, CZA did not work because the single effect of separation achieved in the first column is eliminated in the second.

With suitably chosen sets of parameters, the purification effect takes place within one half cycle and a "pulsation" effect can occur with successive purification and enrichment. The equilibrium model gives the main criteria for the selection of separation conditions (parameters) in a CZA system and envisages three possible scenarios.

Case 1—There is no multiplication of single effect of separation in a CZA or no separation occurs:

$$N = \frac{L}{\dfrac{v\, t_{1/2}}{\varepsilon + \Gamma(T_1)} + \dfrac{v\, t_{1/2}}{\varepsilon + \Gamma(T_2)}} \tag{5.52}$$

where N is a maximal number of one type of bands in one zone. This criterion also means that in each zone with the length L, N pairs of coupled bands can be located, with each pair having a total sublength equal to the value of the denominator of Equation (5.52).

Case 2—CZA works and the degree of separation increases from zone to zone:

$$N < \frac{L}{\dfrac{v\, t_{1/2}}{\varepsilon + \Gamma(T_1)} + \dfrac{v\, t_{1/2}}{\varepsilon + \Gamma(T_2)}} \tag{5.53}$$

The maximal degree of separation at M stages is defined by:

$$SF = 2\left(\frac{\varepsilon + \Gamma(T_1)}{\varepsilon + \Gamma(T_2)} \right)^n - 1 \tag{5.54}$$

Case 3—There is no regular situation (pulsation takes place with arising concentration waves at the outlet of the system) and the following condition is fulfilled.

$$\left\{ L - N\left[\frac{v\, t_{1/2}}{\varepsilon + \Gamma(T_1)} + \frac{v\, t_{1/2}}{\varepsilon + \Gamma(T_2)} \right] \right\} < \frac{v\, t_{1/2}}{\varepsilon + \Gamma(T_1)} \tag{5.55} \text{ or}$$

$$\left\{ L - N\left[\frac{v\, t_{1/2}}{\varepsilon + \Gamma(T_1)} + \frac{v\, t_{1/2}}{\varepsilon + \Gamma(T_2)} \right] \right\} > \frac{v\, t_{1/2}}{\varepsilon + \Gamma(T_2)} \tag{5.56}$$

CZA, like PP, has limited capabilities for resolving concentration problems, in particular, for many-fold concentrations of microcomponents from multispecies solutions like natural waters. The limitation is that the product of separation is the entire volume of solution.

As noted earlier, multiplication of a single dual temperature effect can also be achieved with the help of special methods based on the compression of the concentration peak of the target component, like the dual temperature countercurrent ion exchange method[1,97] and its analogues on stationary beds of sorbents.[10,11] These methods manage such compression effects in sorption beds with the simultaneous accumulation of the component mass in narrow zones and small fractions of processed solutions. A detailed description of this group of dual temperature methods of separation is given below.

5.5 SEPARATION METHODS USING COMPRESSION OF CONCENTRATION PEAK OF TARGET COMPONENT

5.5.1 DUAL TEMPERATURE COUNTERCURRENT METHOD

This method was originally developed in the U.S. and former Soviet Union in the 1940s and 1950s in relation to hydrogen isotope separation and heavy water production for systems of (1) liquid water (with dispersed platinum as a catalyzer) and gaseous hydrogen and (2) liquid water and gaseous hydrogen sulfide.[8,9,146,187,188] Large scale production of heavy water with the second method was achieved in the former Soviet Union in 1947. This technology has proven effective and continues to be used in a number of countries. The process is performed in a countercurrent column that consists of two sections maintained at different temperatures. The dual temperature method of separation of electrolyte mixtures in countercurrent columns was developed by Gorshkov and colleagues.[97,189–195]

Separations conducted at the expense of regulation of temperature, pH values, total concentration, and other parameters were carried out with a technique similar to the method for separating hydrogen isotopes that Gorshkov called separation according to a dual temperature scheme.[97] Figure 5.18 shows this scheme utilizing countercurrent ion exchange. The ion exchange column consists of two sections at different temperatures (T_1 in the upper section and T_2 in the lower). The original solution with a concentration of target component $\tilde{n}_{0,i}$ is fed into the inlet point of the column at its bottom at flow rate v_0 and moves upward. The exhausted (depleted of target component) solution exits the top of the column. The ion exchanger with initial concentration of the component $\bar{c}_{0,i}$ is fed at speed $\bar{v}_0$ to the top of the column and moves downward. The exhausted ion exchanger exits the column from the bottom.

Where $\alpha(T_1) < \alpha(T_2)$, desorption of some part of the target component from the ion exchanger into the solution occurs in the lower section, with most of the component transferred with the solution from the bottom to the top. In the upper section, the inverse sorption of some part of the target component from solution by ion exchange takes place, and most of the component returns to the column with the sorbent moving downward. Thus, at the expense of changing the direction of interphase transfer at the boundary of two sections with different temperatures, the target component gradually accumulates in the area of this boundary. After sufficient accumulation of the component in this area of the column and the achievement of a sufficiently

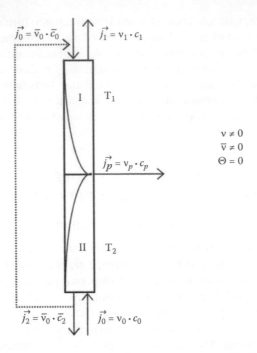

FIGURE 5.18   Dual temperature separation by countercurrent method.

high concentration peak, drawing off the product solution with the concentration c_p can be started at speed v_p. For a steady state regime, a balance equation for fluxes (Figure 5.18) can be written:

$$j_0 + \bar{j}_0 = j_1 + \bar{j}_2 + j_p \tag{5.57}$$

where indices i for the target component are omitted for simplicity; $j_0 = v_0 c_0$ and $\bar{j}_0 = \bar{v}_0 \bar{c}_0$ are the incoming fluxes of the component with moving liquid and sorbent phases; $j_1 = v_1 c$ and $\bar{j}_2 = \bar{v}_2 \bar{c}_2$ are the corresponding depleted fluxes exiting; and $j_p = v_p c_p$ is the product.

The theory of separation via a dual temperature countercurrent scheme was developed by Rosen,[146,196] Bier,[197] and Safonov and Gorshkov.[97] Theoretically, the steady state concentration of the target component at the boundary of sections depends strongly on the ratio of solution and ion exchanger fluxes. Some asymptotic proportions that may serve as bases for preliminary evaluation and choice of parameters for the process can be found in other works.[150,152] In particular, requirements for efficient concentrating of microcomponents of the ionic mixture in the dual temperature column (Figure 5.18) can be expressed, as:

$$\alpha_2 < \frac{v_0 c_{0,\Sigma}}{\bar{v}_0 \bar{c}_{0,\Sigma}} < \alpha_1 \tag{5.58}$$

where $c_{0,\Sigma}$ and $\bar{c}_{0,\Sigma}$ are total concentrations of all components corresponding to the initial feed solution and feed ion exchanger.

The application of countercurrent ion exchange with a dual temperature scheme is successful for reagent-free separation of calcium and potassium ions using the total concentration of solution as a parameter by influencing separation factor α (at the expense of electroselectivity). It is also useful for separation of cesium–rubidium and rubidium–potassium ionic mixtures through pH-driven changes of the selectivity of a polyfunctional resin of the phenol–formaldehyde type.[156,189–191,198] Direct dual temperature separation of calcium–potassium and calcium–iron ionic mixtures on polystyrene sulfonic resin at the expense of temperature-driven change of selectivity was performed by Bailly and Tondeur.[199,200]

5.5.2 CYCLIC COLUMN METHOD

Andreev, Boreskov, and Katalnikov[10,11] proposed this method in 1961 as a process that simulates separation according to a countercurrent scheme and using a stationary bed of ion exchanger. They performed the first dual temperature reagent-free separation on ion exchangers. They studied methods to separate Li^+–NH_4^+ and Cs^+–Na^+ mixtures in a cyclic column with a strong acidic polystyrene sulfonic cation exchange resin and a moving boundary between two temperature zones.

They explained that separation similar to the countercurrent scheme is possible if one of three parameters (speed of solution motion in column v, speed of sorbent motion $\bar{v}$, and speed of temperature front θ) is equal to zero while the other two differ from it. Figure 5.18 shows that the classical dual temperature countercurrent scheme corresponds to a combination: $\theta = 0$, $v \neq 0$, $\bar{v} \neq 0$. Let us also look into the other option. The solution of mixture to be separated is passed through a series of n columns with jackets for external heating or cooling (or a line of sections of one long column, provided with separate thermostatic jackets) with stationary beds of ion exchanger ($\bar{v} = 0$).

At the same time, at a certain combination of speeds $v \neq 0$ and $\theta \neq 0$, it is possible to create conditions that at any time passing the solution from the $(n-1)$-th column at temperature T_2 to the n-th column at T_1, the change of the direction of interphase transfer of the component will be similar to that in the countercurrent ion exchange column. The system has a concentration peak—a wave of target component becoming stronger as it moves through the system. Figure 5.19a illustrates the process. Generally, it is appropriate to use a limited number of columns (sections) and pass the solution through the columns arranged in a circular path with provision for external feeding with a calculated amount of flow of initial solution containing the target component. This cyclic column method shown in Figure 5.19b represents another novel approach proposed by the authors cited above. The flow of solution passes through the cyclic column. From the top, the column is washed by a slowly rotating, umbrella-like spraying machine with hot and cold halves.

A close analogy of the cyclic column method is a variant of chromatography proposed by Zhukhovitsky and co-authors[101,102] in which a chromatographic column is inserted into a series of cylindrical heaters and coolers moving along the column in the direction of mixture pass for the purpose of separation.[103] The theoretical

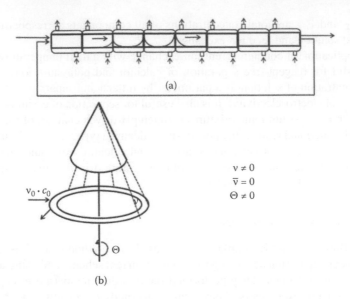

FIGURE 5.19   Dual temperature separation by cyclic column methods. a. Using set of stationary beds. B. Rotating system.

description of the separation method of the cyclic column is similar to that for separation by the dual temperature scheme in a countercurrent column using "fake" speeds of liquid flows and sorbent phases: $v^* = v - \theta$ and $\bar{v}^* = \theta$. In this case, the stationary bed in the dual temperature cyclic column is considered the simulated moving one.

It is important to note that the cyclic method may be analogous to the dual temperature countercurrent method, as it reminds us of CZA and can be studied on the basis of theoretical CZA. In fact, the CZA scheme can be used for multiple concentrations if we do not monitor the total volume of solution coming from the system and focus on the continuously increasing concentration peaks from corresponding columns and select a part of this peak as the target product.

5.5.3 SWINGING WAVE METHOD

Tikhonov theoretically demonstrated the possibility of a new dynamic sorption technique for dual temperature separation of solutes, designated the swinging wave method.[17,18] The processed solution plays the role of heat carrier. Theoretical investigation followed a linear approach. Later, Fokina, Khamizov, and Tikhonov expanded this concept and experimentally identified a number of variants that allow continuous concentration and separation of dissolved substances.[19–22,201–204] They proposed an open mode of the swinging wave method convenient for large scale application in pilot installations for zero-discharge processing of seawater.[205–210]

Swinging wave dual temperature separation is based on the effect of compression of the concentration band during its transfer from a zone with less selectivity of sorbent to a zone of higher selectivity. "Swinging band" would be a more appropriate name, but we will use the "swinging wave" terminology cited in the literature.[20,21] The simplest variant of the method (Figure 5.20) can be realized through a set-up consisting of two identical sorption columns with an ion exchanger, the selectivity of which to the target component depends on temperature. For example, let the selectivity decrease with temperature. At the beginning of the process, the sorbent in both columns is equilibrated with the original solution at low temperature T_1. One column is heated to T_2 and the concentration of target component in the solution in the "hot" column increases in comparison to the solution in the "cold" column due to a decrease of sorbent selectivity. Later, columns are joined with a closed circle of liquid.

The following process consists of repeating the cycles of three stages: (1) substance (species) redistribution through the columns with a special circulating solution; (2) changes of temperatures in columns, achieving substance redistribution between the phases; and (3) the sorption of substance from the original solution. From one cycle to another, the flow direction of the circulating solution is changed.

The cycle begins with the transfer of substance from the hot column to the cold one with the circulating flow of the solution (Figure 5.20a). The quantity of this target component in the hot column before the cycle may be called the accumulation value m (at the beginning, $m \geq 0$). The liquid from the hot column flows into the cold one through a countercurrent heat exchanger and heats the flow of liquid withdrawal from the cold column. The heat losses are compensated by additional heating at point (a) and cooling at point (b). While the flow of solution with accumulation m transfers to the cold column, synchronous compression of the zone of accumulation and the rise of concentration in this zone take place.

The second stage of the cycle is the changing of temperature of two columns. The heat exchanger is turned off (Figure 5.20b) and the circulating solution passes until the hot column becomes cold and vice versa. At this stage, the redistribution of the solution and sorbent in the columns occurs. The second column from the right becomes the hot one. Partial desorption takes place and the solution gains an additional quantity, Δm, of the target component. Accumulation of the target component in solution becomes greater than that in solution phase in column 1 before the first stage.

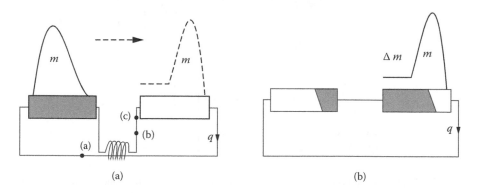

FIGURE 5.20 Two-column swinging wave method. (a) Transfer of enriched zone from hot column to cold column. (b) Change of temperature in column by circulating flow.

The third stage requires feeding the set-up with the target component from the external original solution. The exhausted cold column 1 with reduced concentration of target component is excluded from the system and the cold original solution passes through it until equilibrium is attained. The absorption Δm of the target component from the initial solution takes place, and this column is included in the system and becomes ready for the next cycle in the opposite direction of the flow of circulating solution and with the opposite transfer of enriched zone from column 2 to column 1. At this point, the concentration band of the target component moves in the opposite direction.

When these cycles are repeated several times, the concentration band of the target component swings from left to right and vice versa, gradually accumulating the component and increasing its concentration. When the necessary degree of enrichment is achieved, the preparation phase is complete. In the working phase involving the transfer of the accumulation zone from one column to another, the product solution is picked up from the system at the moment of "pass" of the peak of concentration band of the component through the pick-up point (c). The product solution is withdrawn in recurring mode with repetitive characteristics.

Swinging wave can be realized as a continuous process[203] by including a third column into the system. Figure 5.21 illustrates the set-up. While the stages of redistribution and change of temperature are carried out in two columns, feeding with the

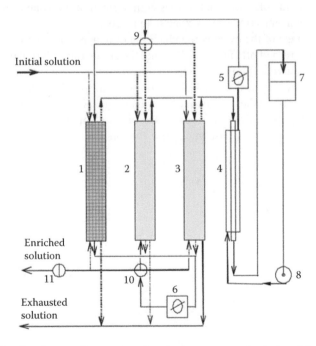

FIGURE 5.21 Continuous (three-column) swinging wave method. 1 and 2 = hot and cold columns undergoing transfer of concentration wave. 3 = column fed with target component from initial solution. 4 = heat exchanger. 5 = heater. 6 = cooler. 7 = tank for circulating solution. 8 = pump. 9 to 11 = valves.

target component takes place in the third column. As a result, the original solution is continuously fed into the system and exhausted solution is withdrawn. The extent of component extraction is defined by the proportion of equilibrium parameters at the selected temperatures T_1 and T_2.

Figure 5.22 shows the results of bromide concentration from seawater in a three-column set-up filled with a polystyrene matrix strong base anion exchanger. During successive operating cycles, the concentration peak of bromide grew, and after 30 cycles its maximal concentration exceeded its original concentration in seawater 80 times ($c/c_0 = 80$). The higher number of cycles represents greater accumulation of microcomponent in the preparatory stage from its original content. The next stage is withdrawal of the concentrated bromide solution that may continue infinitely. At steady state, the average flow rates of feed seawater and final concentrate solution, v_0 and v_p, and the average concentrations of bromide in these flows, c_0 and c_p, are related to the balance of fluxes: $v_p \cdot c_p = v_0 \cdot c_0$.

To reduce the laborious laboratory studies of this method, an alternative is numerical investigation via computer simulation of an appropriate mathematical

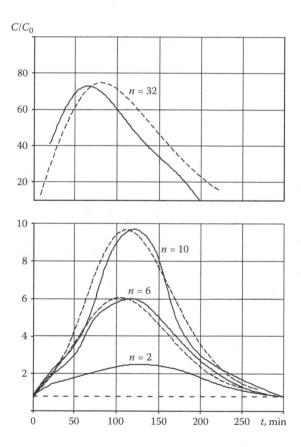

FIGURE 5.22 Experimental (strong) and theoretical breakthrough curves of Br⁻ concentration from seawater by swinging wave method. n = number of treatment cycles.

model of the process for a wide range of variables, and its optimization followed by validation of experiments in calculated regimes. Computer simulation was performed using the following conditions and model parameters obtained from studies of the equilibrium and kinetics of the exchange of Br⁻ and Cl⁻ on strong base anion exchangers:[18,19]

1. The equilibrium was described by Langmuir isotherm with equilibrium coefficients dependent on concentrations and temperature according to Equations (5.4), (5.5), and (5.20).

2. Kinetics of exchange were described by the following model for mixed diffusion:

$$\frac{1}{\gamma^*} = \frac{\Gamma}{\beta} + \frac{1}{\gamma} \tag{5.59}$$

$$\beta = \frac{\beta_0 v^{1/2}}{R^{3/2}} \varphi(\overline{T}), \quad \gamma = \frac{\pi^2 \overline{D}_0}{R^2} \phi(\overline{T}) \tag{5.60}$$

$$\left(\frac{\partial ln\gamma^*}{\partial \theta} \right) = \frac{E}{R\overline{T}^2} \tag{5.61}$$

$$\left(\frac{\partial ln\tilde{K}}{\partial \theta} \right) = \frac{\Delta H_n}{R\overline{T}^2} \tag{5.62}$$

where: $\overline{T}$ is the temperature of sorbent phase that may differ from that for liquid phase; $\overline{D}_0$ represents diffusivity in the sorbent phase; β and γ are the mass transfer coefficients for the solution and sorbent phases, and E is the corresponding activation energy for mass transfer. The following experimental values of parameters were used: $\tilde{K}(\overline{T}) \approx \tilde{K}(\overline{T})|_{298K} = 4.27$, $\Delta H_n = -5.7 \times 10^3$ J/eqv, $\gamma^*(298) = 6 \times 10^{-2}$ s⁻¹; $\overline{D}_0 (298) = 9 \times 10^{-7}$ cm²/s; $\gamma(298) = 7 \times 10^{-3}$ s⁻¹; $E(\gamma^*) \approx 8.3 \times 10^3$ J/eqv; and $E(\gamma) = 16.9 \times 10^3$ J/eqv.

3. The description of the dynamics of ion exchange enrichment of solution with the target component under nonisothermal conditions[20,21] was realized on the basis of the approximate models noted above that use the linear kinetic equations for mass and heat transfer. The dynamic model included the following equations of mass transfer in the column and sorbent phases and for heat transfer:

$$\varepsilon \frac{\partial c_i}{\partial t} + v \frac{\partial c_i}{\partial x} + \frac{\partial \overline{c}_i}{\partial t} = 0, \quad 0 \le x \le L, \quad i = 1,2 \tag{5.63}$$

$$\frac{\partial \overline{c}_i}{\partial t} = \beta \ (\overline{c}_{i,\infty} - \overline{c}_i), \quad \beta = \beta(c, \overline{T}) \tag{5.64}$$

$$\varepsilon \frac{\partial T}{\partial t} + v \frac{\partial T}{\partial x} + \vartheta \frac{\partial \bar{T}}{\partial t} = 0 \tag{5.65}$$

$$\frac{\partial \bar{T}}{\partial t} = \beta^T \left(T - \bar{T} \right) \tag{5.66}$$

$$\frac{c_{01}}{\bar{c}_{1,\infty}} = \tilde{\bar{K}} \left(\bar{T} \right) \frac{c_{02}}{\bar{c}_{2,\infty}} , \quad \sum_{i=1}^{2} \bar{c}_i = \bar{c}_\Sigma \tag{5.67}$$

where $\bar{c}_\Sigma$ is the total exchange capacity per unit of sorbent bed volume; $\bar{c}_i(x,t)$ is the concentration of i–th component in the ion exchanger; t is time ($t = 0$ indicates the beginning of the cycle); x is the linear coordinate in the column ($x = 0$ corresponds to column input; $x = L$ output); β^T is the kinetic coefficient of heat transfer in sorbent (s^{-1}); and ϑ is the coefficient of heat conductivity of sorbent phase per bed unit ($J/cm^{-3} \cdot K$).

4. The mathematical model includes a set of initial and border requirements (boundary conditions) for two columns according to the above descriptions.

This method was evaluated for concentrating bromide from a model solution of seawater using a three-column set-up (Figure 5.21) with the following parameters: volume of bed of strong basic polystyrene-type anion exchanger AV-17 × 8 in column $W = 15$ cm^3; length of ion exchanger in column $L = 20$ cm; $v/L = 2.6$ h^{-1}; $T_1 = 276$ K; and $T_2 = 351$ K. Initial concentrations of chloride and bromide in the model solution were $c_{Cl} = 0.5$ meqv/L and $c_{Br} = 8 \times 10^{-4}$ meqv/l. Isothermal conditions were maintained during one cycle in each column because the full adiabatic process of swinging wave cannot be scaled to smaller dimensions. Nevertheless, the simulation results show that the speed of temperature wave transfer is much higher than the speed of concentration wave transfer and the isothermal column variant simulates the real process quite efficiently. Figure 5.22 compares the experimental concentration profiles of Br⁻ with the calculated results of simulation according to Equations (5.56) through (5.60). The results of the comparison confirmed the validity of the model and efficiency of the swing wave method. Evaluation of some technological parameters indicates that a plant charged with 1 ton of sorbent corresponds to annual production of solution containing 1.2 tons of bromine with energy consumption of 10^8 kJ and heat losses ~15%. The proposed method can be applied to environmentally friendly reagent-free concentration of different components from natural waters.

The swinging wave method can be called a "closed" mode because it uses a special circulating solution for transferring a concentration band from one column to another. The stages of transfer of wave and "feeding" with the target component from the original solution are separated. At small concentrations of target component in the accumulation zone (corresponding well to linear isotherms), it is possible to combine these stages while the original solution passes through the system. Figure 5.23 presents a comparison of calculated results for

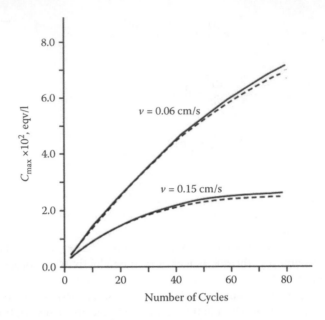

FIGURE 5.23   Results of computation of bromide concentration from seawater by open and closed swinging wave.

bromide concentration from seawater by open and closed methods of swinging wave obtained using Equations (5.56) through (5.60). Both methods yielded almost identical concentration profiles with the Henry's law approximation. Some lower target component concentrations in open mode method, corresponding to large numbers of cycles, are connected with smaller accumulation of component in the cold column.

One technological advantage of the open mode is the ability to achieve a nonstop cyclic process with two columns without using a recycling solution. Figure 5.24 illustrates the module for bromide concentration from seawater by open mode swinging wave. A set-up based on this scheme was installed in a pilot plant for zero-discharge complex processing of seawater.[25,72,197–210] The process was run in an adiabatic regime using the waste heat of the evaporating module of the plant, with additional heating and cooling of columns via heat exchangers as shown in the figure. About 85% heat recuperation was achieved by a set of countercurrent heat exchangers.

5.6   CONCLUSIONS

The dual temperature method (DTM) is an efficient approach for environmentally friendly ion exchange technologies. The investigation of the influence of temperature on ion exchange processes involves three important principles that regulate all known methods of separation in columns based on temperature-regulated selectivity of sorbent:

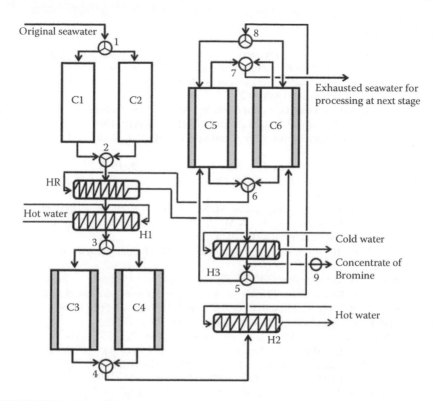

FIGURE 5.24 Bromide concentration module installed in pilot plant for zero-discharge seawater processing. (C1–C2), (C3–C4), (C5–C6) = columns with natural zeolite (for filtering), zeolite of A-type (softening), and anion exchanger (Br⁻ concentration). H1 to H3 = heat exchangers. HR = heat recuperator. 1 through 9 = automatic valves.

1. As the temperature changes, the equilibrium distribution of components between sorbent and liquid phase also changes (distribution rule).
2. As the temperature changes, the speed of motion of the concentration profile in the sorption column changes: essentially speed and selectivity change in opposite manners (rule of change of speed of transfer of concentration wave with temperature in isothermal column).
3. The concentration peak of a compound in a sorbent bed (concentration band) compresses at its transfer from a temperature zone with less selectivity of sorbent to a zone of higher selectivity (rule of change of width of concentration band).

The theoretical basis developed for DTM allows computer simulation, elaboration, and optimization of dynamics of complex processes. A number of strong base anion, weak base anion, and weak acid cation exchangers were found promising for solving several problems of ion exchange concentration and separation without the need for chemical reactants.

REFERENCES

1. Wankat P.C. In *Percolation Processes: Theory and Applications*. Rodrigues A.E. and Tondeur D., Eds. Sijthoff & Noordhoff, Rockville, MD, 1981, 443.
2. Grevillot G. In *Handbook of Heat and Mass Transfer*, Vol. 2. Cheremisinoff N.P., Ed. Gulf Publishing, Houston, 1986, 1429.
3. Tondeur D. and Grevillot G. In *Ion Exchange Science and Technology*. Vol. 2, NATO ACI Series, Rodriguez A.E., Ed. Martinus Nijhoff, Dordrecht, 1986, 369.
4. Ivanov V.A., Timofeevskaya V.D., Gorshkov V.I. et al. *J. Radioanal. Nucl. Chem.* 208, 23 (1996).
5. Gorshkov V.I. and Ivanov V.A. *Solv. Extr. Ion Exch.* 17, 695 (1999).
6. Ivanov V.A., Timofeevskaya V.D., Gorshkov V.I. et al. *Theory Pract. Sorption Proc.* 25, 21 (1999); in Russian.
7. Muraviev D., Noguerol J., and Valiente M. *Solv. Extr. Ion Exch.* 17, 767 (1999).
8. Sakodynsky K.I. and Zhavoronkov N.M. *Uspekhi Khimii* 29, 1112 (1960), Russian.
9. Andreev B.M. *Sep. Sci. Technol.* 36, 1949 (2001).
10. Andreev B.M., Boreskov G.K., and Katal'nikov S.G. *Khimich. Proyshl.* 6, 389 (1961), in Russian.
11. Andreev B.M. and Boreskov G.K. *Zh. Fizich. Khimii* 38, 115 (1964), in Russian.
12. Wilhelm R.H., Rice A.W., and Bendeluis A.R. *Ind. Eng. Chem. Fundam.* 5, 141 (1966).
13. Weiss D.E., Bolto B.A., McNeill R. et al. *Austral. J. Chem.* 19, 561 (1966).
14. Chen H.T., Rak J.L., Stokes J.D. et al. *AIChE J.* 18, 356 (1972).
15. Knaebel K.S. and Pigford R.L. *Ind. Eng. Chem. Fundam.* 22, 336 (1983).
16. Hsu T.B. and Pigford R.L. *Ind. Eng. Chem. Res.* 30, 1067 (1991).
17. Tikhonov N.A. *Zh. Fizich. Khimii.* 68, 856 (1994), in Russian.
18. Poezd A.D. and Tikhonov N.A. *Russ. J. Phys. Chem.* 69, 451 (1995).
19. Fokina O.V. and Khamizov R.Kh. *Russ. Chem. Bull.* 43, 1965 (1994).
20. Tikhonov N.A., Khamizov R.Kh., Fokina O.V. et al. *Dokl. Khimii* 354, 87 (1997), in Russian.
21. Tikhonov N.A., Fokina O.V., Sokolsky D.A. et al. *Russ. Chem. Bull.* 46, 2053 (1997), in Russian.
22. Khamizov R.Kh. and Fokina O.V. *Nauk. Proiz.* 2, 20 (1998), in Russian.
23. Zagorodni A.A. and Muhammed M. In *Ion Exchange: Developments and Applications*. Greig J.A., Ed. Royal Society of Chemists, Cambridge, 1996, 446.
24. Tikhonov N.A. and Zagorodni A.A. *Sep. Sci. Technol.*, 33, 633 (1998).
25. Khamizov R., Muraviev D., and Warshawsky A. In *Ion Exchange and Solvent Extraction*, Vol. 12, Advanced Chemistry Series. Marcel Dekker, New York, 1995, 93.
26. Muraviev D., Noguerol J., and Valiente M. *React. Funct. Polymers* 28, 111 (1996).
27. Muraviev D., Noguerol J., and Valiente M. *Environ. Sci. Technol.* 31, 379 (1997).
28. Muraviev D., Gonzalo A., and Valiente M. *Anal. Chem.* 67, 3028 (1995).
29. Muraviev D., Noguerol J, and Valiente M. *Hydrometallurgy* 44, 331 (1997).
30. Nikolaev N.P., Muraviev D.N., and Muhamed M. *Sep. Sci. Technol.* 32, 849 (1997).
31. Muraviev D., Gonzalo A., and Valiente M. *J. Chromatogr.* 802, 251 (1998).
32. Ivanov V.A., Timofeevskaya V.D., Gorshkov V.I. et al. *Zh. Fizich. Khimii.* 63, 1867 (1989), in Russian.
33. Zagorodni A.A. *Ion Exchange Materials: Properties and Applications*, Elsevier, Amsterdam, 2006.
34. Harjula R. and Lehto J. *React. Funct. Polymers* 27,147 (1995).
35. Lehto J. and Harjula R. In *Ion Exchange: Developments and Applications*. Grieg J.A., Ed. Royal Society of Chemists, Cambridge, 1996.
36. Grant S. and Fletcher P. In *Ion Exchange and Solvent Extraction*, Vol. 11, Advanced Chemistry Series, Marcel Dekker, New York, 1992.

37. Soldatov V.S. In *Ion Exchangers*. Walter deGruyter, Berlin, 1992.
38. Soldatov V.S. *React. Funct. Polymers* 27, 95 (1995).
39. Atkins P.W. and Paula J. *Atkins' Physical Chemistry*. Oxford, University Press, 2001.
40. Gaines G.L. and Thomas H.C. *J. Chem. Phys.* 21, 714 (1953).
41. Tolmachev A.M. and Gorshkov V.I., *Zh. Fizich. Khimii* 40, 1924 (1966), in Russian.
42. Ivanov V.A., Timofeevskaja V.D., and Gorshkov V.I. *Russ. J. Phys. Chem.* 74, 637 (2000), in Russian.
43. Krichevskii I.R. *Concepts and Fundamentals of Thermodynamics*. Khimia, Moscow, 1970, in Russian.
44. Helfferich F. *Ionenaustauscher*. Verlag Chemie, Weinheim, 1959.
45. Tremiyon B. *Separation by Ion Exchanging Resins* (translation). Mir, Moscow, 1967, in Russian.
46. Samuelson O. *Separation by Ion Exchange in Analytical Chemistry* (translation). Khimia, Leningrad, 1967, in Russian.
47. Kokotov Y.A. and Pasechnik V.A. *Equilibrium and Kinetics of Ion Exchange*. Khimia, Leningrad, 1970, in Russian.
48. Soldatov V.S. *Simple Ion Exchange Equilibria*, Nauka, Minsk, 1972, in Russian.
49. Breck D.W. *Zeolite Molecular Sieves*. Wiley, New York, 1974.
50. Timofeevskaya V.D., Ivanov V.A., and Gorshkov V.I. *Zh. Fizich. Khimii*. 62, 2531 (1988), in Russian.
51. Timofeevskaya V.D. Candidate Dissertation. MGU, Moscow, 1990, in Russian.
52. Ivanov V.A., Timofeevskaya V.D., Gorshkov V.I. et al. *Zh. Fizich. Khimii* 65, 2455 (1991), in Russian.
53. Ivanov V.A., Timofeevskaya V.D., and Gorshkov V.I. *React. Polym.* 17, 101 (1992).
54. Ivanov V.A. Candidate Dissertation. MGU, Moscow, 2000, in Russian.
55. Drozdova N.V. Candidate Dissertation, MGU, Moscow, 1996, in Russian.
56. Ivanov V.A., Gorshkov V.I., Timofeevskaya V.D. et al. *React. Funct. Polym.* 38, 205 (1998).
57. Ivanov V.A., Drozdova N.V., Gorshkov V.I. et al. In *Progress in Ion Exchange: Advances and Applications*. Royal Society of Chemists, Cambridge, 1997, 307.
58. Staina I.V. Candidate Dissertation, MGU, Moscow, 1994, in Russian.
59. Ivanov V.A., Gorshkov V.I. Staina I.V. et al. *Zh. Fizich. Khimii* 65, 2184 (1991), in Russian.
60. Ivanov V.A., Timofeevskaya V.D., Gorshkov V.I. et al. *Russ. J. Phys. Chem.* 74, 812 (2000).
61. Klein G. And Villena-Blanco V.T. *Ind. Eng. Chem.* 3, 280 (1964).
62. Soldatov V.S., Novitskaya L.P., Bespalko M.S. et al. In *Synthesis and the Properties of Ion Exchange materials*. Nauka, Moscow, 1968, 216, in Russian.
63. *Lewatit Product Information*. Bayer, Leverkusen, 1993.
64. Muraviev D., Gonzalo A., and Valiente M. In *Ion Exchange: Developments and Applications*. Greig J.A., Ed. Royal Society of Chemists, Cambridge, 1996. 516.
65. Cortina J.L. and Warshawsky A. In *Ion Exchange and Solvent Extraction*. Vol. 13, Advanced Chemistry Series. Marcel Dekker, New York, 1997, Chap. 5.
66. Muraviev D. *Solv. Extr. Ion Exch.* 16, 381 (1998),
67. Juang R.S. and Su J.Y. *Ind. Eng. Chem. Res.* 31, 2774 (1992).
68. Muraviev D., Noguerol J., and Valiente M. In *Ion Exchange: Highlights of Russian Science*. Marcel Dekker, New York, 1999, 767.
69. Muraviev D., Oleinikova M., and Valiente M. *Langmuir*. 13, 4915 (1997).
70. Kuznetsov I.A., Kondorsky A.E., Ivanov V.A. et al. *Molek. Biol.* 18, 457 (1984), in Russian.
71. Kuznetsov I.A., Gorshkov V.I, Ivanov V.A. et al. *React. Polymers* 3, 37 (1984).
72. Khamizov R.Kh. Candidate Dissertation, Geokhi, Moscow, 1998, in Russian.

73. Khamizov R.Kh, Mironova L.I., Tikhonov N.A. et al. *Sep. Sci. Technol.* 31, 1 (1996).
74. Khamizov R.Kh, Mironova L.I., Novitsky E.G. et al. *Tekhnik. Mashinostr.* 4, 112 (1996), in Russian.
75. Khamizov R.Kh, Novikova V.A., and Melikhov S.A. In *Natural Zeolites of Russia*, SO Ran, Novosibirsk, 160, in Russian.
76. Khamizov R.Kh, Novikova V.A., and Melikhov S.A. Russian Federation Patent 2,006,495 (1994).
77. Tchelischev N.F., Volodin V.F., and Kryukov V.L. *Ion Exchange Properties of Zeolites.* Mir, Moscow, 1988, in Russian.
78. Fokina O.V. Candidate Dissertation, Geokhi RAN, Moscow, 1998, in Russian.
79. Khamizov R.Kh., Fokina O.V., and Senyavin M.M. In Abstracts of Lectures and Posters of Sixth Symposium on Ion Exchange, Balatonfured, Hungary, 1991, 140.
80. Nozhov A.M., Kosobryukhova O.M., and Khamizov R.Kh. *Sorbtsion. Khromatograf. Prots.* 3, 159 (2003), in Russian.
81. Reichenberg D. In. *Ion Exchange.* Mir, Moscow, 1968, 104, in Russian.
82. Amelin A.N. and Leikin Y.A. *Calorimetry of Ion Exchange Processes*, VGU, Voronezh, 1991, in Russian.
83. Ivanov V.A., Timofeevskaya V.D., Drozdova N.V. et al. *Russ. J. Phys. Chem.* 74, 641 (2000).
84. Kopylova V.D. *Solv. Extr. Ion Exch.* 16, 267 (1998).
85. Kopylova V.D. In *Ion Exchange: Highlights of Russian Science.* Marcel Dekker, New York, 1999, 270.
86. Kraus K.A. and Raridon R.J. *J. Phys. Chem.* 65, 1901 (1959).
87. Ivanov V.A., Gorshkov V.I., Gavlina O.T. et al. *Russ. J. Phys. Chem.* 80, 1826 (2006).
88. Ivanov V.A., Gorshkov V.I., Gavlina O.T. et al. *Russ. J. Phys. Chem.* 81, 1582 (2007).
89. Kuznetsova E.M. *Zh. Fizich. Khimii* 66, 2688 (1992), in Russian.
90. Kuznetsova E.M. *Zh. Fizich. Khimii* 69, 2092 (1995), in Russian.
91. Kuznetsova E.M. *Russ. J. Phys. Chem.* 70, 473 (1996).
92. Helfferich F.G. and Hwang Y.L. In *Ion Exchangers.* Walter deGruyter,Berlin, 1991, 1288.
93. Dorfner K. In *Ion Exchangers.* Walter deGruyter, Berlin, 1991, 90.
94. Helfferich F.G. *React. Polymers*,13, 191 (1990).
95. Gantman A.I. *Zh. Fizich. Khimii* 69 (1816) 1995, in Russian.
96. Gantman A.I. *Zh. Fizich. Khimii* 69 (2089) 1995, in Russian.
97. Gorshkov V.I., Safonov M.S., and Voskresensky N.M. *Ion Exchange in Counter Current Columns.* Nauka, Moscow, 1981, in Russian.
98. Helfferich F. and Klein G. *Multicomponent Chromatography: Theory of Interference.* Marcel Dekker, New York, 1970.
99. Senyavin M.M., Rubinshtein R.N., Venitsianov E.V. et al. *The Bases for Simulation and Optimization of Ion Exchange Processes.* Nauka, Moscow, 1972, in Russian.
100. Venitsianov E.V. and Rubinshtein R.N. *Dynamics of Sorption from Liquid Media.* Nauka, Moscow, 1982, in Russian.
101. Zhukhovitsky A.A., Zolotarev P.P., Sokolov V.A. et al. *Dokl. AN SSSR* 77, 435 (1951), in Russian.
102. Zhukhovitsky A.A. and Turkeltaub N.M. In *Gas Chromatography.* Nauka, Moscow, 1960, 107, in Russian.
103. Zhukhovitsky A.A. In *Developments in Chromatography.* Nauka, Moscow, 1972, 163, in Russian.
104. Ivanov V.A., Timofeevskaya V.D., Gorshkov V.I. et al. *Vysokoch. Veschest.* 4,133 (1990), in Russian.
105. Ivanov V.A., Gorshkov V.I., Drozdova N.V. et al. *Vysokoch. Veschest.* 10, 10 (1996), in Russian.

106. Gerasimova T.A., Sumina O.A., and Khamizov R.Kh., *Sorbts. Khromatograf. Prots.* 3, 169 (2003), in Russian.
107. Gerasimova T.A., Sumina O.A., and Khamizov R.Kh., *Sorbts. Khromatograf. Prots.* 3, 399 (2003), in Russian.
108. Suzuki T., Sugita Sh., and Miyake M. In Proceedings of Seventh Symposium on Salt Production. 1993, 29.
109. Suzuki T., Jyoshida T., Sugita Sh. Et al. *Nip. Kais. Gakk.* 48, 107 (1994), in Japanese.
110. Lu Zh., Juan J., Zhao J. et al. In Proceedings of Seventh Symposium on Salt Production, 1993, 29.
111. Liu J., Jin S., Ji P. et al. *Qind. Huag. Xuey. Xueb.* 15, 256 (1994), in Chinese.
112. Khamizov R.Kh. and Krachak A.N. Russian Federation Patent 2,201,414 (1998).
113. Khamizov R.Kh., Krachak A.N., and Fokina O.V. Russian Federation Patent 2,115,301 (1998).
114. Beruashvili T.A., Khamizov R.Kh., and Sidamonidze S.I. Georgia Patent U 802 (2001).
115. Steinberg S.L., Ming D.W., Henderson K.E. et al. *Agron J.* 92, 353 (2000).
116. Beruashvili T.A., Kheladze T.A., Takaishvili N.B. et al. *Sorbts. Khromatograf. Prots.* 8, 869 (2008), in Russian.
117. Okudzhava N.G., Beruashvili T.A., and Mamukashvili N.S. *Sorbts. Khromatograf. Prots.* 8, 876 (2008), in Russian.
118. Gotsiridze R.S., Isparyan A.G., Meparishvili N.A. et al. *Sorbts. Khromatograf. Prots.* 9, 175 (2009), in Russian.
119. Leavitt F.W. U.S. Patent 5,681,477 (1997).
120. Bolto B.A. and Weiss D.E. In *Ion Exchange and Solvent Extraction.* Marcel Dekker, New York, 1977, 221.
121. Bolto B.A. *Chemtech.* 5, 303 (1975).
122. Bolto B.A. and Pawlowski L. *Wastewater Treatment by Ion Exchange.* E. & F.N. Spon, London, 1987.
123. Hamman S.D., *Austral. J. Chem.* 24, 1979 (1971).
124. Chada M., Sarkar A., and Modak J.M. *J. Appl. Polymer Sci.* 93, 883 (2004).
125. LeVan M.D. and Vermeulen T. *J. Phys. Chem.* 85, 3247 (1981).
126. Butler J.A. and Ockrent C.,*J. Phys. Chem.* 34, 2841 (1930).
127. Matsuda H., Yamamoto T., Goto S. et al. *Sep. Sci. Technol.* 16, 31 (1981).
128. Butler G.B. *Cyclopolymerization and Cyclocopolymerization*, CRC Press, Boca Raton, FL, 1992.
129. Eppinger K.H. U.S. Patent 4,115,297(1996).
130. France Patent 2,255,334 (1995).
131. Baker B. and Pigford R.L. *Ind. Eng. Chem. Fundam.* 10, 283 (1971).
132. Chada M., Sarkar A.,and Modak J.M., *J. Polym. Materials*, 21, 1 (2004).
133. Chada M., Pillay S.A., Sarkar A. et al. *J. Appl. Polymer Sci.* 111, 2741 (2009).
134 Foo, S.C. *Ind. Eng. Chem. Fundam.* 20, 150 (1981).
135. Baker B. and Pigford R.L. *Ind. Eng. Chem. Fundam.* 10, 283 (1971).
136. Swinton E.A., Bolto B.A., Eldridge R.J. et al. In *Ion Exchange Technology*, Society of Chemical Industries, London, 1984. 542.
137. German Patent 2,822,280. *Chem.Abstr.* 90, 127371k (1979).
138. Dabby S.S. and Zaganiaris E.J. U.S. Patent 4,184,948 (1980).
139. Rousseau R.W. *Handbook of Separation Process Technology.* Wiley-IEEE, 1987.
140. Mattheus F., Goosen, A. and Shayya, W.H. *Water Management, Purification and Conservation in Arid Climates.* CRC Press, Boca Raton, FL, 1999.
141. Comyns A.E. *Encyclopedic Dictionary of Named Processes in Chemical Technology.* CRC Press, Boca Raton, FL, 2007.

142. Zagarodny A.A. and Muhammed M. In *Progress in Ion Exchange: Advances and Application*. Royal Society of Chemists, Cambridge, 1997, 349.
143. Zagarodny A.A., Muraviev D.N., and Muhammed M. *Sep. Sci. Technol.*32, 413 (1997).
144. Muraviev D., Noguerol J., and Valiente M. In *Progress in Ion Exchange: Advances and Applications*. Royal Society of Chemists, Cambridge, 1997, 359.
145. Cohen K., *Theory of Isotope Separation as Applied to the Large Scale Production of U-235*, McGraw-Hill, New York, 1951.
146. Rozen A.M. *Theory of Separation of Isotopes in Columns*. Atomizdat, Moscow, 1960, in Russian.
147. Wancat P.C., Middelton A.R., and Hudson B.L. *Ind. Eng. Chem. Fundam.* 15, 309, (1976).
148. Martin, A.J.P. *Discuss. Faraday Soc.*7, 332 (1949).
149. Sussman, M.V. and Rathore,R.N.S. *Chromatographia* 8, 55 (1975).
150. Sussman, M.V. *Chem. Technol.* 6, 260 (1976).
151. Muraviev D. Noguerol J., and Valiente M. *Anal. Chem.* 69, 4234 (1997).
152. Selemenev V.F., Candidate Dissertation. VGU, Voronezh, 1993, in Russian.
153. Pigford R.L., Baker B., and Blum D.E. *Ind. Eng. Chem. Fundam.* 8, 144 (1969).
154. Wilhelm R.H. and Sweed N.H. *Science* 159, 522 (1968).
155. Sweed N.H. and Wilhelm R.H. *Ind. Eng. Chem. Fundam.* 8, 221 (1969).
156. Chen H.T. and Hill F.B. *Sep. Sci.* 6, 411 (1971).
157. Grevillot G. and Tondeur D. *AIChE J.* 23, 840 (1977).
158. Rolke R.W. and Wilhelm R.H. *Ind. Eng. Chem. Fundam.* 8, 235 (1969).
159. Wankat P.C. *Chem. Eng. Sci.* 33, 723 (1976).
160. Chen H.T., Reiss E.H., Stokes J.D. et al. *AIChE J.* 19, 589 (1973).
161. Rice R.G., Mackenzie M. *Ind. Eng. Chem., Fundam.* 12, 486 (1972).
162. Zhongmin W. and Zhenhua Y. *J. Chem. Indus. Eng.* 22, 206 (1993).
163. Ferreira L.M. and Rodrigues A.E. *Adsorption*, 1, 213 (1995).
164. Simon G., Hanak L., Szanya T. et al. *J. Cleaner Prod.* 6. 329 (1998).
165. Davesac R.R., Pinto L.T., Da Silva F.A. et al. *Chem. Eng. J.* 76, 115 (2000).
166. Hanak L. Candidate Thesis. University of Vesprem, 2005.
167. Gupta R. and Sweed N.H. *Ind. Eng. Chem. Fundam.* 12, 335 (1973).
168. Simon G., Hanak L., Grevillot G. et al. *J. Chromatogr.* 664B, 17 (1995).
169. Simon G., Hanak L., Grevillot G. et al. *J. Chromatogr.* 732A, 1 (1996).
170. Simon G., Grevillot G., Hanak L. et al. *Chem. Eng. Sci.* 52, 467 (1997).
171. Simon G., Grevillot G., Hanak L. et al. *Chem. Eng. J.*, 70, 71 (1998).
172. Watari S. and Hayashi H. In Proceedings of International Conference on Ion Exchange. Takamatsu, 1995, 499.
173. Japan Patent 06 31 109, *Chem. Abstr.* 121, 12988h (1994).
174. Rice R.G. *Ind. Eng. Chem., Fundam.* 14, 362 (1975).
175. Ivanov V.A., Timofeevskaja V.D., Gavlina O.T. et al. *Microporous Mesoporous Mater.* 65, 257 (2003).
176. Otero M., Zabkova M. and Rodrigues A.E. *Chem. Eng. J.* 110, 101 (2005).
177. Otero M., Zabkova M., and Rodrigues A.E. *Water Res.* 39, 3467 (2005).
178. Otero M., Zabkova M., and Rodrigues A.E. *J. Intl. Adsorption Soc.* 11, 887 (2005).
179. Mishra P.K., Nanagre D.M., and Yadav V.L. *J. Sci. Industr. Res.* 66,79 (2007).
180. Huang C.R. and Holleln H.C. In *Handbook of Separation Techniques for Chemical Engineers*, 3rd ed. McGraw Hill, New York, 1997, 637.
181. Richardson J.F., Coulson J.M., Harker J.H. et al. *Chemical Engineering: Particle Technology and Separation Processes*. Butterworth-Heinemann, London, 2002.
182. Couper J.R., Penney W.R., Fair J.R. et al. *Chemical Process Equipment: Selection and Design*. Gulf Publishing, Houston, 2004.

183. Knaebel K.S. and Pigford R.L. *Ind. Eng. Chem. Fundam.* 22, 336 (1983).
184. Gupta R. and Sweed N.H. *Ind. Eng. Chem. Fundam.* 10, 280 (1971).
185. Baker B. and Pigford R.L. *Ind. Eng. Chem. Fundam.* 10, 283 (1971).
186. Foo S.C., Bergsman K.H., and Wancat P.C. *Ind. Eng. Chem. Fundam.* 19, 86 (1980).
187. Murphy J.M., Urey G.S., and Kirshenbaum, I., Eds. *Production of Heavy Water.* McGraw Hill, New York, 1955.
188. Andreev B.M., Zelvensky Y.D., and Katalnikov S.G. *Heavy Isotopes of Hydrogen in Nuclear Techniques*, 2nd ed. AT, Moscow, 2000, in Russian.
189. Gorshkov V.I., Kurbanov A.M., and Apolonik N.V. *Zh. Fizich. Khimii* 45, 2969 (1971), in Russian.
190. Gorshkov V.I., Kurbanov A.M., and Ivanova M.V. *Zh. Fizich. Khimii* 48, 2392 (1974), in Russian.
191. Gorshkov V.I., Kurbanov A.M., and Ivanova M.V. *Zh. Fizich. Khimii* 49, 1276 (1975), in Russian.
192. Gorshkov V.I., Ivanova M.V., Kurbanov A.M. et al. *Vestnik MGU Ser. Chem.* 18, 535 (1977), in Russian.
193. Gorshkov V., Muraviev D., and Warshawsky A. *Solv. Extr. Ion Exch.* 16, 1 (1998).
194. Gorshkov V., Muraviev D., and Warshawsky A. In *Ion Exchange: Highlights of Russian Science*. Marcel Dekker, New York, 1999, 1.
195. Ivanov V.A., Gorshkov V.I., and Timofeevskaya V.D. In *Proceedings of Eighth International Conference on Polymer Technology*. Israel, 1998, 42.
196. Rozen A.M. *Dokl. AN SSSR* 108, 122 (1956), in Russian.
197. Bier K. *Chem. Eng. Technol.* 28, 625 (1956).
198. Tikhonov N.A., Timofeevskaya V.D., Kiryusshin A.A. et al. *Russ. J. Phys. Chem.* 71, 2038 (1997).
199. Bailly M. and Tondeur D. *Inst. Chem. Eng. Symp. Ser.* 54, 111 (1978).
200. Bailly M. and Tondeur D. *J. Chromatogr.* 201, 343 (1980).
201. Khamizov R.Kh. et al. Russian Federation Patent 1,726,387 (1992).
202. Khamizov R.Kh.et al. Russian Federation Patent 1,728,133 (1992).
203. Khamizov R.Kh.et al. Russian Federation Patent 2,034,651(1996).
204. Myasoedov B.F., Senyavin M.M., Khamizov R.Kh. et al. In *Prospective Approaches to the Creation of Technologies for Processing Mineral Raw Materials*. Mekhanobr, Leningrad, 1991, 218, in Russian.
205. Muraviev D, Khamizov R., and Tikhonov N. *Solv. Extr. Ion Exch.* 16, 151 (1998).
206. Muraviev D, Khamizov R., and Tikhonov N. In *Ion Exchange. Highlights of Russian Science*. Marcel Dekker, New York, 1999, 270.
207, Khamizov R.Kh.et al. U.S. Patent 5,814,224 (1998).
208. Khamizov R.Kh.et al. Japan Patent 3,045,378 (2000).
209. Khamizov R.Kh.et al. Israel Patent 119,083 (2000).
210. Khamizov R. and Muraviev D. In *Ion Exchange and Solvent Extraction*. Vol. 16, Advanced Chemistry Series. Marcel Dekker, New York, 2004, 119.

6 Separations by Multicomponent Ionic Systems Based on Natural and Synthetic Polycations

Ecaterina Stela Dragan, Maria Valentina Dinu, and Marcela Mihai

CONTENTS

ABSTRACT

Heavy metals, fine suspensions, and organic matter emitted by industrial processes such as paper processing, paint and pigment production, textile manufacturing, food and cosmetics operations, electronics fabrication, mines, and ceramic production are very dangerous pollutants. Synthetic and, more increasingly, polymers from renewable sources are used alone or in complex systems to enhance the efficiency of separation processes. Multicomponent ionic systems may present a better approach. Their enhanced mechanical, thermal, and adsorption properties are comparable to those of their components alone. This chapter summarizes our investigations of novel materials based on chitosan: (1) composites with clinoptilolite (a common natural zeolite) and their applications for removing heavy metal ions (Cu^{2+}, Co^{2+}, and Ni^{2+}); and (2) preparation of positively charged nonstoichiometric interpolyelectrolyte complexes as colloidal dispersions and their efficiency in solid and liquid separations in comparison to chitosan alone. The adsorption capacity of the chitosan–clinoptilolite composite abruptly increases up to a zeolite content of ~20 wt%. The maximum adsorption capacities are 9.04 mmol/g for Cu^{2+}, 6.4 mmol/g for Ni^{2+}, and 3.4 mmol/g for Co^{2+}. One advantage of using nonstoichiometric interpolyelectrolyte complexes as specialized flocculants is the broadness of the flocculation window—more than double at an optimum dose lower than that of chitosan. Some synthetic polycations were also used to prepare novel multicomponent ionic systems with better efficiency in separation processes than other polycations.

6.1   INTRODUCTION

Environmental protection is based on the principles and strategic elements of sustainable development of society, trying to answer to the present requirements with constant care to ensure that future generations will have ample resources to fulfill their needs. Environmental protection through sustainable development is a central concern of the worldwide scientific community. On a global level, the exchanges of scientific information could improve the understanding and implementation of the sustainable development process. The discovery of the most effective separation processes represents one of the "sine qua non" conditions.

Heavy metals, fine suspensions, and organic matter resulting from industrial processes like paper processing, paints and pigments, textiles, food and cosmetics, fabrication of electronics, mines, and ceramic production are recognized as very dangerous pollutants. Improving the separation processes is critical. Therefore, finding novel and more efficient separation processes is the central task of many groups of researchers.

To solve separation problems, natural polymers are preferred due to their biocompatibility and origins from unlimited renewable resources. However, they do not have consistent structures; their structures depend on their sources. On the other hand, synthetic polymers used in water purification may be designed with controlled structures, molar masses, and functionalities; many of them exhibit good chemical and biological stabilities. These are some of the reasons that our investigations of separation processes and technologies look to natural and synthetic polycations as complementary materials and not as total replacements for polycations coming from nature. Our investigations focused on the preparation of novel materials based on chitosan such as (1) composites with natural zeolites and their application in the removal of heavy metals, and (2) preparation of positively charged nonstoichiometric interpolyelectrolyte complexes and their efficiency in solid and liquid separations compared to chitosan. Some synthetic polycations were also used to prepare multicomponent ionic systems with enhanced efficiency in separation processes compared to polycations.

6.2 COMPOSITE MATERIALS BASED ON CHITOSAN

The presence of heavy metal ions in wastewaters generates huge problems for the environment and for living organisms because of their high toxicity and non-biodegradability. The sources of heavy metal ions in wastewaters include mines, paint and pigment industries, metal fabrication, batteries, corrosion control, fertilizers, textiles, and others. Heavy metal ions are known to cause severe diseases like nervous system damage and even cancer. Conventional methods used to remove heavy metal ions from industrial effluents usually include chemical precipitation, membrane separation, ion exchange, evaporation, and electrolysis, but they are often ineffective, especially in removing the ions from dilute solutions. Among the common conventional techniques for removing heavy metals from wastewaters, the adsorption process is perhaps the most widely utilized, preferred mainly when the enrichment of trace metals or high selectivity for a certain metal is required.[1-3]

Strong interest recently focused on biosorbents derived from polysaccharides as alternatives to existing adsorbents like activated carbon and synthetic ion exchangers.[4-8] Among biopolymers, chitosan (CS)—a linear cationic polysaccharide composed of β-(1→4)-2-amino-2-deoxy-D-glucopyranose and β-(1→4)-2-acetamido-2-deoxy-D-glucopyranose units randomly distributed along a polymer chain—has attracted numerous scientists due to its outstanding biological properties like biodegradability, biocompatibility, and antibacterial activity.[6,7]

The introduction of selective functional groups into the matrix of CS enhances its interaction with a variety of metal ions, improving its selectivity and specificity, and thus increasing its adsorption capacity. Chemical modifications by cross-linking agents increase the chemical stability of the sorbent in acid media, decreasing its solubility in most mineral and organic acids. Its chelating properties for Cu^{2+} cations were recently investigated and correlated with its physical and chemical modifications.[5-8]

Other naturally occurring low cost materials such as zeolites have undergone intensive investigation recently as adsorbents for heavy metals.[9,10] Clinoptilolite (CPL), one of the natural zeolites, is a hydrated alumina silicate member of the heulandite group, found in zeolitic volcanic tuffs in many countries.[11] CPL is characterized by infinite three-dimensional frameworks of aluminum, silicon, and oxygen. Its adsorption capacity for heavy metal cations, and organic pollutants has been demonstrated.[9,10]

A less investigated aspect of research on the sorbents of heavy metals concerns CS-based composite materials with chelating and ion exchange properties.[12,13] The preparation of novel composites based on CS and CPL and the investigation of the relationship of their structures and sorption performances for heavy metal ions was of interest for our group.[14–16] Novel ionic composites were prepared by embedding CPL microparticles in a matrix of cross-linked CS. The ionic composites as beads, stabilized by a "tandem" ionic–covalent cross-linking, were characterized by scanning electron microscopy (SEM), x-ray diffraction (XRD), Fourier transform infrared spectroscopy (FTIR), and thermogravimetric analysis (TGA). The adsorption capacities of Cu^{2+}, Co^{2+}, and Ni^{2+} by the CS–CPL composites were compared with those of cross-linked CS microspheres without zeolite and two synthetic chelating resins bearing iminodiacetate groups.

6.2.1 PREPARATION AND CHARACTERIZATION OF CHITOSAN-BASED COMPOSITES

CS powder was purchased from Fluka. Its ash content was <1% and it was used without further purification. The viscometric average molar mass of CS[17] was estimated:

$$[\eta] = 1.38 \times 10^{-4} \ M_v^{0.85} \tag{6.1}$$

The intrinsic viscosity of a CS solution of 0.3 M CH_3COOH and 0.2 M CH_3COONa (1:1, v/v) was measured with an Ubbelohde viscometer at $25 \pm 0.1°C$. The sample used in this study had an M_v value of 334 kDa. The degree of acetylation (DA) of CS was evaluated by infrared spectroscopy with a Vertex 70 Bruker FTIR spectrometer. Transmission spectra were recorded in KBr pellets. For DA determination, Equation (6.2) was used, treating the 1420 cm^{-1} band as a reference; the characteristic band for N-acetylglucosamine was 1320 cm^{-1}.[18]

$$A_{1320} \big/ A_{1420} = 0.3822 + 0.03133 \ DA \tag{6.2}$$

An average value of DA = 17.6% resulted from three measurements. The natural CPL sample used in this study came from volcanic tuffs containing 60 to 70% CPL, cropped out in the Măcicaş area (Cluj County, Romania), and exhibited an ideal composition: $(NaKCa_{0.5})_{5.4}(Al_{5.4}Si_{30.6}O_{72}).20H_2O$ (Si:Al = 5.7).[11] The CPL sample was ground and moved through a range of sieves. Only the particles that passed through 0.05 mm and remained on 0.032 mm sieves were used to prepare the composites. The H^+ clinoptilolite (CPL-H^+) was prepared by the contact of CPL powder with 1 M HCl (1:10, v/v), for 24 h, followed by washing with distilled water to a neutral pH. Finally, the CPL-H^+ was dried for 2 h at 105°C and 24 h at 40°C.

Scheme 6.1 is an idealized representation of the structures of CS–CPL composites. In the first step, the tripolyphosphate (TPP) multivalent anion forms ionic cross links by the interaction with amino groups of CS, locking the CPL microparticles in the composite microspheres. However, the composites stabilized only by ionotrop gelation do not preserve their integrity when used in an acidic environment. Therefore, covalent cross-linking with epichlorohydrin (ECH) was used to prepare CS–CPL composites as beads with a high chemical stability. ECH was preferred as a cross-linker because further hydrophilic (OH) groups are generated and may contribute to increased chelating performance of the whole system. The proportions of CS and CPL used to prepare the composites are summarized in Table 6.1. Figure 6.1 shows optical images of the composite beads corresponding to the sample $CS_{10}CPL_1$ in swollen and dry states. The narrow distribution of bead sizes is visible in the figure.

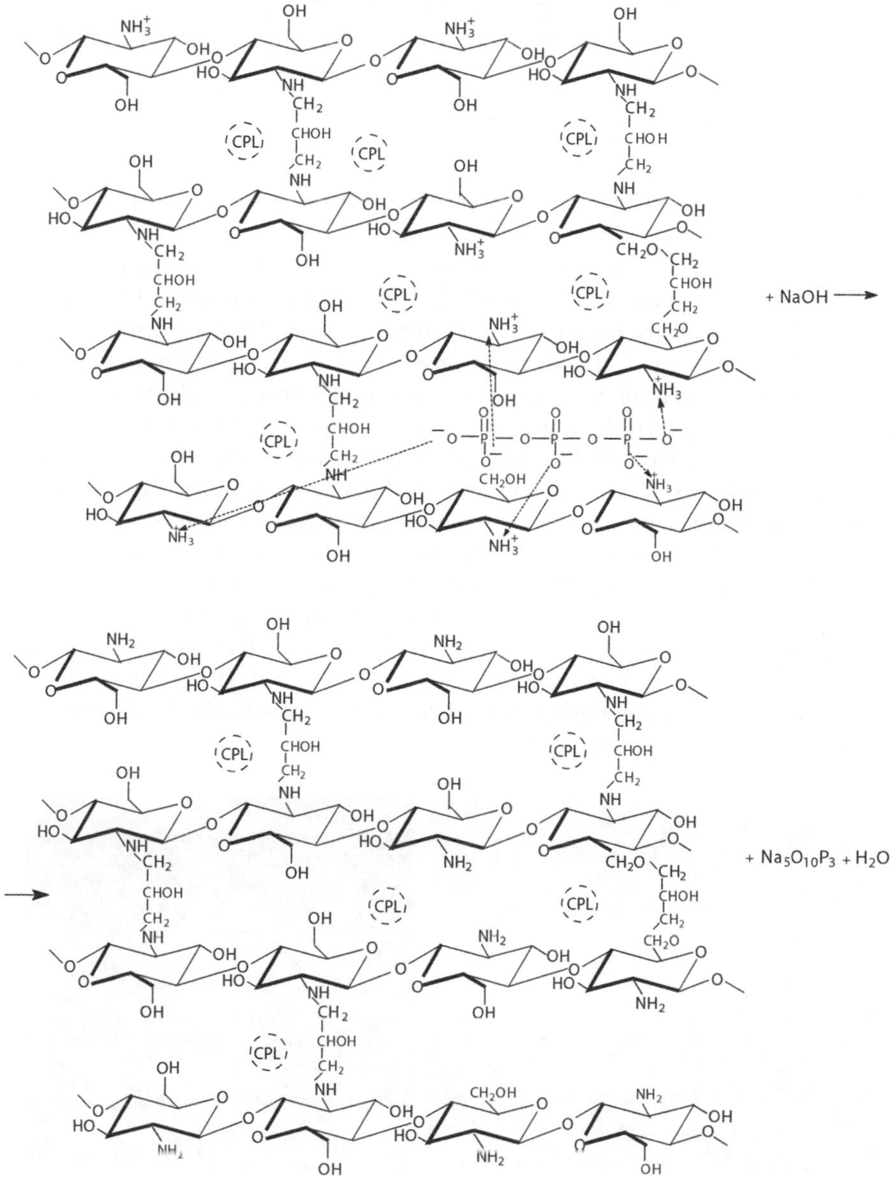

SCHEME 6.1 Ionic and covalent cross linked chitosan–clinoptilolite (CS–CPL) composites and regeneration reaction with 0.1 M NaOH.

TABLE 6.1

Feed Compositions of Chitosan (CS) and Clinoptilolite (CPL) Composites

Sample[a]	Cross-Linked CS	$CS_{10}CPL_1$	CS_5CPL_1	CS_4CPL_1	CS_2CPL_1
CPL-H$^+$, g[b]	0	0.06	0.12	0.15	0.3
CPL-H$^+$, wt.%	0	9.09	16.67	20	33.3

[a] The first subscript shows parts of chitosan; the second indicates parts of clinoptilolite.
[b] CPL was dispersed in 10 mL of water and mixed with 20 g of CS solution in 1 vol.% acetic acid at a concentration of 3%.

The initial data indicating the presence of CPL in different proportions in the CS–CPL composites was given by SEM. Typical cross sectional micrographs of the cross-linked CS beads and the CS–CPL composites are presented in Figure 6.2. A dense and uniform morphology without pores can be seen in Figure 6.2a for the cross-linked CS, in agreement with the report in the literature.[5] The morphology of the CS–CPL composites was different. The presence of CPL appeared as discrete zones in Figure 6.2b for the sample with the lowest CPL content. Some agglomerations of CPL microparticles were observed in the CS_2CPL_1 sample, i.e., for composites with the highest contents of CPL.

Energy dispersive x-ray analysis (EDX) was performed to determine the elemental composition of the CS–CPL composite. SEM-EDX analysis of the $CS_{10}CPL_1$ composite beads is presented in Figure 6.3. EDX spectra were consistent with the presence of Al and Si from CPL and C, N, and O from CS. The weight percentages in selected composites were 0.42, 2.06, 57.4, 5.59, and 28.31%, for Al, Si, C, N, and O, respectively.

X-ray diffractograms of cross-linked CS and CPL showed low crystallinity for the cross-linked CS and a high degree ($\chi = 74\%$) for CPL (Figure 6.4). The crystallinity and number of definite peaks increased with the increase of CPL percentage loaded in composites when CS and CPL were mixed (Table 6.1) to prepare CS–CPL composites.

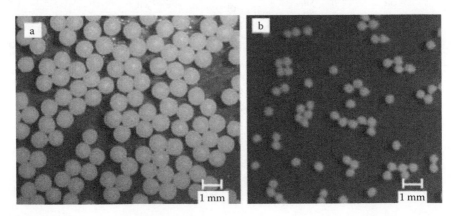

FIGURE 6.1 Optical microscope images of $CS_{10}CPL_1$ composite after synthesis: (a) swollen state; (b) dry state.

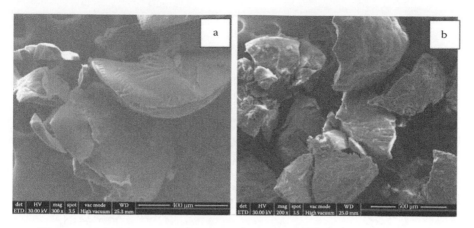

FIGURE 6.2 SEM images: (a) cross-linked CS beads; (b) $CS_{10}CPL_1$ composite.

For a small percentage of CPL, 9.09% in the sample $CS_{10}CPL_1$, only two clear peaks were registered, corresponding to the most intense peaks of CPL: $2\theta = 9.92°$ and 22.5°, respectively, for d = 8.90 Å and 3.94 Å (the second with $I:I_0 = 83.5\%$). The sample CS_2CPL_1, containing the highest percentage (33.3%) of CPL, showed more than 25 well-defined peaks in the same positions as in CPL—which supports a two-phase system and no chemical reactions between CS and CPL. The numerical registration data indicates a smooth left shift in the peaks' position of about 0.1°, which corresponds to an increase in the interplanar distances. This small change would

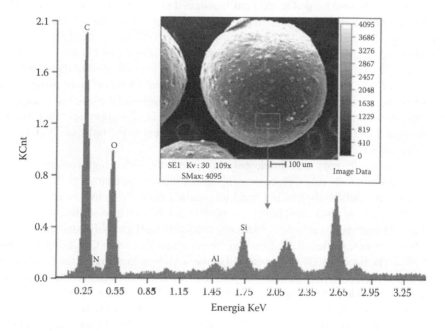

FIGURE 6.3 EDX spectrum obtained by surface analysis of $CS_{10}CPL_1$ composite beads.

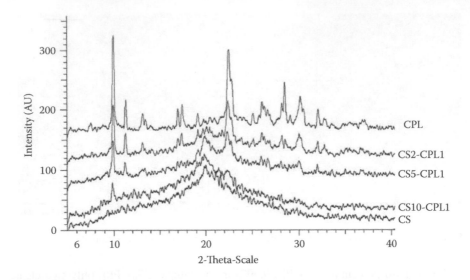

FIGURE 6.4 X-ray diffractograms of CPL, cross-linked CS, and some CS–CPL composites.

support a physical interaction between the continuous phase of the cross-linked CS and dispersed phase constituting CPL microparticles.

FTIR spectroscopy was performed to identify any changes in the structure of CS–CPL composites compared with the cross-linked CS (Figure 6.5). The characteristic bands of CS were located at 1654 cm^{-1}, assigned to the stretching vibrations of the C=O bonds in acetamide groups, amide I band, 1562 cm^{-1}, stretching vibration of N-H bond, amide II band, and at 1322 cm^{-1}, characteristic band for N-acetylglucosamine. The main characteristic bands of CPL were found at wavelengths of 467 cm^{-1}, 608 cm^{-1}, 795 cm^{-1}, 1069 cm^{-1}, and 1635 cm^{-1}. The band at 467 cm^{-1} resulted from the stretching vibrations of Al-O bonds, and the bands at 795 cm^{-1} and 1069 cm^{-1} were assigned to Si-O-Si bonds. The peak at 608 cm^{-1}, assigned to the vibration of the external linkage of the tetrahedral, and the peak at 467 cm^{-1} were observed in all composites. Their intensity increased with the increase of the loaded CPL content in the order of $CS_{10}CPL_1 < CS_5CPL_1 < CS_2CPL_1$.

The band at 795 cm^{-1} was visible only in composites with a high content of CPL (CS_5CPL_1 and CS_2CPL_1). As CPL content increased, some of the characteristic CS bands were either red-shifted, the amide I band was shifted from 1654 cm^{-1}, in cross-linked CS without CPL, to 1643 cm^{-1}, in the composite CS_2CPL_1, or blue-shifted, the intense band at 3418 cm^{-1} assigned to the stretching vibration of O-H and N-H bonds, as well as to hydrogen bonds, shifted to 3435 cm^{-1} in the composites richer in CPL (CS_5CPL_1 and CS_2CPL_1). The band at 1032 cm^{-1}, assigned to the stretching vibration of the C-O bonds in the anhydroglucose ring, diminished with the increase of CPL content and became completely invisible in the composite with the highest CPL content (CS_2CPL_1).

Figure 6.6 summarizes the thermal behavior of two CS–CPL composites compared with CPL and cross-linked CS. The water loss of the CPL-H$^+$ sample was about 10%. The thermo-oxidative behavior of the CS–CPL composites indicates a strong

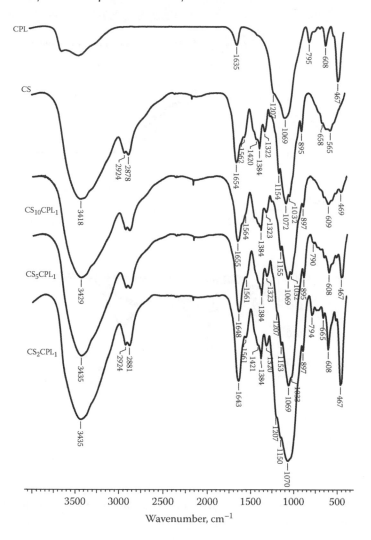

FIGURE 6.5 FTIR spectra of clinoptilolite (CPL), cross-linked chitosan (CS), and three CS–CPL composites (CS$_{10}$CPL$_1$, CS$_5$CPL$_1$, and CS$_2$CPL$_1$).

dependence of the char percentage on the CPL loaded in the composite; the lowest content was found for CS$_{10}$CPL$_1$ and the highest for CS$_2$CPL$_1$, supporting the results obtained by XRD and FTIR. TG and DTG curves revealed different degradation patterns of CS in composites compared with cross-linked CS.

Thus, a high CPL content (CS$_2$CPL$_1$) led to a decrease of temperature corresponding to the first and second degradation steps. The total weight loss was lower than in the case of CS without zeolite: stage 1, maximum at 256°C, weight loss 23.3%; stage 2, maximum at 512°C, weight loss 39%, when compared with cross-linked CS characterized by stage 1, with a maximum at 290°C, weight loss at 35%, and stage 2, with a maximum at 562°C, weight loss at 45%.

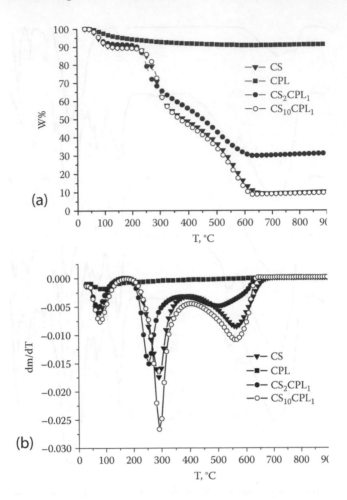

FIGURE 6.6 TG (a) and DTG (b) curves of CPL, cross-linked CS, and two CS–CPL composites, in air at 10°C/min.

6.2.2 ADSORPTION OF METAL IONS BY CHITOSAN-BASED COMPOSITES

Cu^{2+}, Co^{2+}, and Ni^{2+} were selected as heavy metal ions for adsorption experiments on CS–CPL composites compared with two synthetic chelating resins bearing iminodiacetate groups. CR-10 synthesized by our group was derived from acrylonitrile–divinylbenzene copolymers.[2] The other was the Amberlite IRC-748 commercial resin based on a macroporous styrene–divinylbenzene matrix.

6.2.2.1 Sorption of Cu^{2+}

Solution pH can influence metal ion retention by modification of both metal ion concentration and level of ionization of CS. Figure 6.7 shows the retention capacity of Cu^{2+} at equilibrium as a function of pH for cross-linked CS and four samples of composites. The pH values ranged from 2 to 6. The figure indicates that the

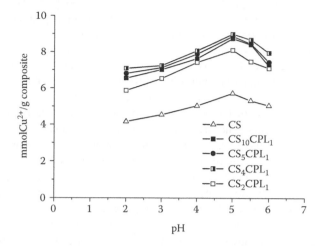

FIGURE 6.7 Cu²⁺ retention on CS–CPL composites as a function of pH. Contact time = 24 h. Initial metal concentration = 0.07 mol/L.

amount of Cu^{2+} adsorbed by the cross-linked CS and CS–CPL composites slowly increased with the increase of pH. Optimum adsorption pH occurred at 5. At pH >5, Cu^{2+} retention dramatically decreased because small amounts started to deposit as $Cu(OH)_2$. This also supports the chelation of Cu^{2+} on all sorbents.

The values of Cu^{2+} retention on cross-linked CS and some CS–CPL composite microspheres as a function of CPL content loaded in a composite for four cycles of adsorption are plotted in Figure 6.8. In the first cycle of adsorption, an abrupt increase of adsorption capacity can be observed for the CS–CPL composites compared with cross-linked CS, starting at the lowest content of CPL loaded in the composite ($CS_{10}CPL_1$), followed by a plateau up to ~20% CPL. The increase of the adsorption capacity of the

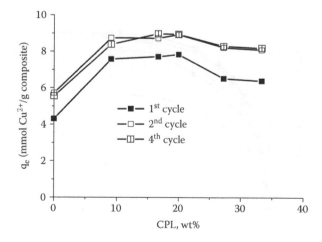

FIGURE 6.8 Cu²⁺ retention on CS–CPL beads at pH 5 as a function of CPL content loaded in CS–CPL composites, for four cycles of adsorption (contact time = 24 h).

CS–CPL composites compared with cross-linked CS may be explained by a synergy of both components; the presence of CPL microparticles led to an increase of accessibility at the functional groups of the CS network. The decrease of sorption capacity and further increase of the CPL loaded in composites was attributed to a decrease of CS content in the whole mass of composite, with CS as the main component contributing to the ability of the functional groups to bind copper ions.

A significant increase of the adsorption capacity of Cu^{2+} was found in the second cycle for all sorbents. The increase during the second cycle, i.e., after sorbent regeneration, is attributed to the increased number of amine groups available for Cu^{2+} binding by the complete removal of TPP anions involved in the ionic gelation step (see regeneration step in Scheme 6.1). The stabilization of the composite by the covalent cross links is demonstrated also by the values of the adsorption capacity— almost constant even after the fourth cycle of adsorption.

The effect of contact time on Cu^{2+} retention capacity of the cross-linked CS and CS–CPL composites is shown in Figure 6.9. The contact time varied from 0 to 36 h, and the initial metal concentration was fixed at 0.07 mol/L; all sorbents were used after regeneration with 0.1 M NaOH. As Figure 6.9 shows, the time required to achieve equilibrium at pH = 5 was about 24 h for all samples. To examine the controlling mechanisms of biosorption, such as mass transfer and chemical reaction, kinetic models were used to fit the experimental data. The kinetics of metal ion adsorption on the CS–CPL composites was determined with three kinetic models: the pseudo first order, pseudo second order, and intraparticle diffusion models. The mathematic equations are presented in Table 6.2.[5,19–22]

The pseudo first order rate equation of Lagergren is one of the most widely used equations for the sorption of solute from a liquid solution.[19,20] The slopes and intercepts of plots of log $(q_e - q_t)$ versus t (Figure 6.10) were used to determine the pseudo first order rate constant k_1 and q_e. The values obtained are presented in Table 6.3. The theoretical q_e values estimated from the first order kinetic model gave significantly

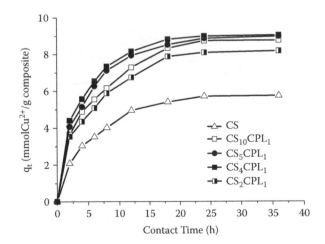

FIGURE 6.9 Cu^{2+} retention on CS–CPS composites at pH 5 as a function of contact time; initial metal concentration = 0.07 mol/L.

TABLE 6.2

Mathematic Equations of Kinetic Models Used in Study

Kinetic Model	Nonlinear Form	Linear Form	Equation
Pseudo first order	$\dfrac{dq_t}{dt} = k_1(q_e - q_t)$	$log(q_e - q_t) = log(q_e) - \dfrac{k_1}{2.303}t$	(6.3)
Pseudo second order	$\dfrac{dq_t}{dt} = k_2(q_e - q_t)^2$	$\dfrac{t}{q_t} = \dfrac{1}{k_2 q_e^2} + \dfrac{1}{q_e}t$	(6.4)
Weber and Morris intraparticle diffusionl	-	$q_t = k_{id}t^{0.5}$	(6.5)

Note: q_e and q_t are the amounts of metal ions adsorbed at equilibrium (mmol/g) and at time t, respectively; k_1 is the rate constant of pseudo first order sorption (min); k_2 is the rate constant of pseudo second order sorption (g/mmol $\times$ min); k_{id} is the intraparticle diffusion rate constant (mmol/(g $\times$ min$^{0.5}$)).

different values from experimental values, and the correlation coefficients were also lower (Table 6.3). These results show that the first order kinetic model is not adequate for describing these sorption systems.

The pseudo second order model is based on the assumption that the rate-determining step may be a chemical sorption involving valence forces through sharing or exchanging electrons between adsorbent and sorbate.[21] The values of q_e and k_2 were obtained from the slope and intercept of a straight line obtained by plotting t/q_t

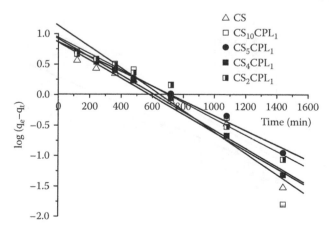

FIGURE 6.10 Pseudo first order model fitted for adsorption of Cu^{2+} on cross-linked CS and CS–CPL composites.

TABLE 6.3

Pseudo First Order Kinetics Parameters for Adsorption of Cu^{2+} on Cross-Linked CS and CS–CPL Composites at pH 5

Sample	$q_{e,exp}$ (mmol/g)	Parameters		
		$q_{e,calc}$ (mmol/g)	k_1 (g/mmol × min)	R^2
CS	5.7446	7.3333	3.408×10^{-3}	0.973
$CS_{10}CPL_1$	8.7591	13.74	4.03×10^{-3}	0.947
CS_5CPL_1	8.978	7.357	2.78×10^{-3}	0.995
CS_4CPL_1	9.0384	8.356	3.45×10^{-3}	0.996
CS_2CPL_1	8.1836	8.7922	3.086×10^{-3}	0.988

against t (Figure 6.11) and compiled in Table 6.4. As the table shows, the correlation factors are high for the cross-linked CS and CS–CPL composites and this indicates the validity of the pseudo second order model for Cu^{2+} adsorption on cross-linked CS and the composites. The pseudo second order kinetics confirm that chemisorption would be the rate-determining step controlling the adsorption of Cu^{2+}.

To reasonably assess the nature of the diffusion process for the adsorption of metal ions onto a composite, pore diffusion coefficients were calculated. The intraparticle diffusion model proposed by Weber and Morris[22] was used to calculate the initial rate of intraparticle diffusion by linearization of the curve $q_t = f(t^{0.5})$ [Equation (6.5) in Table 6.2]. k_{id} was determined by taking into account only the initial period. The amounts of Cu^{2+} adsorbed, q_t versus $t^{0.5}$, for cross-linked CS and CS–CPL composites are plotted in Figure 6.12. All the plots exhibit similar general features described by two distinct linear relationships. The initial linear portion was attributed to intraparticle diffusion. The values of the intraparticle diffusion rate constant, k_{id}, are presented in Table 6.5. The results indicate that the

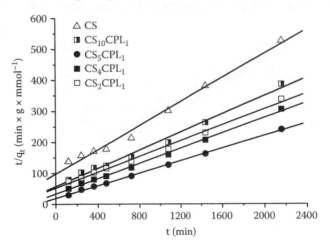

FIGURE 6.11 Pseudo second order model fitted for adsorption of Cu^{2+} on cross-linked CS and CS–CPL composites.

TABLE 6.4
Pseudo Second Order Kinetics for Absorption of Cu²⁺ on Cross-Linked CS and CS–CPL Composites

Sample	$q_{e,exp}$ (mmol/g)	Pseudo Second Order Constants		
		k_2 (g/mmol × min)	$q_{e,calc}$ (mmol/g)	R^2
CS	5.7446	5.59×10^{-4}	6.5924	0.9985
CS_2CPL_1	8.1836	4.358×10^{-4}	9.26	0.9980
CS_4CPL_1	9.0384	6.405×10^{-4}	9.8328	0.9994
CS_5CPL_1	8.978	5.41×10^{-4}	9.862	0.9995
$CS_{10}CPL_1$	8.7591	4.275×10^{-4}	9.7181	0.9984

metal ions diffused quickly through the sorbents at the beginning of the adsorption process, and then intraparticle diffusion slowed and stabilized. If the regression of q_t versus $t^{0.5}$ is linear and passes through the origin, then intraparticle diffusion is the sole rate-limiting step. However, the deviation of straight lines from the origin indicates that intraparticle transport is not the rate-limiting step for the sorbents under study.

The influence of temperature on the adsorption capacities of the CS–CPL composites at an initial concentration of 0.07 mol/L is presented in Figure 6.13a. To evaluate the change of enthalpy, the plot of ln (q/C_e), where q is defined as the equilibrium adsorption capacity of Cu²⁺ on the adsorbent (mmol/g), and C_e is the equilibrium Cu²⁺ concentration in solution (mmol/L), against $1/T$, based on Equation 6.3 is presented in Figure 6.13b.

$$\ln \frac{q}{C_e} = -\frac{\Delta H}{RT} + \frac{\Delta S}{R} \tag{6.3}$$

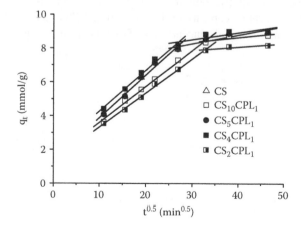

FIGURE 6.12 Weber and Morris intraparticle diffusion model fitted for adsorption of Cu²⁺ on cross-linked CS and CS–CPL composites.

TABLE 6.5

Weber and Morris Intraparticle Diffusion Parameters for Adsorption of Cu²⁺ on Cross-Linked CS and CS–CPL Composites

Sample	Parameters	
	k_{id} (mmol)/(g × min$^{0.5}$)	R^2
CS	0.1401	0.928
$CS_{10}CPL_1$	0.152	0.953
CS_5CPL_1	0.1401	0.928
CS_4CPL_1	0.132	0.920
CS_2CPL_1	0.142	0.953

The changes of enthalpy (ΔH) for cross-linked CS and three composites, determined from the slope (−ΔH/R), were: 3.132, 1.951, 1.474, and 1.797 kJ/mol for cross-linked CS, $CS_{10}CPL_1$, CS_2CPL_1, and CS_4CPL_1, respectively. The positive values of enthalpy show that the adsorption process is endothermic in nature.

Another important characteristic of chelating sorbents is the rate of desorption of the metal adsorbed. The results obtained from desorption of Cu²⁺ from the CS_4CPL_1 composite with 0.1 M HCl as a function of time are compared with the results of cross-linked CS in Figure 6.14. The figure shows that the composite produced faster desorption of Cu²⁺, approximately 20 min, compared with cross-linked CS beads, when desorption was ready in about 60 min.

6.2.2.2 Sorption of Co²⁺ and Ni²⁺

The adsorption capacity of CS for metal ions is related to the pH value in aqueous solution.[23–25] The batch adsorption run in this study at different pH levels showed poor affinity of the cross-linked CS and CS–CPL composites for Co²⁺ and Ni²⁺ ions in an acidic medium. The adsorbents exhibited a decrease in Co²⁺ and Ni²⁺ ion affinity at low pH due to competition with hydronium ions. Figures 6.15 and 6.16 show the effects of pH values on the adsorption capacity of the cross-linked CS and CS–CPL composites for Co²⁺ and Ni²⁺ ions; the initial concentration of metal ions in the aqueous solution was

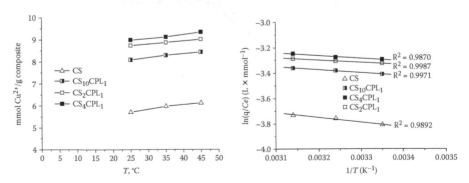

FIGURE 6.13 Effect of temperature on adsorption of Cu²⁺ by CS–CPL composites at pH 5 and initial metal ion concentration of 0.07 mol/L.

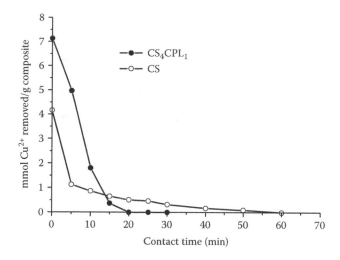

FIGURE 6.14 Release of Cu^{2+} with 0.1 M HCl from CS_4CPL_1 composite compared with cross-linked CS, as a function of contact time.

0.07 mol/L. It is obvious that the optimum adsorption pH is 5, as already observed for Cu^{2+}, at which the maximum uptake capacity (3.4 mmol Co^{2+}/g composite and 6.4 mmol Ni^{2+}/g composite), were obtained. At pH >5, the metal retention decreased because the metal ions started to deposit as $M(OH)_2$. Furthermore, the CS–CPL composites had higher retention capacities than the cross-linked CS, irrespective of pH.

The effect of contact time on Co^{2+} and Ni^{2+} retention capacity of the cross-linked CS and two CS–CPL composites is shown in Figures 6.17 and 6.18. The contact time varied from 0 to 36 h, and the initial metal concentration was fixed at 0.07 mol/L. All sorbents were used after regeneration with 0.1 M NaOH.

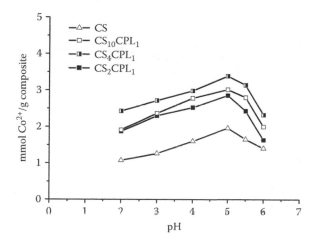

FIGURE 6.15 Co^{2+} retention on CS–CPL composites as a function of pH. Contact time = 24 h. Initial metal concentration = 0.07 mol/L.

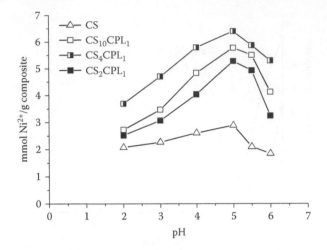

FIGURE 6.16 Ni^{2+} retention on CS–CPL composites as a function of pH. Contact time = 24 h. Initial metal concentration = 0.07 mol/L.

As Figures 6.17 and 6.18 show, the time required to achieve equilibrium at pH = 5 was about 24 h for both metals. The adsorption data were treated according to pseudo second order kinetics [Equation (6.4) in Table 6.2] because they were already shown to be more likely to predict the behavior over the whole range of adsorption, based on the assumption that the rate-determining step may be chemical sorption.[5,9,10,16] The values of q_e and k_2 were obtained from the slope and intercept of the straight lines obtained by plotting t/q_t against t (Figures 6.19 and 6.20) and compiled in Table 6.6. As shown in the table, the correlation factors are high for the cross-linked CS and CS–CPL composites, indicating the validity of the pseudo second order model for

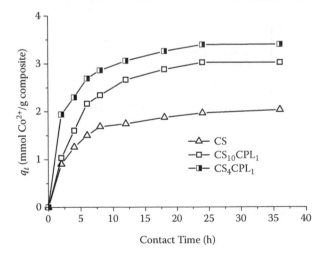

FIGURE 6.17 Co^{2+} retention as a function of contact time, at pH 5 and initial metal concentration of 0.07 mol/L.

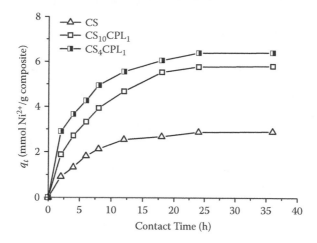

FIGURE 6.18 Ni^{2+} retention as a function of contact time, at pH 5 and initial metal concentration of 0.07 mol/L.

Co^{2+} and Ni^{2+} adsorption on cross-linked CS and the composites. The pseudo second order kinetics confirm that chemisorption is the rate-determining step controlling the adsorption of Co^{2+} and Ni^{2+}, as already shown for the adsorption of Cu^{2+} (Figure 6.11 and Table 6.4).

Desorption of the loaded metal ions was performed with 0.1 M HCl, as for Cu^{2+}. The results obtained by desorption of Co^{2+} and Ni^{2+} from cross-linked CS and two CS–CPL composites as a function of time showed that both composites exhibited faster desorption of Co^{2+} and Ni^{2+}: 20 min for the composite CS_4CPL_1, compared with cross-linked CS, when the desorption was ready in about 60 min.

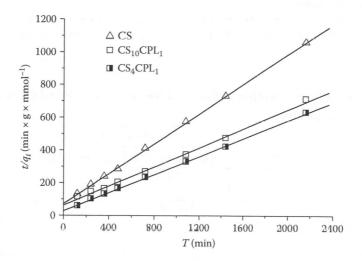

FIGURE 6.19 Pseudo second order model fitted for adsorption of Co^{2+} on cross-linked CS and two CS–CPL composites.

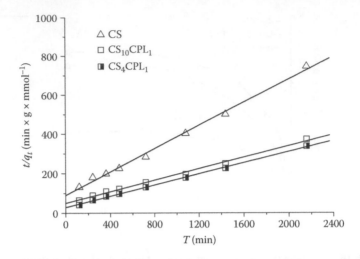

FIGURE 6.20   Pseudo second order model fitted for adsorption of Ni^{2+} on cross-linked CS and two CS–CPL composites.

Figure 6.21 presents results of the uptake of metal ions on synthetic chelating resins. The retention capacities for the metal ions were in the order of Cu^{2+} > Ni^{2+} > Co^{2+}. The affinity of the commercial Amberlite IRC-748 resin for Ni^{2+} and Co^{2+} was lower than that of the CR-10 resin, but higher for Cu^{2+}. However, the metal uptakes of both synthetic chelating resins containing iminodiacetate groups were significant lower than that for CS–CPL composites. Table 6.7 summarizes comparative adsorption capacities for Cu^{2+}, Co^{2+}, and Ni^{2+} based on the results in the literature and our results obtained by the adsorption of these metal ions on CS–CPL composites.

Novel composites with enhanced adsorption capacities for Cu^{2+}, Co^{2+}, and Ni^{2+} could be prepared by loading CPL microparticles in a matrix of cross-linked CS.

TABLE 6.6

Kinetic data for adsorption of Co^{2+} and Ni^{2+} on Cross-Linked CS and CS–CPL composites

Sample	$q_{e,exp}$ (mmol/g)	Pseudo Second Order Constants		
		k_2 (g/mmol × min)	$q_{e,calc}$ (mmol/g)	R^2
Co²⁺				
CS	2.032	2.701×10^{-3}	2.1891	0.9998
$CS_{10}CPL_1$	3.0214	1.253×10^{-3}	3.42	0.9985
CS_4CPL_1	3.4014	2.27×10^{-3}	3.6153	0.9997
Ni²⁺				
CS	2.9012	0.972×10^{-3}	3.389	0.9976
$CS_{10}CPL_1$	5.8034	0.425×10^{-3}	6.896	0.9974
CS_4CPL_1	6.4052	0.692×10^{-3}	7.1022	0.9991

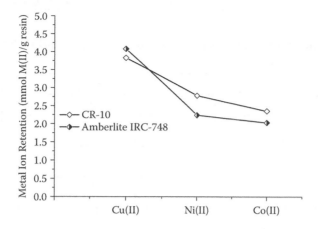

FIGURE 6.21 The effect of the structure of the CRs on the metal ion retention at pH = 5, temperature 25°C, 0.07 mol/L metal ions, contact time 8 h.

A "tandem" ionic–covalent cross-linking was used to stabilize the whole system. The presence of CPL in the composites strongly affected their characteristics and their adsorption capacities for metal ions compared with cross-linked CS without zeolite. The adsorption capacity abruptly increased up to a CPL content of ~20 wt%, and decreased with further increases of CPL content. The maximum adsorption capacities were 9.04 mmol/g for Cu^{2+}, 6.4 mmol/g for Ni^{2+}, and 3.4 mmol/g for Co^{2+}. The adsorption processes of all metal ions obeyed pseudo second order kinetics, confirming that chemisorption was the rate-determining step.

6.3 SEPARATIONS BY COMPOSITE MEMBRANES

A membrane is a natural or artificial interface that separates two fluid phases or volumes of different composition, but enables a selective mass transfer between the two phases. The driving forces that determine mass transfer through membranes are (1) the pressure difference Δp, (2) the concentration difference Δc, (3) the difference in chemical potential $\Delta\mu$, and (4) the electrical potential difference ΔE (depending on the separation process).[33] A composite membrane commonly consist of a thin and highly selective surface layer, and a thick substrate layer that is highly porous and provides mechanical strength. The membrane surface that plays an important role in membrane performance can be modified by physical, chemical, and bulk modification methods. Chemical modification is a way to generate new covalently bound reactive groups that can improve membrane properties such as wettability, antifouling, and the ability to bind other interesting moieties. A prerequisite condition for this modification is the presence of reactive functional groups.

Surface modification by physical methods is mainly based on ionic, H-bond, and hydrophobic interactions. One of the most versatile and convenient surface modification techniques is layer-by-layer (LbL) self-assembly of oppositely charged polyelectrolytes. This method has been used to prepare composite membranes with dense separating layers with thicknesses tailored at nanometer scale.[34] The most

TABLE 6.7

Comparison of Adsorption Capacities for Cu^{2+}, Ni^{2+}, and Co^{2+} for Different Sorbents

Sorbent	pH	q_{max}, mmol/g			Ref.
		Cu^{2+}	Co^{2+}	Ni^{2+}	
CS beads cross-linked CS (epichlorohydrin)	6	0.59 0.55	–	–	5
Magnetic CS nanoparticles	5.5	8.771	–	–	13
Cross-linked CS (glutaraldehyde)	5–6	2.58	1.68	2.34	23
Diazacrown ether, cross-linked CS (epichlorohydrin)	4	0.5385	0.112	0.139	24
Cross-linked CS (tripolyphosphate)	5	3.125	–	–	25
CS beads Cross-linked CS (glutaraldehyde) Cross-linked CS (epichlorohydrin) Cross-linked CS (EGDE)	6	1.26 0.93 0.97 0.72	–	–	26
Cross-linked CS (epichlorohydrin)	5	0.976	–	–	27
CPL treated with $2M$ HCl and $2M$ NH_4Cl	–	0.4056	–	0.2208	28
Alumina–CS composite	5–6	3.2	–	–	29
CS–alginate beads	4.5	1.065	–	–	30
CS–cellulose hydrogel beads	6	0.7874	–	–	31
Cross-linked CS (epichlorohydrin)	6	0.619	–	–	32
CS CS_2CPL_1 CS_4CPL_1 CS_5CPL_1 $CS_{10}CPL_1$	5	5.7446 8.1836 9.0384 8.978 8.7591	3.3931	6.396	This study

substantial advantages of the LbL self-assembly technique are simplicity of the deposition procedure that allows thin film construction on substrates with planar and other shapes, and strict control of the average thickness and nanoarchitecture of oppositely charged species layers.[34,35]

6.3.1 COMPOSITE MEMBRANES DESIGNED BY LbL TECHNIQUE

A charged substrate is dipped into a dilute solution of an oppositely charged polyelectrolyte so that the polymer is adsorbed at the substrate surface and the surface charge is reversed. After washing, the coated substrate is dipped into an aqueous solution of an oppositely charged polyelectrolyte and the surface charge is reversed again. A multilayer assembly is obtained after repeating the adsorption steps. The film thickness is controlled by the number of dipping cycles, pH, and ionic strength of the polyelectrolyte solutions.[35] The charge of the outer layer determines the sign of the membrane surface potential. Thus, positively or negatively charged membrane surfaces can be produced employing the same oppositely charged polyelectrolytes by variation of the number (odd or even) of dipping cycles.

It is known that hydrophobic surfaces adsorb biological materials like proteins during separation processes and membrane fouling occurs. The LbL technique has often been applied to prepare composite membranes because the charged top layer of LbL thin film deposited on the ultrafiltration membranes protects them against fouling during protein filtration and allows the control of membrane performance in other separation processes. This type of membrane has been extensively investigated in pervaporation separation of alcohol and water mixtures.[36,37]

6.3.2 LIQUID SEPARATIONS BY COMPOSITE MEMBRANES PREPARED BY LbL DEPOSITION OF OPPOSITELY CHARGED POLYIONS

The performance of a porous membrane is defined by its selectivity, separation factor (α), and permeate flux (J). Several studies of alcohol–water separation and dehydration of organic solvents by composite membranes have been reported. The water fluxes and separation capabilities of membranes are controlled mainly by charge density (ρ_c). For example, a membrane consisting of poly(vinyl amine) and poly(vinyl sulfate) having an improved separation factor for ethanol–water mixtures compared with poly(allylamine hydrochloride) and poly(styrene sulfonate) (PSS) membranes has been reported.[36] Membranes modified with two weak polyelectrolytes with high charge densities, such as poly(ethyleneimine) and poly(acrylic acid), gave the best separation properties. Membranes modified with strong polyelectrolytes [PSS and poly(diallyldimethylammonium chloride) (PDADMAC)] led to poorer separations.[37]

The objective of our investigations was to modify the surface properties of a microporous composite PAN membrane by physical methods to control the surface charge and swelling degree of the membrane in water and some pure alcohols.[38] Physical modification of PAN membranes was performed by LbL deposition of PSS and one strong polycation containing about 95 mol-% of N,N-dimethyl-2-hydroxypropyleneammonium chloride units in the backbone (PCA$_5$), synthesized according to Dragan and Ghimici.[39] The amount of the polyions adsorbed on the membrane was controlled by the nature and concentration of the supporting electrolyte in polyelectrolyte deposition solutions. After surface modification, the composite membranes were characterized by FTIR spectroscopy and swelling degree (SD) in water and some pure alcohols and compared with the starting PAN membrane.

An asymmetrical PAN supporting membrane reinforced with a polyester fabric (made in Germany), previously treated with oxygen plasma, had an average pore diameter of 12 nm and an average thickness of ~60 μm. The structures of the poly-ions used in the multilayer construction are presented in Scheme 6.2. PSS with a viscometric molar mass of 69,000 g/mol was used in all experiments.

PCA$_5$ has been characterized by intrinsic viscosity in 1 M NaCl at 25°C, $[\eta] =$ 0.46 dL/g. Polyion concentration in salt-free aqueous solutions and in the presence of neutral salts used in the multilayer construction was 10^{-2} mol/L. The protocol for the LbL multilayer build-up consisted of several steps. The PAN microporous membrane was previously treated with 1 wt-% NaHCO$_3$ aqueous solution followed by intensive washing with distilled water to reach neutral pH. Twenty minutes after every polyion adsorption step, each membrane was rinsed three times in distilled water for 1 min. The LbL-modified PAN composite membranes were kept under water and not dried between adsorption steps. The dry steps were necessary only to monitor the adsorbed polymer amount in dependence on the layer pair number. The number of polyions deposited after a certain number of adsorbed layer pairs (P) was determined as follows:[38]

$$P = (m_n - m_i)/m_i \qquad (6.4)$$

where m_n and m_i are the weights of the membrane after n adsorbed layer pairs and before adsorption, respectively. The value of m_n corresponds to the constant weight after drying of the membrane in air (about 4 days). Four samples were averaged.

FTIR spectra of the PAN membrane before and after surface modifications were recorded on a Digilab Scimitar Series spectrometer (made in U.S., resolution of 4 cm^{-1}). Potassium bromide pellets were made with the material scratched on the glossy faces of the membranes. Atomic force microscopy (AFM) was used to inves-tigate the morphology of the PAN membrane surface after LbL adsorption with SPM Solver PRO-M AFM (NT-MTD Co., Moscow, Russia). The images were obtained in tapping mode at room temperature and examined by Gwyddion 2.15 software to measure the surface average roughness (R_a).

Equilibrium swelling of the membranes in water and pure alcohols was investigated by monitoring the weight gained by the membrane immersed in an excess amount of a certain solvent at ambient temperature after 48 h. The mass of the wet membrane

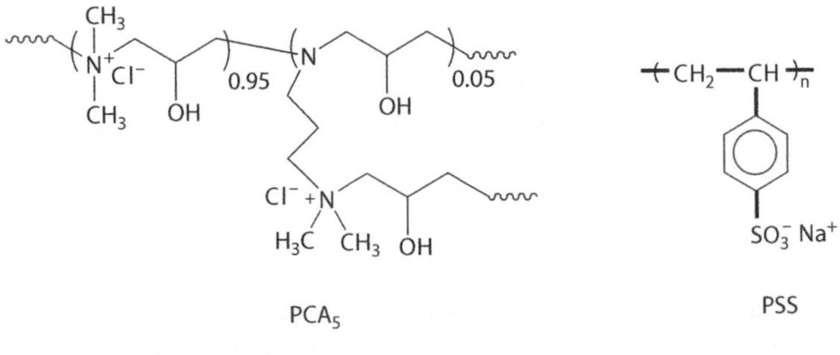

SCHEME 6.2   Structures of polyions used in LbL film construction on PAN membrane.

was obtained after removal of the solvent in excess by wiping the membrane surface carefully. Swelling degree of the membranes (SD, mL/g) was calculated:

$$SD = (m - m_0)/\rho_s \cdot m_0 \qquad (6.5)$$

where m_0 and m are the weights of the dry and the swollen membranes, respectively, and ρ_s is the solvent density. Four samples were measured for each specimen of the composite membrane.

As stated, composite membranes were prepared by the alternate adsorption of PCA_5 and PSS under different conditions on microporous PAN membranes previously treated with $NaHCO_3$. Thus, COOH groups were transformed in COO^-, which electrostatically interacted with PCA_5 as the first adsorbed layer. Figure 6.22 shows variations of the polymer amount adsorbed (P) on the PAN membranes in two low molar mass salt solutions plotted against layer pair number. Three polycation–polyanion double layers were previously adsorbed from aqueous solution as a background to achieve a uniform charge distribution on membrane surfaces. This sample was considered the initial membrane for polyion adsorption in the presence of a supporting electrolyte.

The first consequence of the presence excess counterions was a decrease of the free charges of the macro ions available for the neutralization of the opposite charges of the substrate. A higher amount of oppositely charged polyion is necessary to overcompensate for the support charges. The influence of the counterion can be observed by comparison of the polyion quantities adsorbed from 1 M KBr and 1 M NaBr aqueous solution. A greater number of polyions were adsorbed from 1 M KBr than from 1 M NaBr aqueous solution. As the counterion for PCA_5 was the same (Br^-), the difference is explained by the specific interaction of PSS and the counterions (Na^+ and K^+). Thus, our results are in agreement with the previous data for counterion binding

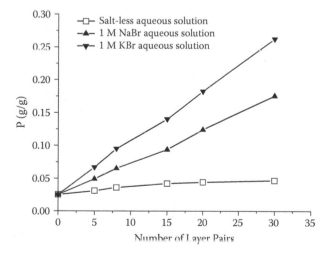

FIGURE 6.22 Polymer (P) adsorbed as a function of the double layer number and nature of the supporting electrolyte in polyion adsorption solutions on starting PAN composite membranes after preliminary treatment with $NaHCO_3$ aqueous solution.

order on polyanions containing sulfonate groups: Li$^+$ < Na$^+$ < K$^+$.[40] An almost linear variation of the polyion adsorbed amount compared with layer pair number was observed under all adsorption conditions. This indicates that regular growth of the PCA$_5$–PSS multilayer took place as observed for other systems.[35]

Multilayer growth was also supported by FTIR spectra performed on a composite membrane having a multilayer of PCA$_5$–PSS$_{20}$, with polyions adsorbed from 1 M NaBr aqueous solutions, compared with a starting PAN membrane (Figure 6.23). The presence of PSS is clearly demonstrated by the characteristic bands of this polyanion: 1162 cm^{-1}, 1121 cm^{-1}, 1028 cm^{-1}, and 1001 cm^{-1} assigned to SO$_3^-$ groups and 1578 cm^{-1}, 1538 cm^{-1}, and 833 cm^{-1} assigned to the aromatic CH stretch of p-disubstituted benzene. Furthermore, in the composite membrane having a multilayer of PCA$_5$–PSS$_{20}$, with polyions adsorbed from 1 M NaBr aqueous solution, the characteristic bands of PAN and PET were almost completely screened.

The morphological changes on the PAN membrane surface modified by LbL, depending on the numbers of double layers, with polyions adsorbed from 1 M KBr or 1 M NaBr aqueous solutions, was followed by AFM. Figure 6.24 shows micrographs of composite membrane samples, including the values of average roughness (R_a). Increasing the number of adsorbed double layers from 10 (Figure 6.24a) to 30 (Figure 6.24b), caused the surface morphology to become looser, sustained also by doubling the R_a value. The influence of electrolyte type on membrane surface morphology, with the same number of layers, can be seen by comparing images b and c in Figure 6.24. Thus, thicker and less homogeneous layers were obtained in the presence of KBr, compared with those obtained in the presence of 1 M NaBr, with average surface roughness (R_a) of about 10 nm.

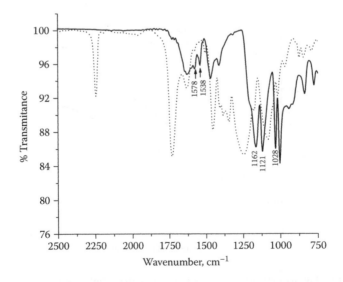

FIGURE 6.23 FTIR spectra of starting PAN membrane (dotted line) and after LbL deposition of 20 double layers of PCA$_5$ and NaPSS adsorbed from 1 M NaBr aqueous solution (solid line).

(a) R_a = 4 nm   (b) R_a = 8 nm   (c) R_a =10 nm

FIGURE 6.24   Amplitude (up) and 3D height (down) tapping mode AFM images of some LbL modified membranes: a) (PCA5/PSS)10 adsorbed in NaBr 1M; b) (PCA5/PSS)30 adsorbed in NaBr 1M; c) (PCA5/PSS)30 adsorbed in KBr 1M.

Figure 6.25 illustrates SD variations in water versus the number of layer pairs for the composite membranes prepared by LbL self-assembly of PCA_5 and PSS with two supporting electrolytes in polyion adsorption solutions. The increase of the SD values versus the number of double layers reflects the same influence of the counterion nature as observed in Figure 6.22 for variation of P values. The highest SD values in water were found for polyion adsorption from 1 M KBr aqueous solution and this shows the highest hydrophilicity of the membrane. The high concentration of salt electrostatically screened the charged groups in the polymer, leading to a film with a more loopy structure. Some polyion charges could remain extrinsically compensated, unlike those directly involved in the interpolyion complex formation in the LbL self-assembled film (intrinsic compensation) that induced greater hydrophilicity.[41]

The SD values of some pure alcohols on a microporous PAN composite membrane modified by the LbL technique with polyions adsorbed from either salt-free aqueous solution or 1 M NaBr are plotted in Figures 6.26 and 6.27, respectively. The SD values in pure water are included for comparison. As Figure 6.26 shows, the SD values decreased significantly after the first three double layers for all solvents and slightly varied after that. The SD values increased with the decrease of solvent polarity (E_T) and the highest values were observed for t-butanol, which had the lowest polarity (43.3 kcal/mol compared with 55.5 for methanol, 50.2 for 1-butanol, and 63.1 for water). This shows that the SD values were influenced by both the pore volume of the modified membrane, which decreased by pore filling with a polyelectrolyte complex, and also by the polar interactions.

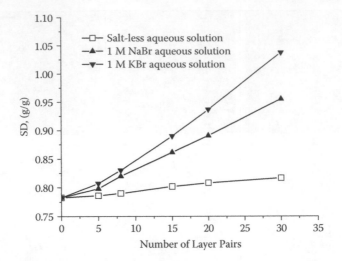

FIGURE 6.25 Swelling degree (SD) in water as a function of number of layer pairs and concentration of supporting electrolyte in polyion adsorption solutions on PAN membranes after preliminary treatment with $NaHCO_3$ aqueous solution.

Further comments are necessary concerning swelling in water. The lowest SD values were found after adsorption of the first three polyion double layers, followed by a slight but continuous increase of SD with the increase of the double layer number. This shows a slight increase of hydrophilicity of the multilayer and may indicate an increase of the hydrophilic group content (OH) brought about by the PCA_5 chains. Figure 6.27 shows that the order of SD values in pure alcohols

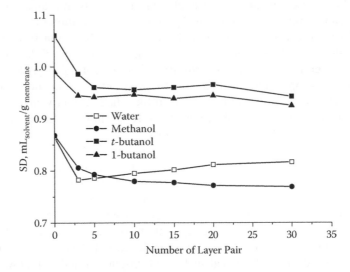

FIGURE 6.26 Swelling degree (SD) in water and pure alcohols as a function of number of layer pairs with both polyions adsorbed from salt-free aqueous solutions on a starting PAN composite membrane after preliminary treatment with $NaHCO_3$ aqueous solution.

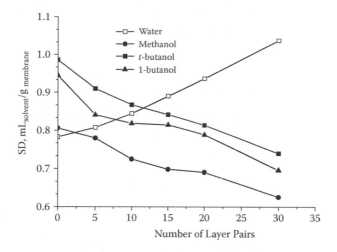

FIGURE 6.27 Swelling degree (SD) in water and pure alcohols as a function of number of layer pairs with both polyions adsorbed from 1 M NaBr on a starting PAN membrane after preliminary treatment with NaHCO$_3$ aqueous solution.

remained the same for the membrane modified by the LbL self-assembly of the polyions adsorbed from 1 M NaBr, but continuously decreased with the number of polyion double layers.

The decrease of the SD values for pure alcohols can be correlated with the decrease of pore volume by their filling with the interpolyion complex in the multilayer construction. The number of polyions (P) adsorbed from 1 M NaBr was much higher than the number adsorbed from a salt-free solution (Figure 6.22). The fast and almost linear increase of SD in water shows an increase of membrane hydrophilicity because a higher number of charges could remain compensated by small counterions in such a film. This also indicates an increase of membrane selectivity for water molecules.

Composite membranes resulting from LbL self-assembly of PCA$_5$ and PSS on a microporous PAN membrane demonstrated the possibility of using this pair of polyions for modulation of membrane separation properties. The adsorbed polyion quantity was controlled by the nature and concentration of the supporting electrolytes in polyion adsorption solutions. The SDs in water and some pure alcohols depended on both the modification of the PAN composite membrane and the solvent polarity. Thus, the order of SD values was t-butanol > 1-butanol > water ≥ methanol when a PAN membrane was modified by LbL self-assembly of polyions adsorbed from salt-free aqueous solutions. The SD values in pure alcohols continuously decreased with the number of double layers when PCA$_5$ and PSS were adsorbed from 1 M NaBr aqueous solutions, but SD values in water abruptly increased with the increase of the layer pair number. Thus, the composite membranes presented above could serve as alternatives for the separation of water from alcohols.

6.4 SEPARATIONS BY NONSTOICHIOMETRIC COMPLEX NANOPARTICLES BASED ON CHITOSAN

Interpolyelectrolyte complexes (IPECs) with different structures and properties can arise from a very fast ionic exchange reaction between two oppositely charged polyelectrolytes, accompanied by the release of corresponding small counterions, leading to an increase of system entropy and complex stability.[42–44] The formation and properties of IPECs based on synthesis conditions are still under investigation.

Scheme 6.3 presents a representation of IPEC formation. Based on their properties, IPECs can be divided into three main types: (1) insoluble and amorphous precipitates (polysalts),[42–50] (2) soluble IPECs produced mainly when complementary polyelectrolytes have significant different molar masses in dilute aqueous solutions and nonstoichiometric systems (the structural differences between oppositely charged polyions and low amounts of NaCl are also necessary),[43,44,51,52] and (3) IPECs as stable colloidal dispersions.[53–65] Both soluble IPECs and IPECs as colloidal dispersions bearing free charges in excess are nonstoichiometric complexes (NIPECs).

NIPECs as stable colloidal systems bearing positive charges in excess have been used as flocculants for cellulose and clay dispersions and organic compounds, as well as for surface modifications of different solid substrates.[53,54] Preparation of NIPECs as nanoparticles and their applications for destabilization of solid and liquid systems is one of our directions of work.[55–65]

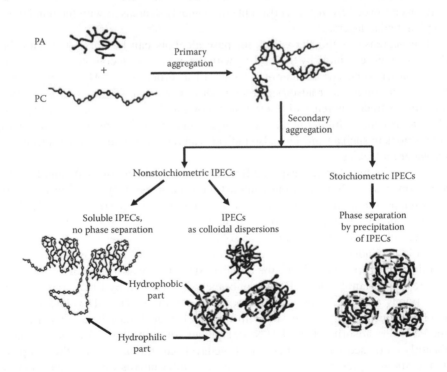

SCHEME 6.3   IPEC formation.

It is well known that the efficiency of a certain flocculant in separation processes is evaluated as a function of two main parameters: the optimum flocculation concentration, which should be as low as possible, and the flocculation window, which must be as large as possible.[66] The most widely used polymeric flocculants are polycations because most solid particles in suspensions have negative charges.[44,64,67–69] However, polycations have a main drawback: narrow flocculation windows.

An improvement of this aspect was achieved by using NIPECs as colloidal dispersions bearing free positive charges in excess.[53,54,58] The optimum concentration of the complex required for flocculation was much higher than the optimum concentration of the polycation alone, even if the flocculation window was much larger with NIPECs. Therefore, a study of factors that allow tuning of complex properties to simultaneously fulfill both conditions was highly desirable. Furthermore, most synthetic flocculants are obtained from petroleum-based raw materials by processing chemistry, which is not always environmentally friendly because many contaminants appear in water from residual unreacted monomers.

In this context, the ionic polymers coming from renewable resources as bioflocculants are gaining more attention. Chitosan (CS) has been extensively used in wastewater remediation, both alone[69-71] and in IPECs formed with natural polyanions.[72] Data about the formation and applications of NIPECs between CS and synthetic polyanions in separation processes are lacking. Thus, finding the formation conditions of novel CS-based IPECs as colloidal dispersions and tailoring their properties for certain applications are still challenges to be met. Ionic and nonionic copolymers are involved in tuning complex properties through both their controlled charge densities and hydrophilic–hydrophobic balances. Therefore, our interest has focused on the formation and properties of novel NIPECs as colloidal dispersions based on CS with different molar masses and ionic and nonionic random copolymers of 2-acrylamido-2-methylpropanesulfonate (AMPS) with t-butyl acrylamide (TBA), along with their efficiency in solid–liquid separations.[73–75]

6.4.1 Formation and Stability of Complex Nanoparticles Based on Chitosan

The chemical structures of the polyelectrolytes used in this study are presented in Scheme 6.4 and some of their characteristics are summarized in Table 6.8. CS samples

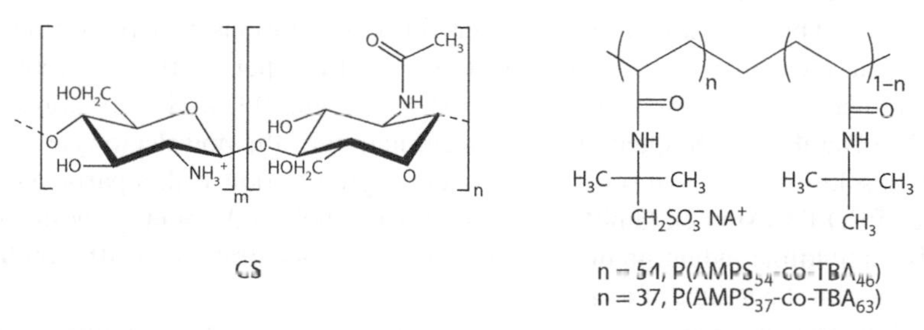

CS

n = 54, P(AMPS$_{54}$-co-TBA$_{46}$)
n = 37, P(AMPS$_{37}$-co-TBA$_{63}$)

SCHEME 6.4 Chemical structures of oppositely charged polyions used to prepare IPEC nanoparticles based on chitosan.

TABLE 6.8

Characteristics of Oppositely Charged Polyions Used to Prepare Complex Nanoparticles Based on Chitosan

Sample	η (mPas)	[η] (dL/g)	M_v (kDa)	M_u^a (g/charge)	b^b (nm)	CD^c
CS I	400	9.12^d	470	196.8	0.643	0.8
CS II	800	12.33^d	670	196.8	0.643	0.8
CS III	1000	14.07^d	780	196.8	0.643	0.8
P(AMPS$_{54}$-co-TBA$_{46}$)^e	–	0.33^f	175	337.0	0.463	0.54
P(AMPS$_{37}$-co-TBA$_{63}$)^{el}	–	0.47^f	–	440.6	0.667	0.37

a   Mass per charge; $M_u = [M_{IM} \times f_{IM} + M_{NM} \times (1 - f_{IM})]/f_{IM}$ where M_{IM} and M_{NM} are molar masses of ionic and the nonionic monomers, respectively and f_{IM} is the molar fraction of ionic monomer.

b   Distance between charges.

c   Linear charge density.

d   In 0.3 M CH$_3$COOH and 0.2 M CH$_3$COONa (1:1 v/v) at 25°C.

e   Synthesized according to Vadivelan, V. and Kumar, K.V. *J. Colloid Interface Sci.* 286, 90–100, 2005.

f   In 1M NaCl, at 25°C.

were purchased from Heppe GmbH as flakes (ash content <1%) and used without further purification. The viscometric average molar masses of CS samples were estimated using Equation (6.1) and are included in Table 6.8. Degree of acetylation (DA) was evaluated by infrared spectroscopy with a Vertex 70 Bruker FTIR spectrometer. For DA determination, Equation (6.2) was used, taking the 1420 cm^{-1} band as a reference; for N-acetylglucosamine, the band chosen was at 1320 cm^{-1}. The medium value of DA = 20% for all CS samples was taken into account.

CS solutions with concentrations of 1 g/L were obtained by dissolving the flakes in 1 vol.-% acetic acid solution with moderate stirring for at least 48 h. To use the CS solution for IPEC preparation, the concentration was adjusted to 0.5 mM by dilution with 1 vol.-% acetic acid solution. Polyanions with concentrations of 5 mM were prepared by adequate dilution of the stock solutions (10 mM) with distilled water.

The CS and polyanion solutions with certain concentrations were prepared taking into account the molar mass of the repeat unit in grams per charge (see Table 6.8). Before use, all solutions were adjusted to pH = 4.0, with 0.1 M HCl for polyanions, and 0.1 M NaOH for CS. Quantitative determination of the polyelectrolyte concentration in solution was performed by polyelectrolyte titration with a particle charge detector PCD 03 (Müteck GmbH, Germany) using poly(sodium ethylenesulfonate) or poly(diallyldimethylammonium chloride) at a concentration of 10^{-3} mol/L in dependence on the nature of charges.

IPEC dispersions were prepared at room temperature (~25°C) by mixing aqueous solutions of oppositely charged polyelectrolytes. The amount of CS was kept constant while the amount of polyanion was varied according to the desired mixing molar ratio between charges, $n^-:n^+$. It was already demonstrated that titrant addition rate is a valuable parameter for controlling particle size, shape, and polydispersity.[62–65] A constant addition rate of 3.8 mL polyanion/mL CS/h was used because this was the optimum rate in our previous investigations, taking into account the complex morphology and colloidal stability. After mixing, the formed dispersions were stirred for 60 min and characterized after 24 h or 1 wk of storage.

Particle sizes and their polydispersities were evaluated by dynamic light scattering (DLS) using a Zetasizer 3000 (Malvern Instruments, U.K.) equipped with a 10 mW He-Ne laser operating at $\lambda = 633$ nm at a scattering angle of 90°. The reported results represent the average of two independent DLS measurements. AFM was used to characterize particle sizes and shapes in the dry state. Prior to use, the silicon wafer substrates were cleaned in two steps: (1) in "piranha solution" followed by intensive rinsing with deionized water, and (2) with a mixture of NH_4OH, H_2O_2, and deionized water, at 70°C in an ultrasonic bath, followed by intensive rinsing with water and drying under a nitrogen flow.

The clean silicon wafers were immersed in IPEC dispersions identical to those used for DLS for 20 min, then immersed for 1 min three times in distilled water, and finally air dried for 48 h at room temperature. The shapes of IPEC particles were examined with a Nanoscope IIIa Dimension 3100 SPM (Digital Instruments, Veeco Metrology Group, Woodbury, NY, USA). The topographic images were obtained in tapping mode and repeated on different areas of the same sample. The sizes of about 50 individual IPEC particles adsorbed on the silicon substrate were measured directly from AFM topographic images, by the particle analysis function of the device software.

The physicochemical and biological properties of CS depend mainly on the fraction and distribution of the two kinds of repeating units along the chains (glucosamine and N-acetylglucosamine), the pH of solution, and the degree of polymerization. In our experiments, the initial solutions of polyelectrolytes were adjusted at pH = 4.0. Due to the constant pH, the charge density of CS remained unchanged, whereas the charge density of polyanions was adjusted in the synthesis process by the ratio between the ionic and nonionic comonomers. As shown in Figure 6.28a, the particle sizes and their hydrodynamic diameters (D_h) were strongly influenced by the characteristics of the oppositely charged polyions. As a common trend, the particle sizes slowly decreased up to $n^-:n^+$ of about 0.8, remained almost unmodified up to $n^-:n^+$ of about 1.2, and abruptly increased after that, mainly when CS II and CS III were the starting polyions. The particle sizes were always higher when $P(AMPS_{37}\text{-co-}TBA_{63})$ was the added polyion.

The decrease of particle sizes before stoichiometry suggests that when the polyanion was in default and CS in excess, the addition of polyanion allowed both the formation of new particles and also their rearrangement toward more compact structures. The $n^-:n^+$ values ranging from ~0.8 up to 1.2, characterized by almost constant sizes of the complex particles, were characteristic of the complexes formed between CS and the two random copolymers P(AMPS-co-TBA).[73] As Figure 6.28a shows,

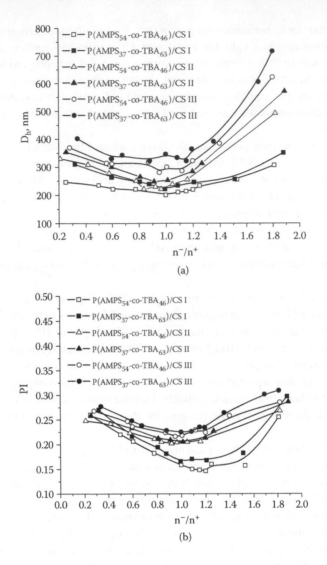

FIGURE 6.28 D_h (a) and PI (b) of IPEC dispersions as a function of ratio between charges $(n^-:n^+)$ for all polyion pairs.

the secondary aggregation of the particles occurred after $n^-:n^+$ of ~1.2, leading to a monotonous increase of complex particle sizes after that.

The polyanion charge density influenced particle sizes, both before and after stoichiometry. Thus, when $P(AMPS_{37}\text{-co-}TBA_{63})$ was used as the polyanion, the higher content of the nonionic and hydrophobic comonomer (TBA) led to the increase of polyanion required for the charge compensation; particle sizes were always larger than the sizes of the IPECs formed with $P(AMPS_{54}\text{-co-}TBA_{46})$ as the polyanion. Figure 6.28b shows that the polydispersity index (PI) followed almost the same trend as the particle sizes: the smallest values of PI were obtained at the ratio between charges, n^-:n^+, ranging from 0.8 to 1.2. A much lower polydispersity of the complex particles was found when synthetic polycations containing quaternary ammonium salt centers in the backbone were used in the complex formation with the P(AMPS-co-TBA) as the polyanion when the PI was lower than 0.1.[63]

The AFM amplitude images obtained in tapping mode of some IPEC nanoparticles are presented in Figures 6.29 and 6.30. The images indicate that before stoichiometry (n^-:n^+ = 0.6), irrespective of the CS molar mass, the adsorbed complexes appeared as small, individual, dispersed particles when the IPECs were formed with P(AMPS$_{54}$-co-TBA$_{46}$) as the added polyion (Figures 6.29a and b). Highly aggregated structures dispersed between individual particles were observed when P(AMPS$_{37}$-co-TBA$_{63}$) was added to CS III (Figure 6.30a). Comparing Figure 6.29b with Figure 6.30a, one may note that the influence of polyanion structure on particle shape was stronger than the influence on their average size. Close to stoichiometry, at n^-:n^+ = 1.2, the particle sizes strongly increased with the increase of CS molar mass

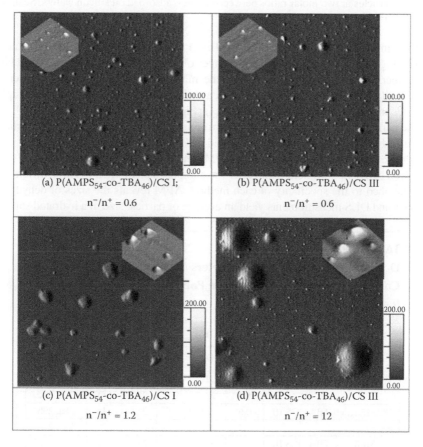

FIGURE 6.29 Tapping mode amplitude AFM images of P(AMPS$_{54}$-co-TBA$_{46}$)–CS complex particles at different molar ratios between charges, adsorbed on silicon wafers. Scan size = 5 × 5 μm² for all images. Insets show three-dimensional height AFM images of selected areas. Scan size = 2 × 2 μm² for all inset images.

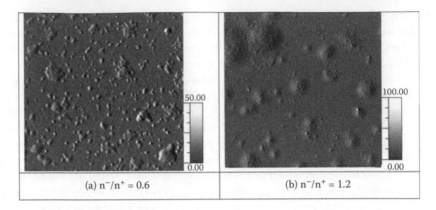

FIGURE 6.30 Tapping mode amplitude AFM images of P(AMPS$_{37}$-co-TBA$_{63}$)–CS III complex particles at two molar ratios between charges, adsorbed on silicon wafers. Scan size = 5 × 5 μm² in both images.

for the same polyanion [P(AMPS$_{54}$-co-TBA$_{46}$) added to CS I (Figure 6.29c) and to CS III (Figure 6.29d)]. Comparing the shapes of IPECs formed with CS III and both copolymers (Figures 6.29d and 6.30b), the main difference consists of numerous very small particles that act like a corona around a condensed core for the complex particles formed with P(AMPS$_{37}$-co-TBA$_{63}$) (Figure 6.30b).

The particle sizes measured by DLS were compared with those of the dried particles measured by AFM. The results are presented in Table 6.9. The medium particle sizes of the complex nanoparticles measured by AFM were always lower than the average value measured by DLS, both before and after the complex stoichiometry. These differences were ascribed to the specificity of each method: AFM reveals the sizes of dehydrated particles and DLS measurements yield an average of particle size in a hydrated state.[73]

TABLE 6.9

D_h Measured by DLS and Diameters Measured by AFM for Complex Particles Prepared from Polyion Pairs and Two Molar Ratios between Charges

Sample	n⁻:n⁺	DLS	AFM			Figures
		D_h	d_{min}	d_{med}	d_{max}	
P(AMPS$_{54}$-co-TBA$_{46}$)CS I	0.6	222	22	158	579	28a, 29a
	1.2	231	27	219	651	28a, 29c
P(AMPS$_{54}$-co-TBA$_{46}$)CS III	0.6	281	22	213	519	28a, 29b
	1.2	272	29	257	877	28a, 29d
P(AMPS$_{37}$-co-TBA$_{63}$)CS III	0.6	337	22	216	611	28a, 30b
	1.2	359	44	315	509	28a, 30b

(d_{min}) = minimum diameter; (d_{med}) = medium diameter; (d_{max}) = maximum diameter.

CS in a dilute solution (0.5 mM in this work) has a worm-like conformation. Synthetic polyanions adopt a coiled conformation inasmuch as their concentration was 10 times higher than that of CS (5 mM). Therefore, in the formation of IPECs between CS in excess and P(AMPS-co-TBA) polyanions, some peculiarities should be considered. The charge density of the polyanions was about 0.5 for P(AMPS$_{54}$-co-TBA$_{46}$) and much lower for P(AMPS$_{37}$-co-TBA$_{63}$) (see Table 6.8). Furthermore, the nonionic TBA comonomer is hydrophobic and some difficulties in complex formation may be expected. Based on DLS and AFM results, the mechanism for the formation of positively charged NIPECs nanoparticles is suggested in Scheme 6.5.[73] According to this mechanism, the added polyanion interacts first with CS chains, leading to the primary aggregates that may contain more CS chains connected by fewer polyanion chains because of the differences in the flexibility of the complementary polyions and the mismatch of charges; such aggregates would exhibit high densities of free positive charges compensated by small counterions (extrinsic compensation) and not by polyanion charges.

Further addition of polyanion (step II, Scheme 6.5) led to the step-by-step neutralization of the positive charges of CS included in the primary aggregates by polyanion charges, and, by rearrangements of the chains, to the formation of more compact particles with lower sizes. This assumption was supported by a monotonous decrease of particle sizes with the increase of the molar ratio n^-:n^+ up to ~0.8 found by DLS measurements (Figure 6.28a). Generally, the level of aggregation of IPECs as colloidal dispersions is mainly controlled by the properties of oppositely charged polyelectrolytes.[55–63] Using CS as the starting polyion and ionic and nonionic copolymers as polyanions added to CS, DLS results (Figure 6.28a) indicated that the particle sizes remained almost constant for n^-:n^+ ranging from ~0.8 up to 1.2 (step III, in Scheme 6.5). The ratio between charges of ~1.2 is critical for these systems, because an abrupt increase of particle sizes and polydispersities was observed after this ratio (Figures 6.28a and b) and supported also by the sizes measured on AFM images (Table 6.4). However, such complex particles formed after stoichiometry are not of interest for separation processes.

A very important characteristic of IPECs as colloidal dispersions tailored for use in separation processes is their colloidal stability during storage. Particle sizes (D_h) measured 24 h after preparation were compared with those measured after 1 wk of storage.[73] Figure 6.31 illustrates the influence of CS molar mass and polyanion structure on the colloidal stabilities of complex nanoparticles. As seen in Figure 6.31a, when CS I was the starting polyion, an increase in particle sizes was observed before stoichiometry, irrespective of the polyanion structure, probably caused by some intraparticle and interparticle rearrangements. After stoichiometry, the D_h

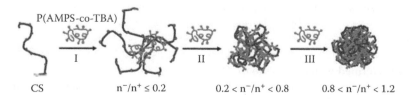

P(AMPS-co-TBA)

I   II   III

CS $n^-/n^+ \leq 0.2$ $0.2 < n^-/n^+ < 0.8$ $0.8 < n^-/n^+ < 1.2$

SCHEME 6.5 Mechanism for formation of NIPEC nanoparticles from chitosan as starting polyion and P(AMPS-co-TBA) as added polyions.

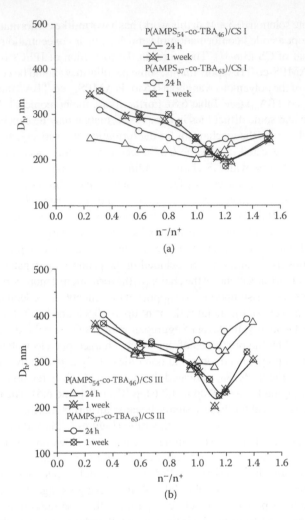

FIGURE 6.31 Storage colloidal stability of IPEC nanoparticles based on CS at room temperature (22°C) without stirring.

decreased for IPECs prepared with both polyanions. With CS III as the starting polyion (Figure 6.31b), the abrupt decrease of particle sizes after 1 wk of storage was observed for both polyanions, mainly at stoichiometry and after that. The lower D_h values show that the bigger particles separated over time. The measurements after 7 days were made by collecting the supernatant without stirring. The poorer colloidal stability of the complex formed with CS III is a consequence of its high molar mass. Highly aggregated structures were observed by AFM (Figure 6.30b), for IPECs formed with this polyion.

During studies of the formation of IPEC nanoparticles as colloidal dispersions from CS (molar masses of 470, 670, and 780 kDa) and the ionic and nonionic random copolymers of AMPS with TBA, valuable information on the mechanisms of complex formation and colloidal stability of NIPEC dispersions was obtained. The particle sizes increased with the increase of CS molar mass and when $P(AMPS_{37}\text{-co-}TBA_{63})$ was the added polyion, i.e., where the polyanion had the highest content of hydrophobic comonomer. According to the suggested mechanism for the formation of NIPEC nanoparticles, the primary aggregates formed at $n^-:n^+ \leq 0.2$ are larger than the particles formed at higher ratios between charges, and would have higher densities of free positive charges not involved in complexation with polyanion; this recommends them for solid–liquid separation processes.

6.4.2 CHITOSAN-BASED NONSTOICHIOMETRIC COMPLEX NANOPARTICLES AS SPECIALIZED FLOCCULANTS

Investigations to date on applications of NIPECs as colloidal dispersions bearing positive charges in excess have concentrated on the use of NIPECs with molar ratios between charges ranging from 0.4 to 0.8. The main advantage in flocculation induced by NIPECs was a substantially wider optimum concentration range for flocculation. As noted earlier, the main drawback of NIPECs as colloidal dispersions is that the optimum concentration required for flocculation was much higher than the optimum concentration for flocculation with polycations, which makes NIPECs as flocculants not very cost effective.

The evaluation of positively charged NIPEC nanoparticles formed by the interaction of CS and two random copolymers of AMPS with TBA prepared at molar ratios between charges ($n^-:n^+$) ranging from 0.15 to 0.5 in the flocculation of kaolin from a model dispersion compared with CS, is presented in this section.[74,75] Kaolin suspension was used as a representative colloidal material because kaolin is a well known and common thickening agent that acts as a key component in various industrial fluid formulations and has attracted great interest in recent years. The kaolin dispersions at a concentration of 1 g/L at pH 6 were prepared by 15 min ultrasonic treatment of kaolin powder (Aldrich, 600-nm particle size) followed by 1 h of vigorous stirring.

Destabilization experiments were conducted at room temperature. Volumes of 50 ml kaolin model suspension were stirred at 120 to 150 rpm in beakers and then different volumes of CS solution or NIPEC dispersion were added. Stirring continued with the same speed for ~2 min, and then decreased to ~50 rpm for 15 min. After a settling time of ~20 min, an OD_{500} reading was performed with a Specord M42. Residual turbidity (RT) was calculated:

$$RT = (OD_{500s}/OD_{500i}) \times 100, \% \qquad (6.6)$$

where $OD_{500s} = OD_{500}$ after the addition of flocculant, and $OD_{500i} = OD_{500}$ of the initial model suspension. The flocculation window broadness was calculated as the difference between the redispersion concentration (highest dose of polymer for which the turbidity of the supernatant OD_{500} was<0.01 a.u.) and the optimum dose.

Some characteristics of the NIPEC nanoparticles selected for separation are presented in Table 6.10. The kaolin flocculation induced by NIPECs based on CS II and ionic and nonionic copolymers of AMPS with TBA prepared at different molar ratios between charges $n^-:n^+$ was compared with flocculation by CS. The variation of RT as a function of the ratio between the amount of polymer per substrate ($C_p:C_s$, mg flocculant/g kaolin) is plotted in Figure 6.32. For flocculation with NIPECs, the polymer mass (C_p) is the

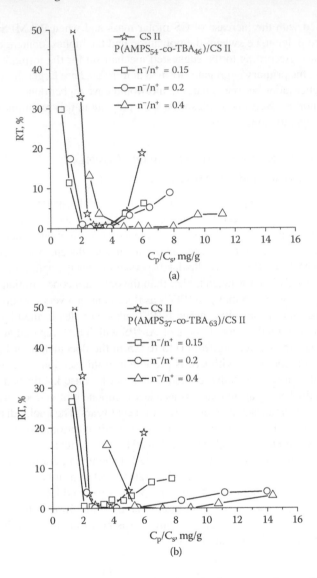

FIGURE 6.32 Residual turbidity (RT) as a function of polymer per substrate (C_p:C_s, mg/g) in kaolin flocculation with NIPECs based on CS II at different molar ratios between charges.

sum of the mass of polycations and polyanions corresponding to a certain molar ratio between charges.

The efficiency of flocculation by polyelectrolytes depends on certain parameters such as the types and densities of charges, polyelectrolyte molar mass, and polymer conformation.[64] Polymer conformation exerts great influence on flocculation efficiency. CS in a dilute solution (1 g/L in this study) has a worm-like conformation. Because of its high molar mass, it can induce bridging by simultaneous interaction with two or more particles.[70]

Figure 6.32 shows that a low dose of CS II was needed for destabilization and sedimentation of kaolin suspension. Thus, the minimum value of RT was achieved at an optimum dose of 2.7 mg flocculant/g kaolin—the broadness of the flocculation window ranged from 2.7 to 4 mg CS/g kaolin. By further increasing the CS dose, RT increased and the system restabilized.

As Figure 6.32 shows, the flocculation of kaolin by CS-based NIPECs was very effective in a broad range of polymer concentrations; the values of optimum dose and flocculation window depended on the polyanion nature and the molar ratio between charges, n^-:n^+. The presence of a hydrophobic nonionic comonomer in the polyanion structure led to increased NIPEC particle sizes (Table 6.10). Thus, the flocs were larger, compared to those formed with CS II, and their sedimentation was faster. A smaller amount of flocculant was enough to achieve the optimum dose. At the same time, the broadness of the flocculation window was enlarged and the NIPEC–kaolin flocs were well protected against redispersion due to flocculant characteristics, namely larger particle size, smaller flexibility, and the hydrophobicity caused by the polyanion. Figure 6.33 summarizes flocculation efficiency (optimum dose and window broadness) as a function of the molar ratio between charges n^-:n^+ for NIPECs as specialized flocculants in comparison with CS II. Three regions are evident in Figure 6.33a: (1) by using NIPECs with n^-:n^+ = 0.15, the optimum dose abruptly decreased, comparative to the optimum dose of CS II (dashed line); (2) for NIPECs prepared at molar ratio between charges ranging from 0.15 to 0.25, the optimum dose increased almost linearly until the values reached that of CS; (3) at a further increase of the molar ratio between charges, the amount of flocculant increased over the limit needed for optimum CS flocculation.

TABLE 6.10

Characteristics of NIPEC Dispersions

Polyion Pair	N^-:n^+	CD^a (meq/L)	$D_h{}^b$ (nm)	PI^c
P(AMPS$_{54}$-co-TBA$_{46}$)/CS II	0.15	0.47	288	0.254
	0.2	0.44	284	0.251
	0.3	0.39	267	0.246
	0.4	0.33	254	0.24
	0.5	0.29	244	0.231
P(AMPS$_{37}$-co-TBA$_{63}$)/CS II	0.15	0.51	326	0.31
	0.2	0.48	321	0.3
	0.3	0.43	311	0.28
	0.4	0.38	306	0.26
	0.5	0.33	302	0.25

[a] Particle charge density determined by PCD03 particle charge detector.
[b] Hydrodynamic diameter measured by DLS.
[c] Polydispersity index measured by DLS.

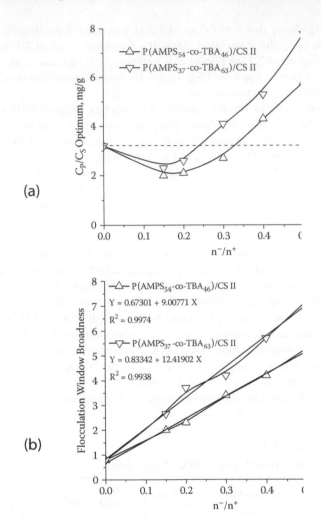

FIGURE 6.33 Influence of molar ratio (n^-/n^+) between charges on the optimum dose. C_P/C_S optimum (a) and flocculation window (b) for kaolin flocculation with different NPECs based on CS II (dashed line = C_P/C_S optimum for flocculation with CS II).

Regarding flocculation window broadness, Figure 6.33b shows an almost linear increase, for all NIPECs with an increase of molar ratio between charges—the most important increase being observed for the complex $P(AMPS_{37}\text{-co-TBA}_{63})/CS$ II over the whole investigated molar ratio domain. As observed by DLS (Table 6.10), the sizes of NIPEC particles decreased with the increase of the molar ratio n^-/n^+. The complex particles used in this study had hydrodynamic diameters (D_h) in the range of 240 to 330 nm. Increasing the molar ratio between charges decreased the charge density and sizes of the complex particles, leading to the adsorption of a higher number of flocculants on the particle surfaces.

The thicker layer of NIPEC particles and the higher hydrophobicity of the flocculant more efficiently protected the particles against redispersal, leading to the increase of the flocculation window. For the same molar ratio, the polyanion nature strongly influenced the flocculation efficiency; both optimum dose and flocculation window broadness increased with polyanion hydrophobicity.

The postulated mechanisms by which polyelectrolytes can cause flocculation are charge neutralization, bridging, and electrostatic patch. Charge neutralization arises from a reduction in the electric double layer repulsion between particles due to the adsorption of highly charged polyelectrolytes on oppositely charged particles. It is generally believed that low molar mass polymers tend to adsorb and neutralize opposite charges on the particles.[68] Bridging flocculation occurs when segments of the same polymer molecule are attached to more than one particle, thereby linking the particles together.

Electrostatic patch flocculation is operative for polyions of very high charge densities that interact with oppositely charged particles of low charge density. The net residual charge of the polymer patch on one particle surface can attach to the bare part of another oppositely charged particle.

Based on the results presented above, a flocculation mechanism with positively charged NIPECs based on CS was proposed in Scheme 6.6. The increase of the flocculation efficiency of NIPECs prepared at a molar ratio $n^-:n^+ = 0.15$ was most certainly caused by improved bridging ability of the NIPECs compared to single CS molecules, mainly due to larger sizes and higher densities of free positive charges (step I, Scheme 6.6 and step I, Scheme 6.5). Increasing the molar ratio between charges decreased the size and charge density of the complex particles; the NIPEC nanoparticles were adsorbed on the kaolin surface by electrostatic interactions, neutralizing the kaolin particle surfaces (step II, Scheme 6.6); the nonattached parts of the NIPECs nanoparticles may interact by an electrostatic patch mechanism with the bare part of another kaolin particle to form aggregates that are very difficult to restabilize.

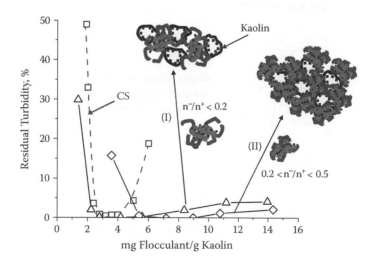

SCHEME 6.6 Residual turbidity of kaolin model suspension as function of ratio of flocculant and kaolin after destabilization with CS and CS-based NIPEC dispersions.

Strong electrolytes are present in most industrial effluents that must be processed by flocculation. It is thus important to evaluate the influence of ionic strength on the efficiency of kaolin flocculation with CS to establish a relationship between salt concentration and performance in the separation processes. It was reported that flocculation efficiency is strongly influenced by the ionic strength of the environment. The optimum dose decreases and the flocculation window broadens with increasing ionic strength.[66] Therefore, some measurements were performed at ionic strengths ranging from 0.01 to 0.05 M and the flocculation behavior was observed and compared for a one-component (CS) system and NIPEC systems.[75] Figure 6.34 shows the influence of ionic strength on the destabilization of the kaolin dispersions with CS II. Even a small concentration of NaCl (0.01 M) significantly influenced flocculation with CS II, the optimum dose sharply decreasing from 2.7 mg flocculant/g kaolin in salt-free solution to 1.3 mg flocculant/g kaolin in 0.01 M NaCl. This behavior may be explained by the greater flexibility of chitosan chains caused by the partial screening of ionic groups on the polymeric chains by excess counterions that may thus attach to more particles, connecting them in bigger flocs than appear in the absence of NaCl.

Flocculation of kaolin with NIPECs based on CS II and both copolymers P(AMPS-co-TBA) used as polyanions and prepared at a molar ratio between charges of $n^-{:}n^+ = 0.15$, in the presence of NaCl at different concentrations is presented in Figure 6.35. As Figure 6.35a shows, NIPEC performance in kaolin flocculation was enhanced when the complex particles were prepared in the presence of NaCl. The optimum dose decreased and the flocculation window broadness increased with the increase of ionic strength. The presence of NaCl has two main consequences: an increase of the chain flexibility of CS in excess and a decrease of complex particle sizes.[48,53] Both effects lead to a more effective interaction with kaolin particles. The level of these effects depends on the characteristics of the complementary polyelectrolytes.

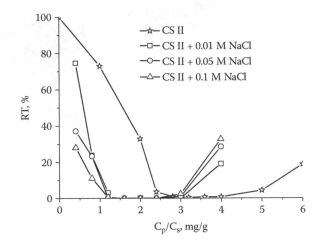

FIGURE 6.34 Effect of NaCl concentration on flocculation of kaolin dispersion by CS II.

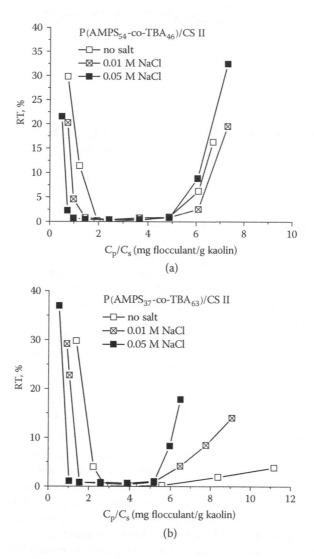

FIGURE 6.35 Residual turbidity (RT) as function of polymer per substrate (C_p/C_s, mg/g) in kaolin flocculation with NIPECs based on CS II and P(AMPS-co-TBA) copolymers at a molar ratio between charges of 0.15 in the presence of NaCl at different concentrations.

Thus, as Figure 6.35b shows, when P(AMPS$_{37}$-co-TBA$_{63}$) is the added polyanion, the NIPEC particles prepared in the presence of NaCl mainly decreased the optimum dose of flocculation. The presence of NaCl may also influence particle-to-particle adhesion, which becomes weaker, increasing the adsorption capacity of the positively charged CS (or positively charged NIPECs) on the negatively charged kaolin faces.

In conclusion, NIPEC nanoparticles were more effective than CS for kaolin separation, especially at low molar ratios between charges when the broadness of the flocculation window more than doubled at an optimum dose lower than that of CS. The main advantage of NIPECs was the increase of critical concentration for kaolin restabilization. The NIPEC particles adsorbed on the kaolin surface protected them more efficiently against redispersion. The efficiency of NIPECs in kaolin flocculation was enhanced when the complex particles were prepared in the presence of NaCl.

6.5 MULTICOMPONENT IONIC SYSTEMS BASED ON SYNTHETIC POLYCATIONS AND THEIR FLOCCULATION EFFICIENCIES

The use of polycations in fields like environmental technologies requires a clear understanding of their behavior in the presence of various organic and inorganic low molar mass compounds. Polyion conformation, a very important characteristic in the adsorption and flocculation by polyelectrolytes, is strongly influenced by their intrinsic characteristics such as molar mass and charge density, hydrophilic–hydrophobic balance, counterion valence, and polarizability.

Among the synthetic polycations, nitrogen-based polycations are important—alone or in specialized systems—for drinking water treatment and coagulation and flocculation processes in wastewater remediation.[62,64,67,68] Because low molar mass salts are often involved in such applications, it is vital to know how they influence the behaviors of polycations in solution. NIPECs based on polycations bearing quaternary ammonium centers in side chains and ionic and nonionic copolymers of AMPS will be also evaluated as flocculants in this section.

6.5.1 SYNTHETIC POLYCATIONS IN PRESENCE OF DIVALENT COUNTERIONS

We know that the interaction between polyelectrolytes and multicharged counterions constitutes a route for preparing ionically cross-linked complex systems. In our previous studies on the interactions of strong polycations and counterions, polycations containing ammonium quaternary centers in their backbones were investigated in the presence of monovalent, divalent, and trivalent counterions.[76,77] The interactions between divalent counterions like sulfate and persulfate and strong polycations bearing ammonium quaternary centers attached to macromolecular chains were recently investigated by our group.[78] Polycations were prepared from poly(N,N-dimethylaminoethylmethacrylate) (PDMAEM) by quaternization with benzyl chloride (BC) (Q_x polycations).

The effects of polyion charge density and the concentrations of $S_2O_8^{2-}$ and SO_4^{2-} anions on the behavior of polycations in aqueous solution were investigated by viscometry and AFM. Information about the morphology of the complexes formed by polycations and $S_2O_8^{2-}$ anions in a dried state was revealed by AFM. The destabilization of kaolin from a model suspension and the removal of Congo Red (CR) with Q_x polycations compared with PCA_5 cations having ammonium quaternary salt centers in their backbones was also investigated in the presence and absence of divalent counterions.

PDMAEM was synthesized by free radical polymerization. PDMAEM with a molar mass M_w = 136,000 g/mol and M_w/M_n = 1.97 was used in this study. Quaternization of PDMAEM with BC was performed as previously shown.[79] The chemical structure of Q_x polycations is presented in Scheme 6.7, and some characteristics are summarized in Table 6.11. The polycation PCA$_5$ (Scheme 6.2) was synthesized and purified according to Dragan and Ghimici.[39]

Polyelectrolyte solutions with different concentrations were prepared in a salt-free aqueous solution or in salt solutions with concentrations corresponding to a certain molar ratio $C_s:C_p$, where C_s is the concentration of salt in meq/L and C_p is the concentration of polycation in meq/L, taking into account the molar mass of the repeat unit in grams per charge (Table 6.11). Na$_2$SO$_4$ (SS), (NH$_4$)$_2$SO$_4$ (AS), and (NH$_4$)$_2$S$_2$O$_8$ (APS) were used as sources of divalent counterions. Viscometric measurements were performed with an Ubbelohde viscometer with internal dilution at 25°C.

Silicon wafer substrates used for AFM measurements were cleaned as described in Section 6.4.1. The drop deposition method was used to prepare the samples (0.02 mL of polycation solutions, with or without divalent counterions), which were deposited on clean silicon wafers that were air-dried at room temperature in a dust-free

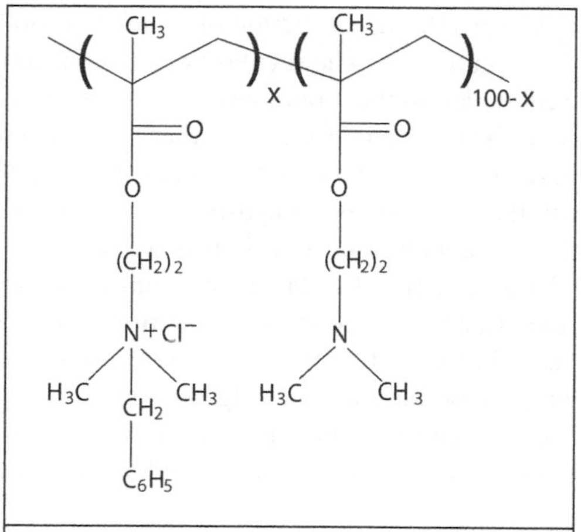

SCHEME 6.7   Structures of Q_x polycations.

TABLE 6.11

Characteristics of Polycations Used in This Study

Polycation	Quaternization Degree (x)	M$_u$ (g/charge)	Linear Charge Density
Q$_{50}$	50	220.5	0.5
Q$_{85}$	80	264.5	0.85
PCA$_5$[a]	–	140.35	0.95

[a] $[\eta] = _{1\ M\ NaCl},\ 25°C = 0.46$ dL/g.

environment for ~48 h. Two silicon wafers were prepared and at least ten images were obtained for each sample. The images were examined by scanning probe microscope software (Gwyddion 2.15) to measure the average roughness (R_a), and maximum height (h_{max}) of the adsorbed species.

Destabilization experiments of kaolin model suspension were conducted at room temperature under the same conditions cited in Section 6.4.2. Residual turbidity (RT) was calculated with Equation (6.6). The UV-Vis spectra of CR in the presence of polycations Q_{85} and PCA_5 were recorded using a Specord M42 spectrophotometer. Increasing volumes of polycation solution at a concentration of 2 meq/L, without or with low molar mass salt, were added to a 50 mL dye solution at a concentration of 2×10^{-4} M under stirring. The stirring continued for 10 min and the residual concentration of CR was determined by measuring the absorbance in the supernatant at the characteristic maximum of CR, 497 nm, after 24 h.

As a consequence of counterion binding, the solubility of polyelectrolyte decreased because the counterions are more strongly bound to the ionic centers. The reduced viscosity (η_{sp}/C_p) profiles as a function of the C_s:C_p ratio for polycations Q_{50} and Q_{85} is illustrated in Figure 6.36. The increase of salt concentration at a constant concentration of polycation led to the monotonous decrease of the reduced viscosity values. The exchange of the C$-$ counterions by other halide counterions in the case of polycations containing cationic centers in their backbones showed that counterion binding increased with the decrease of the hydrated counterion radius.[76]

Conversely, the divalent counterions used in this study ($S_2O_8^{2-}$, SO_4^{2-}) can generate both intramolecular and intermolecular interactions as ionotrop cross links. As Figure 6.36 shows, the strength of the interactions of Q_x polycations and divalent counterions was strongly influenced by the counterion nature and polycation charge density, correlated with the probability as bridges to be formed. The specific interactions between polycation charges and divalent counterions is evidenced by the reduced viscosity as a function of the ratio C_s:C_p, which was lower for $S_2O_8^{2-}$ than

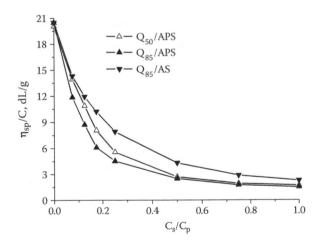

FIGURE 6.36 Reduced viscosity as function of C_s:C_p ratio for two polycations derived from PDMAEM. C_p = 2 meq/L. C_s= salt concentration (meq/L). Temperature = 25°C.

for SO_4^{2-}. Thus, the interaction between $S_2O_8^{2-}$ counterions and quaternary ammonium centers was stronger than in the case of SO_4^{2-}, even if both counterions are divalent.

AFM allowed the investigation at nanoscale level of the polycation morphology in the presence of divalent counterions adsorbed on planar substrates of silicon wafers as a function of polycation concentration and the ratio between salt and polycation ($C_s{:}C_p$). Figure 6.37 shows tapping mode amplitude AFM images of the polycation Q_{85} adsorbed on silicon wafers at two polymer concentrations, 2 meq/L (a and c) and 20 meq/L (b and d), in the presence of $S_2O_8^{2-}$. The ratios $C_s{:}C_p$ were 0.5 (a and b) and 1.0 (c and d). From each height AFM image, the maximum heights h_{max}, of the adsorbed structures and the surface average roughness, R_a (arithmetic average of vertical deviations of a real surface from its ideal form) were determined. The electrostatic interaction between divalent counterions and polycations reduces the charge density of the polyions and leads to a lower intramolecular repulsive interaction and also to ionic cross links. As Figures 6.37a and c show, at low polymer concentrations, the hemispherical domains are compact and small. Increasing the polycation concentration made the adsorbed layer on the surface smoother than that observed at $C_p = 2$ meq/L, irrespective of the salt concentration, with R_a and h_{max} values always lower than those found at low polycation concentrations (Figures 7.37b and d).

The results of the destabilization of a kaolin model suspension with polycation Q_{85} (Figure 6.38a) and with PCA_5 (Figure 6.38b) in salt-free aqueous solution in the presence of APS or SS show the influence of the counterion nature and the $C_s{:}C_p$ ratio on the flocculation efficiencies of both polycations. In the absence of electrolyte, polycation Q_{85} appears more efficient than PCA_5, both at an optimum

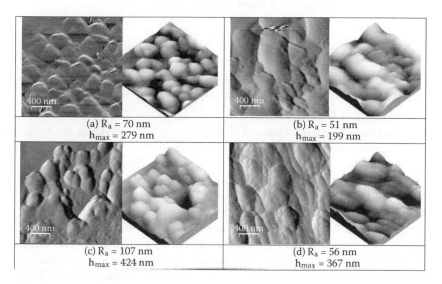

(a) $R_a = 70$ nm $h_{max} = 279$ nm	(b) $R_a = 51$ nm $h_{max} = 199$ nm
(c) $R_a = 107$ nm $h_{max} = 424$ nm	(d) $R_a = 56$ nm $h_{max} = 367$ nm

FIGURE 6.37 Tapping mode amplitude and three-dimensional height AFM images of Q_{85} polycation adsorbed on silicon wafers at two polymer concentrations: 2 meq/L (a, and c) and 20 meq/L (b, and d). $C_s{:}C_p$ ratios = 0.5 (a and b) and 1.0 (c and d).

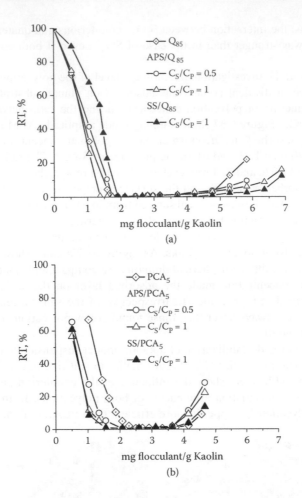

FIGURE 6.38 Residual turbidity (RT) as function of polymer per substrate (mg/g) for kaolin flocculation with polycations Q_{85} (a) and PCA_5 (b) in the absence and presence of salts.

flocculation dose (1.6 mg/g compared to 2.1 mg/g, respectively) and for broadness of the flocculation window (3.2 for Q_{85} and 2.2 for PCA_5). These differences may be attributed to the mobility of the cationic centers in the side chains of Q_{85} and the hydrophobic interactions of the benzyl substituents that help increase the flocculation performances of Q_{85} compared to PCA_5.

Adding APS led to a significant decrease of optimum flocculation dose and enlargement of the flocculation window at a C_s:C_p of 1 when Q_{85} was used as the flocculant. The values of these parameters decreased to 1.35 mg/g and increased to 3.85, respectively (Figure 6.38a). The influence of APS was less important when PCA_5 was used as the flocculant; only a small decrease of the optimum flocculation dose was observed (Figure 6.38b). Conversely, SS at a C_s:C_p of 1 shifted the optimum flocculation dose to a higher value (1.9 mg/g) and enlarged the broadness of the flocculation window up to 4.4 with polycation Q_{85} (Figure 6.38a). With PCA_5, the positive effect of SS at a C_s/C_p ratio of 1 was observed as a decrease of optimum flocculation dose (1.3 mg/g) and enlargement of flocculation window broadness up to 2.9 (Figure 6.38b).

To explain the differences among the polycations selected for the destabilization of kaolin dispersion in the presence of divalent counterions, the structural differences of both the divalent counterions and the counterions should be considered. The meeting of both requirements for better flocculation efficiency observed for Q_{85} in the presence of APS at $C_s/C_p = 1$ and for PCA_5 in the presence of SS at the same ratio was similar to improvements noted for NIPECs formed by CS and ionic and nonionic copolymers P(AMPS-co-TBA), at a ratio between charges up to 0.15.[75] The results obtained for the Q_{85} pendant-type polycations obtained in the presence of APS at a concentration of 2 meq/L may be explained by the formation of an ionically cross-linked system in which more polycation chains are involved in the formation of a loosely cross-linked hydrogel. Such structures were characterized by greater roughness and maximum height compared to those obtained at a higher concentration (20 meq/L), as shown in AFM images (Figure 6.37c and d). Due to the flexibility of the spacer between the backbones and cationic centers and the low concentration, the probability of cationic centers to interact with anionic charges of persulfate anions decreased. These systems contain enough free positive charges to recommend them as effective agents for separating negatively charged matter. For PCA_5 molecules having cationic centers in their backbones, the formation of an ionically intermacromolecular cross-linked gel was most likely with SS.

The removal of CR was followed by a comparison of polycations Q_{85} and PCA_5 in the absence of electrolyte and in the presence of APS or SS at a C_s/C_p ratio of 1. Figure 6.39 shows the variation of relative absorbance (A_c/A_i, where A_c and A_i represent absorbances after and, respectively, before the addition of polycation) as a function of the molar ratio between the polycation and the dye ([P]/[D]). As Figure 6.39a shows, the complex stoichiometry with polycation Q_{85} alone was located at around 2, as expected. The stoichiometry shifted at about 1.6 by the addition of APS, but remained almost unchanged when SS was added. The positive influence of both salts was the greater complex stability reflected in the larger flocculation window, i.e., a higher [P]/[D] molar ratio was necessary to increase the relative absorbance (A_c/A_i) after the complex stoichiometry, compared with polycation Q_{85} alone. The removal of the dye by PCA_5 alone was very difficult, but by adding APS, the complex stoichiometry was evident at about 1.25 (Figure 6.39b). The dye was completely removed in the presence of SS. The difference was the higher [P]/[D] molar ratio corresponding to the stoichiometry in the presence of SS and the higher complex stability observed in the presence of APS.

It was thus demonstrated that the interaction of strong polycations that possess ammonium quaternary centers attached to acrylic macromolecular chains, Q_x polycations, with divalent counterions such as SO_4^{2-} and $S_2O_8^{2-}$, produced novel systems that more effectively removed kaolin in model dispersions than polycations without salts.

A simultaneous decrease of the optimum flocculation dose and enlargement of flocculation window broadness in the destabilization of kaolin model dispersion were observed for Q_{85} in the presence of APS, and for PCA_5 in the presence of SS, both at a $C_s:C_p$ of 1. The best results for CR removal were obtained with both polycations in the presence of APS—the complex stoichiometry located at lower [P]/[D] and the complex stability enhanced compared with polycations in the absence of salt.

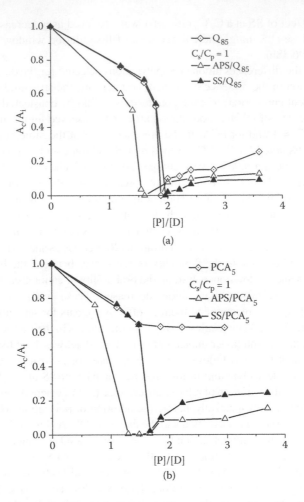

FIGURE 6.39 Removal of Congo Red from aqueous solution with polycations Q_{85} (a) and PCA_5 (b), in the absence and presence of salts.

6.5.1 NONSTOICHIOMETRIC COMPLEX NANOPARTICLES BASED ON SYNTHETIC POLYCATIONS AS FLOCCULANTS

NIPECs as colloidal dispersions bearing positive charges in excess were prepared by the interactions of polycation Q_{85} in excess and two ionic and nonionic copolymers of AMPS with TBA, $P(AMPS_{54}\text{-co-}TBA_{46})$ or methyl methacrylate (MM) $P(AMPS_{52}\text{-co-}MM_{48})$, as polyions in default (added polyions).[80] The characteristics of $P(AMPS_{54}\text{-co-}TBA_{46})$ are presented in Table 6.8. The characteristics of $P(AMPS_{52}\text{-co-}MM_{48})$ were as follows: $M_v = 285$ kDa, $M_u = 322.05$ g/charge, average distance between charges (b) = 0.482 nm, and charge density (CD) = 0.52. The concentration of polycation was adjusted to 0.5 mM and that of polyanions to 5 mM, taking into account the molar mass of the repeat unit in grams per charge (see Tables 6.8 and 6.11).

The molar ratio between charges varied from 0.05 to 0.4. The particle sizes measured by DLS for the NIPEC dispersions as a function of the molar ratio between charges are plotted in Figure 6.40. A small decrease of particle size with the increase of the molar ratio between charges in the range 0.1 to 0.4 is obvious. Particle sizes were lower when $P(AMPS_{52}\text{-co-}MM_{48})$ was the added polyion. However, the difference between the highest size ($n^-{:}n^+ = 0.1$) and the lowest ($n^-{:}n^+ = 0.4$) was around 15 nm—very small compared with the differences found for CS-based NIPECs (Figure 6.28a).

Based on the results of flocculation tests with CS-based NIPECs (Figure 6.32), NIPEC dispersions prepared at $n^-{:}n^+$ molar ratios of 0.05 and 0.1 were selected for flocculation tests with NIPECs based on polycation Q_{85}. To evaluate and compare flocculation performances, an arbitrary RT limit of 5% was fixed; under this value the flocculant was considered effective.

Figure 6.41 shows that, even if the particle sizes were influenced by the polyanion structure, the flocculation efficiency was mainly influenced by the ratio between charges, i.e., by the number of free positive charges available for interaction with kaolin particles. Thus, at $n^-{:}n^+ = 0.05$ (Figure 6.41a) the optimum flocculation dose was almost identical with that of polycation alone (1.7 and 1.6 mg flocculant/g kaolin, respectively), but the broadness of the flocculation window increased from 4.9 to 6.1 mg flocculant/g kaolin when NIPECs were used as flocculants; the values were almost the same for both polyanions. The increase of the ratio between charges up to 0.1 caused the increase of the optimum flocculation dose up to 2.6 mg flocculant/g kaolin and to a spectacular enlargement of the flocculation window broadness up to 9.5 mg flocculant/g kaolin for NIPECs with $P(AMPS_{54}\text{-co-}TBA_{46})$ and 10.5 mg flocculant/g kaolin for NIPECs with $P(AMPS_{52}\text{-co-}MM_{48})$ (Figure 6.41b).

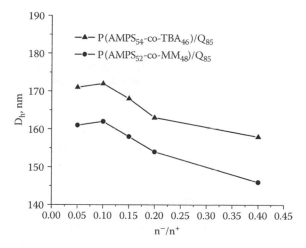

FIGURE 6.40 Particle size (D_h) as function of $n^-{:}n^+$ molar ratio between charges in formation of NIPECs based on polycation Q_{85} in excess and two ionic and nonionic copolymers as added polyanions.

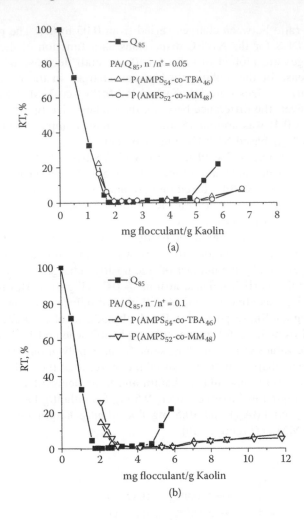

FIGURE 6.41   Residual turbidity (RT) as function of polymer per substrate (C_P/C_S, mg/g) for kaolin flocculation with NIPECs based on polycation Q_{85} and two ionic and nonionic copolymers.

Thus, the NIPEC nanoparticles prepared from polycations bearing cationic centers in side chains (e.g., polycation Q_{85}) with very low n^-/n^+ ratios between charges ranging from 0.05 to 0.1 may be considered potential specialized flocculants.

6.6   CONCLUSIONS AND OUTLOOK

This chapter summarizes recent advances in the field of separation processes based on natural and synthetic polycations in multicomponent ionic systems. Novel CS-based composites with enhanced adsorption capacities for Cu^{2+}, Co^{2+}, and Ni^{2+} may be prepared by loading CPL microparticles in a matrix of cross-linked CS. The assembly is stabilized by a "tandem" ionic–covalent cross linking.

The presence of CPL in the composites strongly affects their adsorption capacities for metal ions compared with cross-linked CS without zeolite. The adsorption capacity abruptly increased up to a CPL content of about 20 wt.-%, and decreased with further increases of CPL content. The maximum adsorption capacities were 9.04 mmol/g for Cu^{2+}, 6.4 mmol/g for Ni^{2+}, and 3.4 mmol/g for Co^{2+}. The adsorptions of all metal ions obeyed pseudo second order kinetics, confirming that chemisorption is the rate-determining step.

CS-based NIPEC nanoparticles were prepared with ionic and nonionic copolymers of AMPS as the added polyions. These materials were more effective than CS alone for separating kaolin from a model dispersion, especially at low molar ratios between charges, when the broadness of the flocculation window more than doubled at an optimum dose lower than that of CS. The main advantage of NIPECs was the increase of critical concentration for kaolin restabilization. The NIPEC particles adsorbed on the kaolin surface protected them more efficiently against redispersion. The efficiency of NIPECs in kaolin flocculation was enhanced when the complex particles were prepared in the presence of NaCl and the optimum dose decreased when the NIPECs were prepared in the presence of NaCl.

The interactions among strong polycations possessing quaternary ammonium centers attached to acrylic macromolecular chains with divalent counterions such as SO_4^{2-} and $S_2O_8^2$ led to the preparation of novel multicomponent ionic systems that were more effective in removing kaolin in model dispersion than polycations without salts. Simultaneous decreases of the optimum flocculation dose and enlargements of flocculation window broadness in the destabilization of model kaolin dispersion were observed for polycation Q_{85} in the presence of APS at a $C_s:C_p$ ratio of 1.

The best results for CR removal were obtained in the presence of APS. The complex stoichiometry occurred at lower values of [P]/[D] and the complex stability was enhanced compared to polycations in the absence of salt. NIPEC nanoparticles prepared from polycations bearing cationic centers in their side chains (e.g., polycation Q_{85}) at a very low (0.05 to 0.1) $n^-:n^+$ ratios between charges may be considered specialized flocculants with potential for removing negatively charged matter. These results reveal many possibilities for enhancing the efficiency of natural and synthetic polycations in separation processes if their behavior in multicomponent systems is determined and controlled. Further investigations of such systems may provide information for developing other cost-effective systems.

REFERENCES

1. Baraka, A., Hall, P.J., and Heslop, M.J. Preparation and characterization of melamine–formaldehyde–DTPA chelating resin and its use as an adsorbent for heavy metals removal from wastewater. *React. Funct. Polym.* 67, 585–600, 2007.
2. Dinu, M.V. and Dragan, E.S. Heavy metals adsorption on some iminodiacetate chelating resins as a function of the adsorption parameters. *React. Funct. Polym.* 68, 1346–1354, 2008.
3 Dinu, M.V., Dragan, E.S., and Trochimczuk, A.W. Sorption of Pb(II), Cd(II) and Zn(II) by iminodiacetate chelating resins in non-competitive and competitive conditions. *Desalination* 249, 374–379, 2009.

4. Chauhan, K., Chauhan, G.S., and Ahn, J.H. Synthesis and characterization of novel guar gum hydrogels and their use as Cu^{2+} sorbents. *Bioresour. Technol.* 100, 3599–3603, 2009.
5. Chen, A.H., Liu, S.C., and Chen, C.-Y. Comparative adsorption of Cu(II), Zn(II), and Pb(II) ions in aqueous solution on the cross linked chitosan with epichlorohydrin, *J. Hazard. Mater.* 154, 184–191, 2008.
6. Crini, G. Recent developments in polysaccharide-based materials used as adsorbents in wastewater treatment. *Prog. Polym. Sci.* 30, 38–70, 2005.
7. Guibal, E. Interactions of metal ions with chitosan-based sorbents: a review. *Sep. Purif. Technol.* 38, 43–74, 2004.
8. Osifo, P.O., Webster, A., van der Merve, H. et al. Influence of the degree of cross-linking on adsorption properties of chitosan beads. *Bioresour. Technol.* 99, 7377–7382, 2008.
9. Argun, M.E. Use of clinoptilolite for the removal of nickel ions from water: Kinetics and thermodynamics. *J. Hazard. Mater.* 150, 587–595, 2008.
10. Günay, A., Arslankaya, E., and Tosun, I. Lead removal from aqueous solution by natural and pretreated clinoptilolite: Adsorption equilibrium and kinetics. *J. Hazard. Mater.* 146, 362–371, 2007.
11. Bedelean, H., Stanca, M., Măicăneanu, A. et al. Zeolitic volcanic tuffs from Măcicaş (Cluj County): Natural raw materials used for NH_4^+ removal from wastewaters. *Stud. Univ. Babes-Bolyai Geol.* 52, 43–49, 2006.
12. Chang, Y.C., Chang, S.W., and Chen, D.H. Magnetic chitosan nanoparticles: Studies on chitosan binding and adsorption of Co (II) ions. *React. Funct. Polym.*, 66, 335–341, 2006.
13. Copello, G.J., Varela, F., Martinez-Vivot, R. et al. Immobilized chitosan as biosorbent for the removal of Cd(II), Cr(III) and Cr(VI) from aqueous solutions, *Bioresour. Technol.* 99, 6538–6544, 2008.
14. Dragan, E.S. and Dinu, M.V. Removal of copper ions from aqueous solution by adsorption on ionic hybrids based on chitosan and clinoptilolite. *Ion Ex. Lett.* 2, 15–18, 2009.
15. Dragan, E.S., Mihai, M., and Dinu, M.V. Separations with composite materials based on natural and synthetic polycations, In *Proceedings of 24th International Symposium on Physicochemical Methods of Separation.* Ars Separatoria 2009, 41–47.
16. Dragan, E.S., Dinu, M.V., and Timpu, D. Preparation and characterization of novel composites based on chitosan and clinoptilolite with enhanced adsorption properties for $Cu^{2+.}$ *Bioresour. Technol.* 101, 812–817, 2010.
17. Gamzazade, A.I., Shimac, V.M., Skljar, A.M. et al. Investigation of the hydrodynamic properties of chitosan solutions. *Acta Polym.* 36, 420–424, 1985.
18. Brugnerotto, J., Lizardi, J., Goycoolea, F.M. et al. An infrared investigation in relation with chitin and chitosan characterization. *Polymer* 42, 3569–3580, 2001.
19. Vadivelan, V. and Kumar, K.V. Equilibrium, kinetics, mechanism, and process design for the sorption of methylene blue onto rice husk. *J. Colloid Interface Sci.* 286, 90–100, 2005.
20. Kumar, K.V. and Sivanesan, S. Selection of optimum sorption kinetics: Comparison of linear and non-linear method, *J. Hazard. Mater.* 134B, 277–279, 2006.
21. Ho, Y.S. Second-order kinetic model for the sorption of cadmium onto tree fern: A comparison of linear and non-linear methods. *Water Res.* 40, 119–125, 2006.
22. Weber, J.W.J. and Morris, J.C. Kinetics of adsorption on carbon from solution. *J. Sanit. Eng. Div. Am. Soc. Civil Eng.* 89, 31–60, 1963.
23. Cao, Z., Ge, H., and Lai, S. Studies on synthesis and adsorption properties of chitosan cross-linked by glutaraldehyde and Cu(II) as template under microwave irradiation. *Eur. Polym. J.* 37, 2141–2143, 2001.
24. Ding, S., Zhang, X., Feng, X. et al. Synthesis of N,N′-diallyl dibenzo 18-crown-6-crown ether cross linked chitosan and their adsorption properties for metal ions. *React. Funct. Polym.* 66, 357–363, 2006.

25. Lee, S.T., Mi, F.L., Shen, Y.J. et al. Equilibrium and kinetic studies of copper(II) ion uptake by chitosan tripolyphosphate chelating resin, *Polymer* 42, 1879–1892, 2001.

26. Wan Ngah, W.S., Endud, C.S., and Mayanar, R. Removal of copper(II) ions from aqueous solution onto chitosan and cross-linked chitosan beads. *React. Funct. Polym.* 50, 181–190, 2002.

27. Navarro, R., Guzman, J., Saucedo, I. et al. Recovery of metal ions by chitosan: sorption mechanisms and influence of metal speciation. *Macromol. Biosci.* 3, 552–561, 2003.

28. Sprynskyy, M., Buszewski, B., Terzyk, A.P. et al. Study of the selection mechanism of heavy metal (Pb^{2+}, Cu^{2+}, Ni^{2+}, and Cd^{2+}) adsorption on clinoptilolite. *J. Colloid Interface Sci.* 304, 21–28, 2006.

29. Steenkamp, G.C., Keizer, K., Noeomagus, H.W. et al. Copper (II) removal from polluted water with alumina–chitosan composite membranes. *J. Membr. Sci.* 197, 147–156, 2002.

30. Wan Ngah, W.S., and Fatinathan, S. Adsorption of Cu(II) ions in aqueous solution using chitosan beads, chitosan–GLA beads and chitosan–alginate beads. *Chem. Eng. J.* 143, 62–72, 2008.

31. Li, N. and Bai, R. Copper adsorption on chitosan–cellulose hydrogel beads: Behaviors and mechanisms. *Sep. Purif. Technol.* 42, 237–247, 2005.

32. Coelho, T.C., Laus, R., Mangrich, A.S. et al. Effect of heparin coating on epichlorohydrin cross-linked chitosan microspheres on the adsorption of copper(II) ions, *React. Funct. Polym.* 67, 468–475, 2007.

33. Warner, M.G., and Hutchison, J.E. In *Synthesis, Functionalization and Surface Treatment of Nanoparticles*. Baraton, M.I., Ed. American Scientific, San Francisco, 2003, 67–89.

34. Decher, G. Fuzzy nanoassemblies: Toward layered polymeric multicomposites. *Science* 277, 1232–1237, 1997.

35. Dragan, E.S. and Bucatariu, F. In *New Trends in Ionic (Co)polymers and Hybrids*. Dragan, E.S., Ed. Nova Science, New York, 2007, 165–234.

36. Toutianoush, A., Krasemann, L., and Tieke, B. Polyelectrolyte multilayer membranes for pervaporation separation of alcohol/water mixtures. *Colloids Surf. A* 198–200, 881–889, 2002.

37. Meier-Haack, J., Lenk, W., Lehmann, D. et al. Pervaporation separation of alcohol/water mixtures using composite membranes based on polyelectrolyte multilayer assemblies. *J. Membr. Sci.* 184, 233–243, 2001.

38. Dragan, E. S., Mihai, M., Schauer, J. et al. PAN composite membrane with different solvent affinities controlled by surface modification methods. *J. Polym. Sci. A Polym. Chem.* 43, 4161–4171, 2005.

39. Dragan, S. and Ghimici, L. Cationic polyelectrolytes XI. Polymers with quaternary N-atoms in the main chain obtained by condensation polymerization of epichlorohydrin with amines. *Angew. Makromol. Chem.* 192, 199–211, 1991.

40. Hara, M., Lee, A.H., and Wu, J. Solution properties of ionomers: Counterion effect. *J. Polym. Sci. Polym. Phys.* 25, 1407–1418, 1987.

41. Schlenoff, J.B., Ly, H., and Li, M. Charge and mass balance in polyelectrolyte multilayers. *J. Am. Chem. Soc.* 120, 7626–7634, 1998.

42. Michaels, A.S. and Miekka, R.G. Polycation–polyanion complexes: Preparation and properties of poly-(vinylbenzyltrimethylammonium) poly-(styrenesulfonate). *J. Phys. Chem.* 65, 1765–1773, 1961.

43. Voycheck, C.L. and Tan, J.S. Ion-containing polymers and their biological interactions. In *Polyelectrolytes: Science and Technology*. Hara, M., Ed. Marcel Dekker, New York, 1992, 299–388.

44. Dautzenberg, H., Jaeger, W., Köetz, J. et al. *Polyelectrolytes: Formation, Characterization and Application*. Karl Hanser Verlag, Munich, 1994, 248–271.

45. Dragan, S., Cristea, M., Luca, C. et al. Polyelectrolyte complexes I. Synthesis and characterization of insoluble polyanion–polycation complexes. *J. Polym. Sci. A Polym. Chem.* 34, 3485–3494, 1996.

46. Dragan, S., Dragan, D., Cristea, M. et al. Polyelectrolyte complexes II. Specific aspects of formation of polycation/dye/polyanion complexes. *J. Polym. Sci. A Polym. Chem.* 37, 409–418, 1999.

47. Dautzenberg, H. Polyelectrolyte complex formation in highly aggregating systems 1. Effect of salt–polyelectrolyte complex formation in the presence of NaCl. *Macromolecules* 30, 7810–7815, 1997.

48. Dragan, S. and Cristea, M. Influence of low-molecular-weight salts on the formation of polyelectrolyte complexes based on polycations with quaternary ammonium salt groups in the main chain and poly(sodium acrylate). *Eur. Polym. J.* 37, 1571–1575, 2001.

49. Dragan, S. and Cristea, M. Polyelectrolyte complexes IV. Interpolyelectrolyte complexes between some polycations with N,N-dimethyl-2-hydroxypropyleneammonium chloride units and poly(sodium styrene sulfonate) in dilute aqueous solution. *Polymer* 43, 55–62, 2002.

50. Dragan, S. and Cristea, M. In *Recent Research Developments in Polymer Science*, Vol. 7. Gayathri, A., Ed. Transworld, Trivandrum, India, 2003, 149–181.

51. Kabanov, V.A. and Zezin, A.B. Soluble interpolymeric complexes as a new class of synthetic polyelectrolytes. *Pure Appl. Chem.* 56, 343–354, 1984.

52. Kabanov, V.A. and Kabanov, A.V. Supramolecular devices for targeting DNA into cells: Fundamentals and perspectives, *Macromol. Symp.* 98, 601–613, 1995.

53. Buchhammer, H.M., Petzold, G., and Lunkwitz, K. Salt effect on formation and properties of interpolyelectrolyte complexes and their interactions with silica particles. *Langmuir* 15, 4306–4310, 1999.

54. Mende, M., Schwarz, S., Petzold, G. et al. Destabilization of model silica dispersions by polyelectrolyte complex particles with different charge excess, hydrophobicity, and particle size. *J. Appl. Polym. Sci.* 103, 3776–3784, 2007.

55. Dragan, S. Interpolyelectrolyte complexes as colloidal dispersions. *Ann. West Univ. Tim. Ser. Chem.* 12, 871–878, 2003.

56. Dragan, S. and Schwarz, S. Polyelectrolyte complexes. VI. Polycation structure, polyanion molar mass and polyion concentration effects on complex nanoparticles based on poly(sodium 2-acrylamido-2-methylpropanesulfonate), *J. Polym. Sci. Part A: Polym. Chem.*, 42, 2495-2505, 2004.

57. Dragan, E.S. and Schwarz, S. Polyelectrolyte complexes VII. Complex nanoparticles based on poly(sodium 2-acrylamido-2-methylpropanesulfonate) tailored by titrant addition rate. *J. Polym. Sci. A Polym. Chem.* 42, 5244–5252, 2004.

58. Schwarz, S. and Dragan, E.S. Nonstoichiometric polyelectrolyte complexes as colloidal dispersions based on NaPAMPS and their interactions with colloidal silica particles. *Macromol. Symp.* 210, 185–191, 2004.

59. Dragan, E.S., Mihai, M., and Schwarz, S. Polyelectrolyte complex dispersions with a high colloidal stability controlled by the polyion structure and titrant addition rate *Colloids Surf. A* 290, 213–221, 2006.

60. Mihai, M. and Dragan, E.S. Formation and colloidal stability of some polyelectrolyte complex dispersions based on random copolymers of AMPS. *Rev. Roum. Chim.* 52, 267–273, 2007.

61. Mihai, M. and Dragan, E.S. Polyelectrolyte complex nanoparticles investigated by dynamic light scattering and atomic force microscopy. In *Progress in Nanoscience and Nanotechnology*, Vol. 11, Series on Micro and Nanoengineering, Academic Press, 2007, 58–67.

62. Dragan, E.S., Ghimici, L., and Mihai, M. Pollutant removal with synthetic polycations and their complexes. *Env. Eng. Mgt. J.* 5, 625–634, 2006.

63. Mihai, M., Dragan, E.S., Schwarz, S. et al. Dependence of particle sizes and colloidal stability of polyelectrolyte complex dispersions, etc. *J. Phys. Chem. B* 111, 8668-8675, 2007.

64. Ghimici, L., Dinu, I.A., and Dragan, E.S. In *New Trends in Ionic (Co)Polymers and Hybrids*. Nova Science, New York, 2007, 31–64.

65. Dragan, E.S. and Mihai, M. Polyanion structure and mixing conditions: Useful tools to tailor characteristics of polyelectrolyte complex particles. *J. Optoel. Adv. Mater.* 9, 3927–3932, 2007.

66. Laue, C. and Hunkeler, D. Chitosan-graft-acrylamide polyelectrolytes: Synthesis, flocculation, and modeling. *J. Appl. Polym. Sci.* 102, 885–896, 2006.

67. Lupascu, T., Dranca, I., Sandu, M. et al. Study of some flocculating processes with cationic polyelectrolytes. *Angew. Makromol. Chem.* 220, 11–19, 1994.

68. Dragan, S., Maftuleac, A., Dranca, I. et al. Flocculation of montmorillonite by some hydrophobically modified polycations containing quaternary ammonium salt groups in the backbone. *J. Appl. Polym. Sci.* 84, 871–876, 2002.

69. Mihai, M., Dragan, E.S., and Ghimici, L. Application of polycations in separation processes. *Env. Eng. Mgt. J.* 6, 416–423, 2007.

70. Renault, F., Sancey, B., Badot, P.M. et al. Chitosan for coagulation and flocculation processes: An eco-friendly approach. *Eur. Polym. J.* 45, 1337–1348, 2009.

71. Bratskaya, S., Avramenko, V., Schwarz, S. et al. Enhanced flocculation of oil-in-water emulsions by hydrophobically modified chitosan derivatives. *Colloids Surf. A* 275, 168–176, 2006.

72. Roussy, J., Van Vooren, M., and Guibal, E. Influence of chitosan characteristics on coagulation and flocculation of organic suspensions. *J. Appl. Polym. Sci.* 98, 2070–2079, 2005.

73. Dragan, E.S., Mihai, M., and Schwarz, S. Complex nanoparticles based on chitosan and ionic/nonionic strong polyanions. *ACS Appl. Mater. Interfaces* 1, 1231–1240, 2009.

74. Mihai, M., Dragan, E.S., Manolova, N. et al. Chitosan and its positive nonstoichiometric polyelectrolyte complexes as flocculants for fine suspensions. *Proceedings of Third Conference of International Congress of Chemistry and Environment*. Kuwait, November 2007, 387–392.

75. Mihai, M. and Dragan, E.S. Chitosan based nonstoichiometric polyelectrolyte complexes as specialized flocculants. *Colloids Surf. A* 346, 39–46, 2009.

76. Ghimici, L., Dragan, S., and Popescu, F. Interaction of low molecular weight salts with cationic polyelectrolytes. *J. Polym. Sci. Polym. Phys.* 35, 2571–2581, 1997.

77. Ghimici, L. and Dragan, S. Behaviour of cationic polyelectrolytes upon binding of electrolytes: effects of polycation structure, counterions and nature of solvent. *Colloid Polym. Sci.* 280, 130–134, 2002.

78. Dragan, E.S., Dinu, I.A., and Mihai, M. Conformational changes of strong polycations in the presence of divalent counterions and their influence upon the flocculation efficiency. *Colloids Surf. A* 348, 282–288, 2009.

79. Dragan, S., Petrariu, I., and Dima, M. Cationic Polyelectrolytes IV. Kinetic study of reactions of acrylic polymers containing tertiary amine groups with halogenated compounds in dipolar aprotic solvents. *J. Polym. Sci. Polym. Chem.* 19, 2881–2894,1981.

80. Dinu, I.A., Mihai, M, and Dragan, E.S. Comparative study on the formation and flocculation properties of polyelectrolyte complex dispersions based on synthetic and natural polycations. *Chem. Eng. J*, under review

7 Polymeric-Inorganic Hybrid Ion Exchangers

Preparation, Characterization, and Environmental Applications

Sudipta Sarkar, Prakhar Prakash, and Arup K. SenGupta

CONTENTS

ABSTRACT

Hybrid ion exchangers are composite materials with two phases. The first is a host containing a functionalized porous polymer and the second is the irreversible dispersion of metal oxide nanoparticles inside the host. The nanoparticles, due to their high surface area-to-volume ratio, offer fast kinetics and enhanced sorption capacity for many contaminants of environmental significance. Encapsulation of nanoparticles within a porous polymeric support improves the mechanical strength and hydraulic properties of the nanoparticles. Dispersal of nanoparticles such as magnetite imparts magnetic properties to polymeric sorbents, rendering them effective for many applications that are conducive only to magnetic separation. By intelligently applying the Donnan membrane effect, it is possible to tailor the enhancement or reduction of the sorption affinities of metal oxide nanoparticles for many environmentally significant ions and ligands. Finally, dispersal of nanoparticles within a polymeric phase makes it amenable to regeneration and reuse for multiple sorption–desorption cycles. This chapter discusses the preparation, characterization, and application of different types of polymeric-inorganic hybrid ion exchangers. Through a combination of the diverse physical and chemical properties and morphologies of the two phases, it is possible to specifically design hybrid ion exchangers for different applications. Commercial hybrid anion exchangers (HAIXs) containing hydrated ferric oxide nanoparticles have been used to remove arsenic from groundwater in many parts of the world. Relevant examples are cited in this chapter.

7.1   INTRODUCTION

Polymeric ion exchange resins are viewed as strongly ionized electrolytes with low dielectric constants. They can be made suitable for various uses including trace contaminant removal in environmental separation systems, hydrometallurgy, catalysis, product purification, and sustainable industrial processes. The science of ion exchange has evolved continuously. New ion exchange materials have been synthesized and new processes are being developed to cater to various applications in industries as diverse as power and water utilities, biotechnology, chemical purification, foods and beverages, agriculture, pharmaceuticals, microelectronics, and water treatment.

Today, many ion exchange resin manufacturers around the world produce hundreds of materials aimed to serve a wide variety of applications in these industries. As different as they may appear, these ion exchangers can be well defined and their specific interactions may be characterized and enhanced by four independent composition variables: matrix, functional groups, cross linking, and pore structure. In addition to the diverse ion exchangers characterized by these four variables, a new class of ion exchangers has been synthesized. In these resins, a dispersed phase of metal or metal oxide nanoparticles is embedded in a polymeric phase to form a synergy not available otherwise. The resulting new materials are heterogeneous even at the level of 10 nanometers (nm) and are called hybrid ion exchangers.

Figure 7.1 illustrates the fundamental composition variables defining polymeric ion exchangers.

7.2 HYBRID ION EXCHANGERS

Due to their extremely high surface area-to-volume ratios, nanoparticles of many metals and metal oxides offer fast kinetics and enhanced sorption capacities for many reactions of environmental significance. For example, (1) hydrated Fe (III) oxides and HFO particles can selectively sorb dissolved heavy metals such as zinc, copper, or metalloids like arsenic oxyacids or oxyanions; (2) Mn(IV) oxides are fairly strong solid phase oxidizing agents; (3) magnetite (Fe_3O_4) crystals are capable of imparting magnetic activities; and (4) elemental Zn^0 and Fe^0 are excellent reducing agents for both inorganic and organic contaminants[1–9].

Figure 7.2 depicts the properties of several inorganic nanoparticles. Synthesis of these nanoparticles and their aggregates is environmentally safe, operationally simple, and inexpensive. However, the nanoparticles are not suitable for direct application in fixed bed columns, in reactive barriers, or in plug flow-like application because of poor durability and excessive pressure drops across columns. Conversely, many commercially available porous polymeric beads are durable and offer excellent hydraulic properties such as reduced pressure drops in fixed bed columns. Novel hybrid polymeric–inorganic materials that combine the excellent hydraulic characteristics of spherical polymer beads with favorable sorption, redox, and magnetic properties of inorganic nanoparticles are worth developing.

It is notable that the host materials (polymer beads) improve hydraulic permeability in flow-through systems alone but are unable to influence or alter the behavior of the metal oxide nanoparticles unless the polymeric phase is functionalized. Therefore, depending on the need, polymeric host materials can be chosen to contain functional groups that provide specific sorption affinities for many environmental contaminants and/or impart the Donnan membrane effect.

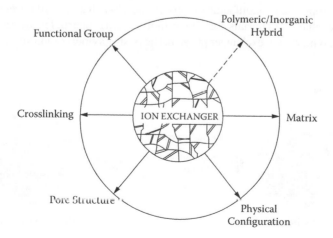

FIGURE 7.1 Fundamental composition variables for polymeric ion exchangers.

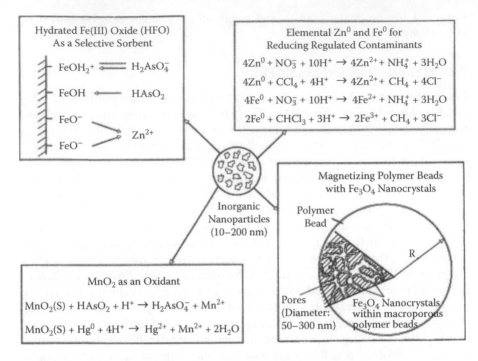

Hydrated Fe(III) Oxide (HFO)
As a Selective Sorbent

$$FeOH_2^+ \rightleftharpoons H_2AsO_4^-$$

$$FeOH \longleftarrow HAsO_2$$

$$FeO^-$$
$$\quad\searrow$$
$$FeO^- \nearrow \quad Zn^{2+}$$

Elemental Zn^0 and Fe^0 for
Reducing Regulated Contaminants

$$4Zn^0 + NO_3^- + 10H^+ \rightarrow 4Zn^{2+} + NH_4^+ + 3H_2O$$
$$4Zn^0 + CCl_4 + 4H^+ \rightarrow 4Zn^{2+} + CH_4 + 4Cl^-$$
$$4Fe^0 + NO_3^- + 10H^+ \rightarrow 4Fe^{2+} + NH_4^+ + 3H_2O$$
$$2Fe^0 + CHCl_3 + 3H^+ \rightarrow 2Fe^{3+} + CH_4 + 3Cl^-$$

Inorganic
Nanoparticles
(10–200 nm)

Magnetizing Polymer Beads
with Fe_3O_4 Nanocrystals

Polymer
Bead

R

MnO_2 as an Oxidant

$$MnO_2(S) + HAsO_2 + H^+ \rightarrow H_2AsO_4^- + Mn^{2+}$$

$$MnO_2(S) + Hg^0 + 4H^+ \rightarrow Hg^{2+} + Mn^{2+} + 2H_2O$$

Pores
(Diameter:
50–300 nm)

Fe_3O_4 Nanocrystals
within macroporous
polymer beads

FIGURE 7.2 Favorable properties of some metal and metal oxide nanoparticles.

A hybrid ion exchanger essentially contains two phases (1) a functionalized polymeric host or ion exchanger and (2) metal or metal oxide nanoparticles dispersed within the polymer phase. Although the nanoparticles can be incorporated during the prepolymerization steps, it is often difficult to control and retain the desired properties of inorganic nanoparticles in the final product[10–15]. Moreover, hybrid materials prepared this way, for example, magnetic polymers, tend to be overly expensive because of proprietary manufacturing processes. This chapter will cover only those hybrid ion exchange materials for which inorganic nanoparticles are incorporated inside the polymeric ion exchanger phase in the post-polymerization stage. Figure 7.3

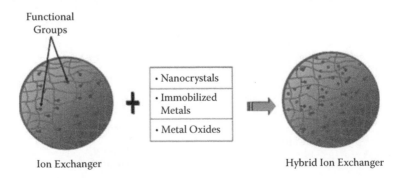

Functional
Groups

• Nanocrystals
• Immobilized Metals
• Metal Oxides

Ion Exchanger   Hybrid Ion Exchanger

FIGURE 7.3 Hybrid ion exchanger.

illustrates hybrid ion exchangers. We will focus on the preparation, characterization, and environmental applications of three classes of hybrid ion exchangers: (1) magnetically active polymeric particles (MAPPs); (2) hybrid nanosorbents for selective removal of ligands; and (3) tunable sorbents. We will also discuss a new class of hybrid sorbents—hybrid ion exchange fibers that offer, along with other property enhancements, better kinetics that benefit many applications.

7.3 MAGNETICALLY ACTIVE POLYMER PARTICLES (MAPPs)

Magnetic polymers have long been pursued by the biomedical, electronics, and materials science worlds for their unique properties and potential for incorporation into novel processes related to protein and biomolecule separation, water and wastewater treatment, color imaging, and information storage. In many areas of biological importance, magnetic polymers are used as carriers for cells and biomolecules such as nucleic acids and proteins. Magnetic polymers offer the advantages of easy manipulation and potential for automation and miniaturization. In addition, fast and cost-efficient separation of magnetic carriers from studied biological mixtures without filtration or centrifugation makes these polymers particularly attractive.

In the fields of environmental separation and control of contaminants, polymeric sorbent materials with specific affinities for heavy metals, metalloids, inorganic and organic ligands, chlorophenols, and pesticides are extensively used for remediation[16-22]. In environmental forensics, the origin of a pollutant of interest in natural waters can be determined using highly selective polymeric adsorbents. Despite their excellent sorption properties for targeted contaminants, use of polymeric sorbents in complex matrices may not be practical because of difficulties in retrieving the saturated sorbents from the matrices. The problem can be overcome if nonmagnetic (i.e., diamagnetic) polymeric sorbents are imparted with magnetic activities to enable these sorbents to be recovered from complex matrices by applying magnetic fields. When magnetized, polymeric sorbent materials can be very effective for sequestering target contaminants and can easily be recovered from complex environmental matrices such as slurries and sediments with high suspended solids content, viscous or radioactive liquids, or media with high concentrations of biomasses. Thus, the introduction of magnetic activity into a wide array of commercially available or tailored polymeric substances makes them fit for unique applications in complex environmental systems.

Magnetically active polymer particles of different morphologies can be prepared to satisfy specific needs. Particles can contain superparamagnetic cores embedded in a polymeric shell, an evenly dispersed magnetic material within the polymer, or a magnetic material heterogeneously dispersed within the sorbent. For core-shell type magnetic polymers, magnetic core materials are encapsulated by a polymer coating that is separately applied to the core by phase inversion or solvent evaporation method[23-26]. Morphological properties of these magnetic polymers such as particle size and particle size distribution depend on the polymer coating fabrication method and the nature of the polymer. When monomers and magnetic particles are mixed intimately and the monomer is polymerized using different polymerization techniques, the magnetic material inside the resultant magnetic polymeric material is evenly distributed within its polymeric matrix[27–30]. The processes for preparing these two types of magnetically active polymers are proprietary in nature and the products are very expensive. Furthermore, the rigorous reactions involved in the polymerization or polymer deposition stages allow very little flexibility in developing a wide variety of magnetically active polymers with specific sorption affinities for different types of environmental contaminants.

For diverse environmental applications, the requirements for effective uses of MAPPs are different and certain requirements apply: (1) the magnetization process should be universally applicable for a wide range of reusable polymeric sorbent particles; (2) the imparted magnetic activity or process should not interfere with the sorption properties (equilibrium and kinetics); and (3) MAPPs should retain magnetic activity over many cycles of operation. To meet these requirements, MAPPs are prepared in situ. In one method, preformed magnetite nanocrystals are deposited inside the polymer phase through swelling. Micron-size polystyrene (PS) particles are swollen in an aqueous solution of N-methyl-2-pyrrolidone (NMP) and then mixed with superparamagnetic iron oxide nanoparticles [31]. The magnetic nanoparticles are able to diffuse into polymer microspheres and become entrapped within the microspheres. The difficulty associated with the process is that without a proper selection of the proportion of solvent, the polymeric support tends to dissolve and is lost in the solution.

Of all the ferromagnetic materials, magnetite is environmentally benign, inexpensive, and chemically stable. The magnetic activity of a polymeric sorbent can be greatly enhanced by irreversibly dispersing nanoscale magnetite particles within the polymeric phase. The challenge in magnetization of polymeric beads lies in controlling the process conditions so that the formation of $Fe_3O_4(s)$ is preferred over the formation of nonmagnetic $Fe(OH)_3(s)$ and $Fe(OH)_2(s)$. In a reducing environment (without oxygen), $Fe(OH)_2(s)$ is the predominant solid phase; in a moderate to highly oxidizing environment, $Fe(OH)_3(s)$ predominates. The environmental conditions under which magnetite crystals are formed are very sensitive to pH and redox conditions. There is a very small envelope in which the ferromagnetic material is preferably formed over the nonmagnetic hydrated Fe(II) or Fe(III) oxides[32]—hence formation is a challenging task. Figure 7.4 shows the stability and predominance at 25°C (E_h–pH) of the pertinent solid phases. Considering oxygen as the sole electron acceptor, the solid phase transition between $Fe_3O_4(s)$ and $Fe(OH)_3(s)$ can be presented as follows:

$$4Fe_3O_{4(S)} + O_{2(aq)} + 18H_2O \rightleftharpoons 12Fe(OH)_{3(S)}$$

The magnetization process demands an extremely low concentration of dissolved oxygen to oxidize Fe^{2+} to Fe_3O_4 without forming essentially nonmagnetic $Fe(OH)_3(s)$. All these factors were taken into consideration while designing a process to form magnetite crystals within polymer beads. Figure 7.5 depicts a scheme for dispersing magnetite nanocrystals within a polymeric bead. This figure exemplifies the formation of magnetite crystals inside a cation exchange resin with sulfonic acid groups. The same procedure can also be applied to other types of functionalized polymers

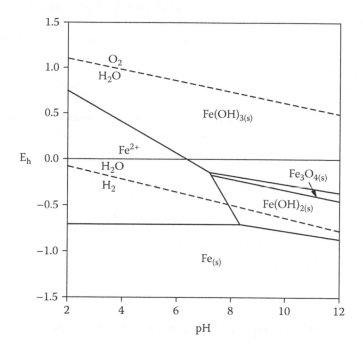

FIGURE 7.4 Predominances of different species of Fe and Fe oxides.

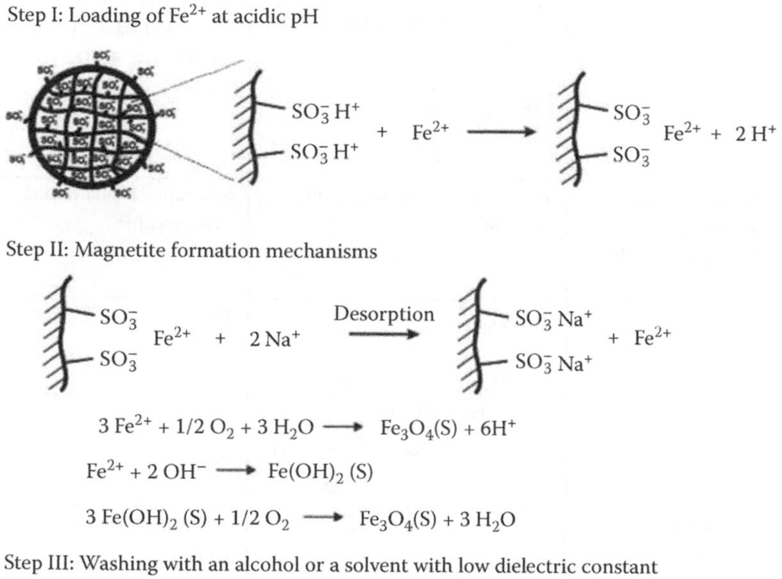

Step I: Loading of Fe^{2+} at acidic pH

$$SO_3^- H^+$$
$$SO_3^- H^+$$ $+$ Fe^{2+} $\longrightarrow$ SO_3^- Fe^{2+} $+$ $2 H^+$
$$SO_3^-$$

Step II: Magnetite formation mechanisms

$$SO_3^-$$ Fe^{2+} $+$ $2 Na^+$ $\xrightarrow{\text{Desorption}}$ $SO_3^- Na^+$ $+$ Fe^{2+}
$$SO_3^-$$ $SO_3^- Na^+$

$$3 Fe^{2+} + 1/2\ O_2 + 3 H_2O \longrightarrow Fe_3O_4(S) + 6H^+$$

$$Fe^{2+} + 2 OH^- \longrightarrow Fe(OH)_2\ (S)$$

$$3 Fe(OH)_2\ (S) + 1/2\ O_2 \longrightarrow Fe_3O_4(S) + 3 H_2O$$

Step III: Washing with an alcohol or a solvent with low dielectric constant

FIGURE 7.5 Steps involved in preparation of MAPPs.

with both macroporous and gel type morphologies[33,34]. Table 7.1 shows the salient properties of different types of ion exchangers for which magnetic activity has been imparted.

Sulfonic acid groups exhibit very low affinities toward hydrogen ions[35]. When a cation exchange resin with sulfonic acid functional groups in hydrogen form is contacted with an acidic solution containing Fe^{2+} ions, the ions are immediately taken up by the cation exchanger in exchange for H^+ ions. The concentrations of Fe(II) species within the ion exchanger are very high—2 N (eq/L) or more. In the next step, as the ion exchanger in Fe(II) form is contacted with high concentrations of Na^+ ions from an alkaline solution of NaCl and NaOH, the Na^+ ions replace the Fe^{2+} ions from the ion exchange sites. The unbound Fe(II) concentration inside the polymer phase is very high and the pH in the phase is also alkaline. If a trace concentration of oxygen is present in the polymer phase, depending on the oxygen concentrations inside, the following reactions can take place to form ferrous hydroxide, ferric hydroxide, and Fe_3O_4. Both the Fe(II) and Fe(III) hydroxides are nonmagnetic while the magnetite crystals are ferromagnetic.

TABLE 7.1
Properties of Different Types of Ion Exchangers Used for The Synthesis of MAPPs.

Characteristic	Composition of the functional group	Manufacturer trade name
High metal–ion affinity	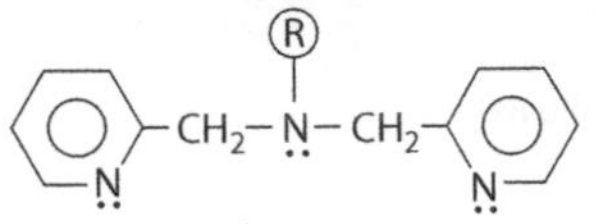 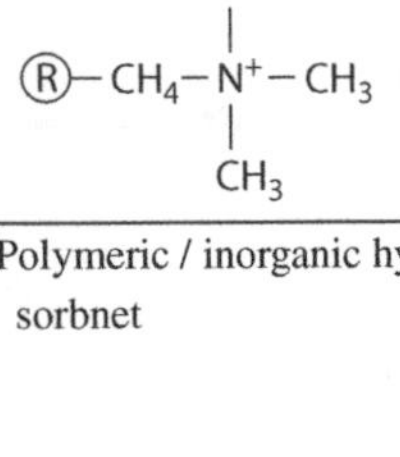 $\langle O \rangle - CH_2 - \overset{R}{\underset{\cdot\cdot}{N}} - CH_2 - \langle O \rangle$ $\underset{\cdot\cdot}{N}$ $\underset{\cdot\cdot}{N}$ $\textcircled{R} - CH_4 - \overset{+}{\underset{CH_3}{N}} - CH_3$	Dow Chemicals Dow SN or XFS 4195 Rohm and Haas Co., IRA-900 Purolite Inc, A-500
High affinity for chromate, benzene sulfonate, pentachlorophenate. High arsenic selectivity	Polymeric / inorganic hybrid sorbnet	SenGupta, Demarco and Greenleaf [24]
Cation exchange resin	$\textcircled{R} - SO_3^-$ $\textcircled{R} - CH_2 - \underset{COO^-}{\overset{3O_3^-}{CH}} - CH_2 - \underset{HO-P-OH}{\overset{}{C}} \\ \qquad\qquad\qquad\qquad \overset{\|}{O}$	Purolite Inc., C-145
Metal–selective multi-functional cation exchanger	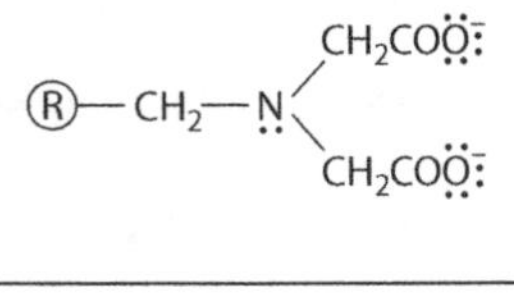 $\textcircled{R} - CH_2 - \overset{\cdot\cdot}{N} \overset{CH_2CO\overset{\cdot\cdot}{O}\underset{\cdot\cdot}{:}}{\underset{CH_2CO\overset{\cdot\cdot}{O}\underset{\cdot\cdot}{:}}{\Big\langle}}$	Eichrome industries, Diphonix
High metal–ion affinity		Rohm and Haas Co., IRC–718

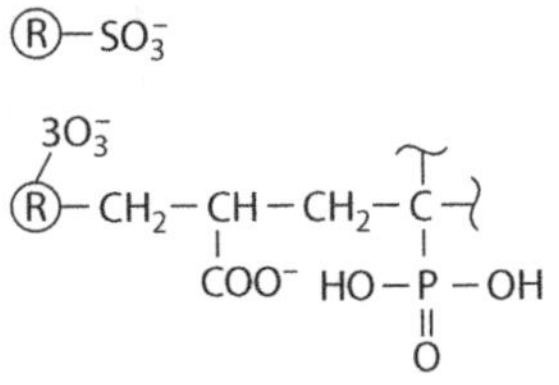

$$Fe^{2+} + 2OH^- \rightarrow Fe(OH)_{2\,(S)}$$

$$4Fe^{2+} + O_2 + 10H_2O \rightarrow 4Fe(OH)_{3(S)} + 8H^+$$

$$3Fe^{2+} + \frac{1}{2}O_2 + 3H_2O \rightarrow Fe_3O_{4(S)} + 6H^+$$

In a laboratory, magnetization for different polymeric sorbent particles was carried out in a batch reactor as shown in Figure 7.6. The specific steps followed were as follows:

1. Prepare 500 ml of ferrous chloride solution (500 mg/L as Fe) inside a reactor vessel at acidic conditions (pH 2.5 to 3.5) under nitrogen pressure only and heat the solution to ~60 to 70°C. Then introduce ~10 g of polymeric sorbent particles inside the reactor in a porous nylon pouch.
2. After 15 to 20 min, slowly raise the pH to ~10 by adding 5% NaCl + 0.5% NaOH solution and bubble ~0.1 to 1% oxygen by volume with carrier nitrogen and stir ~60 min. Formation of magnetite is marked by a dark black color.
3. Stop stirring, switch off the heating plate, and then remove the pouch containing the polymeric particles. Rinse with distilled water and dry the particles in a vacuum furnace for 1 h (rinsing with ethanol helps produce relatively dry particles).
4. Repeat the process for a second cycle as necessary.

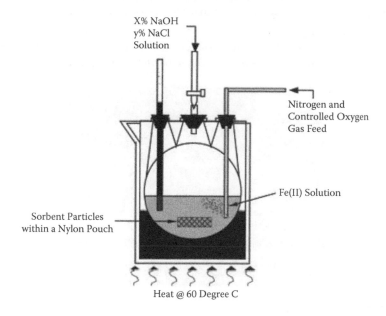

FIGURE 7.6 Batch reactor used for synthesis of MAPPs.

7.3.1 MAPP CHARACTERIZATION

The procedure for dispersing magnetite nanoparticles inside polymers is noninvasive; the chemical reactions involved with dispersal of magnetite nanocrystals within beads did not interfere with or alter the polymeric phase. The morphology of the polymeric phase remained unaltered except for the black color. Figure 7.7 is an enlarged view of a magnetized C-145 cation exchange resin bead. The resin looked exactly the same as the parent resin except for the color change. When both the parent and magnetized resins were cut for a visual observation, both their surfaces appeared similar; a pitch dark magnetic coating was observed only in the MAPP.

Figure 7.8 shows x-ray diffractograms of a sliced MAPP and a pure magnetite crystal used as a reference standard. The similarities of the peaks prove the existence of magnetite crystals inside the bead. The degree of magnetization can be visually determined by comparing responses to a laboratory magnet applied to four different kinds of materials (Figure 7.9). Only the ion exchanger dispersed with magnetite showed significant magnetic activity. The parent ion exchanger, the ion exchanger dispersed with ferric [Fe(III)] hydroxide, and the ion exchanger dispersed with ferrous [Fe(II)] hydroxide did not exhibit any magnetic activity.

7.3.2 FACTORS AFFECTING ACQUIRED MAGNETIC ACTIVITY

A dimensionless parameter known as magnetic susceptibility (χ_m) is used to determine the magnetic behavior of a material. Magnetic susceptibility is defined as the degree of magnetization induced in a material under the influence of one unit of magnetic field. For nonmagnetic particles, magnetic susceptibility is negative. Water is diamagnetic; it has a magnetic susceptibility of -9×10^{-6} whereas the magnetic susceptibility of pure magnetite ranges from 1 to 5.7 [36]. The magnetic susceptibility of the MAPPs prepared in a laboratory at Lehigh University was measured using a susceptibility meter (Sapphire Instruments SI-2). The sample particles were placed within a copper

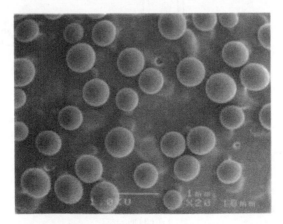

FIGURE 7.7 Enlarged view of magnetized C-145 polymer beads (20× magnification). Physical configurations of the particles were unchanged following magnetization.

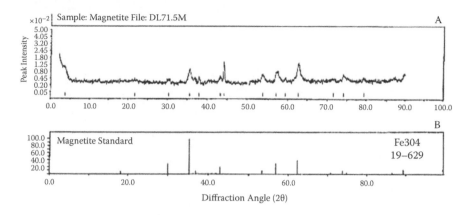

FIGURE 7.8 X-ray diffractograms and characteristic peaks of (A) sliced magnetized polymeric particle (Purolite C-145) and (B) magnetite standard.

coil of the instrument and a magnetic field was applied externally and carried through the coil. A core logging device measures the magnetic susceptibility of samples in terms of dimensionless volume susceptibility.

Figure 7.10 shows the experimentally determined magnetic susceptibilities of MAPPs prepared via the method detailed earlier. Although, their degrees of magnetization were different, all forms of hybrid sorbents acquired magnetic activity, regardless of the physical and chemical natures of the parent polymeric sorbents. The magnetic activity of all the hybrid sorbents increased when the loading of magnetite continued for multiple cycles. A closer look at the acquired magnetic susceptibility values in Figure 7.10 and an examination of the chemical natures of the functional groups of the polymeric sorbents detailed in Table 7.1 reveal that the degree of magnetic susceptibility acquired depends on the nature of the functional groups.

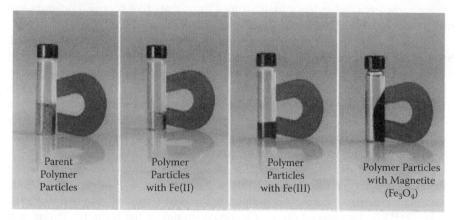

FIGURE 7.9 Comparison of responses to a laboratory magnet for four types of Diphonix Polymer beads: (i) no magnetization and dispersed with (ii) Fe(II) hydroxide, (iii) Fe(III) hydroxide, and (iv) magnetite.

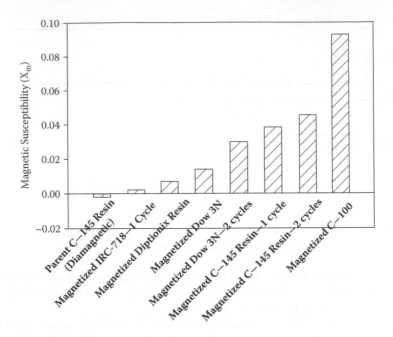

FIGURE 7.10 Experimentally determined magnetic susceptibilities of various magnetized polymeric beads.

For the magnetized C-100, the magnetic susceptibility was the highest after one cycle. The magnetized C-145 and Dow 3N resins were next in the series of decreasing magnetic susceptibilities. Note that C-100 is a strong acid cation exchange resin with sulfonic acid functional groups while C-145 is a weak acid cation exchange resin with a carboxylate functional group. Dow 3N has weakly basic characteristics but has a high affinity for metals like copper. Diphonix is a metal-selective multifunctional ion exchanger.

The more acidic the functional groups are, the more magnetic activity is acquired. Strong acid cation exchangers contain sulfonic acid functional groups that impart a very low affinity toward hydrogen ions, whereas the affinity for hydrogen ions increases as the acidity of the functional groups decreases. During the first step of magnetization, the ion exchanger in hydrogen form is contacted with a solution containing Fe^{2+} ions so that the ion exchanger uptakes Fe^{2+} in preference to hydrogen ions. The uptake of Fe^{2+} is the highest for strong acid cation exchangers compared to resins with low acidity or basic characteristics.

It is estimated that the concentration of Fe^{2+} ions inside a strong acid cation exchanger reaches a concentration of about 2 N or more at this step. As a result, when magnetite nanoparticles are formed in the next step via controlled oxidation of the Fe(II) species sorbed on the functional groups of the resin, the strong acid cation exchange resin acquires a higher degree of magnetization compared to other resins. Although the other functional groups did not provide favorable conditions for acquiring magnetic activity, the degree of magnetic activity of the ion exchangers can always be enhanced by loading more magnetite nanoparticles through repetition of the loading procedure over multiple cycles.

All other conditions remaining identical, greater magnetic susceptibility of a material allows effective separation at even low magnetic fields. Therefore, partial functionalization of any polymeric sorbent with sulfonic acid functional groups leads to a greater degree of magnetization, thus widening the opportunities for using magnetic separation for complex environmental operations. One method of partial functionalization to achieve magnetic activity while retaining other properties of a polymeric sorbent is using dual zone sorbents discussed later in this chapter.

7.3.3 Retention of Magnetic Activity and Sorption Behavior

To be viable when applied to complex environmental separation processes, MAPPs must simultaneously retain both magnetic activity and specific sorption capacity. To assess these capabilities, magnetized Dow 3N ion exchangers were utilized. Previous studies indicated that DOW 3N and XFS 4195 exhibit high selectivity and high capacity for copper ions even at low pH [37,38]. After it is exhausted, the resin can be effectively regenerated by a solution of ammonia [18,39,40]. Magnetized DOW 3N particles were subjected to cyclic adsorption and desorption over 15 cycles to determine the effects of magnetization on the sorption–desorption characteristics of the resin and the retention of magnetic susceptibility over multiple cycles of adsorption–desorption. Each cycle consisted of equilibration with 20 mg/L copper solution at pH 4.0 followed by desorption with 5% ammonia solution.

Figure 7.11 shows experimentally determined copper sorption capabilities over 15 cycles. The copper removal capacity of about 2 meq Cu/g of magnetic DOW 3N was comparable to that of the parent resin reported in a previous study. Figure 7.12 shows specific magnetic susceptibility of the MAPP sorbent over the 15 cycles of use. No change of magnetic activity of the MAPP was observed.

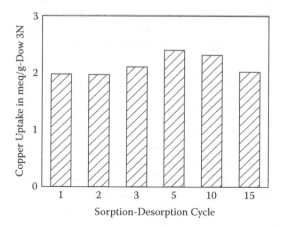

FIGURE 7.11 Copper sorption capacities of magnetized metal-selective Dow 3N polymer beads for 15 consecutive sorption–desorption cycles.

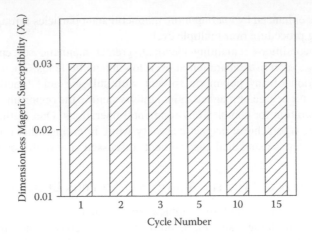

FIGURE 7.12 Specific magnetic susceptibilities of the same DOW 3N polymer beads in Figure 7.11 for 15 cycles.

Thus, the MAPP retained its magnetic activity despite use over many cycles of harsh chemical activities of adsorption at low pH and desorption at high pH using ammonia solution as the regenerant. Since magnetic activity is imparted by the magnetite nanoparticles dispersed within the polymer matrix, retention of magnetic activity implies no loss of magnetite within the MAPP beads even after several adsorption–desorption cycles. This observation confirms that the magnetite particles inside were irreversibly dispersed and stayed within the beads despite external physical forces. Their chemical properties remained unchanged over many cycles of operation.

Figure 7.13 shows the results of kinetic tests of zinc removal using both magnetized and parent Diphonix resin beads. The presence of magnetite did not affect the uptake rates of zinc ions. Under batch conditions, intraparticle diffusion is usually the rate-limiting step. Thus it may be concluded that the presence of fine magnetite particles or the formation of magnetite particles inside the beads did not influence the effective intraparticle diffusivity of polymer particles.

7.4 HYBRID NANOSORBENTS FOR SELECTIVE REMOVAL OF LIGANDS

Oxides of some polyvalent metals such as Al(III), Fe(III), Ti(IV), Zr(IV), and others are environmentally benign and show amphoteric sorption behaviors around neutral pH. They are innocuous, inexpensive, readily available, and chemically stable over a wide range of pH values. Below the pH of zero point charge (pH_{PZC}), a metal oxide behaves like a Lewis acid—as an electron pair acceptor. At this pH, which is below the isoelectric point, such oxides are capable of selectively binding to Lewis bases that donate electron pairs to form coordinate or inner sphere complexes by binding to central metal atoms of the metal oxide. Examples of environmentally significant ligands are inorganic oxyanions and oxyacids of arsenic, phosphorus, and selenium[41–51].

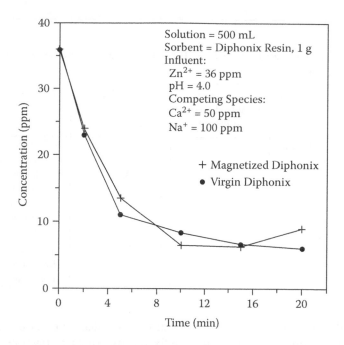

FIGURE 7.13 Comparison of batch kinetic test results for zinc removal with parent and magnetized Diphonix resins.

Investigations involving extended x-ray absorption fine structure (EXAFS) spectroscopy confirmed that both As(III) and As(V) are selectively bound to hydrated ferric oxide surfaces through coordinate bonding [52]. Table 7.2 shows the ligand characteristics or electron pair donating ability for As(III) and As(V) species and for phosphorus oxyanions. Based on their high selectivities for ligands, the metal oxides can selectively adsorb the ligands from a background of other common ions such as chloride, sulfate, nitrate, and others. These anions can form outer sphere complexes with the metal oxide surfaces only through Coulombic interaction. Compared to crystalline forms, amorphous metal oxides exhibit higher surface area per unit mass or specific surface area. Since adsorption sites primarily remain at the surface, amorphous metal oxides show higher adsorption capacities than other forms.

Hydrated ferric oxide (HFO) particulates are popular adsorbents for arsenic and phosphorus species. The sizes of freshly precipitated amorphous particles range from 20 to 100 nm. Although the small size helps achieve a very high specific surface area, it is nearly impossible to use these nanoscale particles in the field to remove trace contaminants from contaminated ground water or wastewater. Fixed bed adsorption is universally acceptable for water and wastewater treatment because it is simple, requires virtually no start-up time, and is forgiving for fluctuation in influent concentrations. The nanoscale adsorbents are not suitable for fixed bed applications because of high pressure drops and poor mechanical strength. Attempts have been made to develop granulated hydrated ferric oxide or ferric hydroxide particles and they were used in fixed bed columns for selective removal of arsenic from contaminated water.

TABLE 7.2

Sorption Properties and Lewis Structures of Oxyacids and Oxyanions of As(V) and As(III)

Parent Oxyacids	PK$_0$ Values	Predominant Dissolved species at PH 6.0	Predominant Dissolved species at PH 8.0	Sorption Interaction
AS(V): H$_3$AsO$_4$	$_p$K01 = 2.2 $_p$K02 = 8.98 $_p$K03 = 11.6			As(V) species or arsenates can undergo Coulomtite (can exchange) as well as Lewis acid-base interaction
AS(III) HASO$_2$	$_p$K21 = 9.2	O=As—OH	O=As—OH	AS(III) species or arsenates can undergo Lewis acid-base interaction

However, the mechanical properties were still weaker than those of polymeric and other inorganic sorbents normally used in fixed beds. Another drawback is that the material is not regenerable and produces voluminous contaminated waste after only one cycle of operation.

To overcome the problems associated with the isolated use of nanosorbents, it is necessary to disperse them irreversibly within a host material that can provide enough mechanical and chemical support for their use in fixed beds. Several attempts were made to use naturally occurring or synthetic (polymeric or inorganic) materials such as alginates, zeolites, activated carbon, chitosan, cellulose, polymeric sorbents, and polymeric cation exchangers as hosts to contain nanosorbents[53–63]. All these support materials improved the permeability and durability of adsorbents in fixed bed columns. However, it is necessary to understand how and to what extent the chemical and physical nature of a host material influences the nature of a hybrid nanosorbent. Obviously, the morphology of the polymeric host materials (e.g., pore size and distributions) influence the size and nature of the nanoparticles dispersed within the pores.

Some host materials such as polymeric cation exchangers have charged surface functional groups; to a varying degree, the nature and charge density of the functional groups influences the process and extent of dispersion of nanoparticles within the polymeric matrix. Most importantly, it is vital to determine whether a charged functional group alters or synergistically enhances the sorption behavior of the hybrid material. Depending on the type of functional group on the polymeric host, three types of hybrid nanosorbents are possible: (1) hybrid anion exchange resins (HAIX); (2) hybrid cation exchange resins (HCIX); and (3) hybrid resins with no ion exchange capability (HNIX). The next section focuses on the first two types and discussion is limited mainly to their preparation, characterization, and use with hydrated ferric oxide nanoparticles.

7.4.1 SYNTHESIS OF HYBRID NANOSORBENTS

Table 7.3 presents salient features of two gel type polymeric supports containing cation exchange and anion exchange functional groups used to synthesize hybrid sorbent materials. Similar exchangers with both gel and macroporous morphologies from other manufacturers were also used. The preparation of the hybrid cation exchanger (HCIX) consisted of three steps:

1. Loading of Fe(III) onto the sulfonic acid sites of the cation exchanger by passing through 4% $FeCl_3$ solution at an approximate pH of 2.0
2. Desorption of Fe(III) and simultaneous precipitation of Fe(III) hydroxides within the gel phase of the exchanger by passage of a solution containing NaCl and NaOH (both at 5% w/v concentration).
3. Rinsing and washing with a 50:50 ethanol:water solution followed by a mild thermal treatment (50 to 60°C) for 60 min.

Figure 7.14 depicts the major steps of the process. Step 2 was twice repeated to achieve greater Fe(III) loading. Experimental observations suggest that at the conclusion of Step 3 both amorphous and crystalline phases were present and submicron or nanoscale HFO particles coalesced to form agglomerates. Use of ethanol lowered the dielectric constant of water and supposedly enhanced the agglomeration of submicron particles through suppression of surface charges. HFO agglomerates were irreversibly encapsulated within the spherical exchanger beads; turbulence and mechanical stirring did not result in any loss of HFO particles. High concentrations of sulfonic acid functional groups allowed

TABLE 7.3
Properties of Parent Polymeric Ion Exchangers Used as Host Materials

Anion and Cation Resin	Purolite A-400 and IRA-900	Purolite C-100
Structure (Repeating Unit)		
Functional Group	Quaternary ammonium	Sulfonic acid
matrix	Polystyrene	Polystyrene
Capacity (meq/c resin)	3.6	4.6 (Dry) 2.5 (Wet)
Manufacturer	Purolite Inc. and Rohm and Haas Co., Philadelphia	Purolite Inc. Philadelphia

Step 1. Loading with $FeCl_3$ solution at pH < 2.0

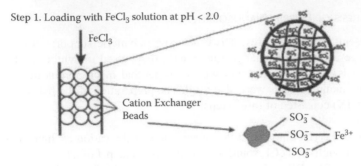

Step 2. Desorption and simultaneous precipitation in the gel phase and pores

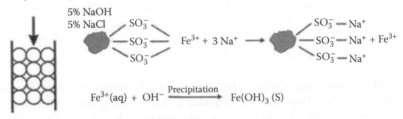

$$Fe^{3+}(aq) + OH^- \xrightarrow{\text{Precipitation}} Fe(OH)_3 \ (S)$$

Step 3. Alcohol wash and mild thermal treatment

$$Fe(OH)_3 \ (S) \xrightleftharpoons{60°C} FeOOH \ (S) \ \text{(Crystalline)}$$

FIGURE 7.14 Three-step procedure to disperse crystalline and amorphous HFO microparticles inside spherical polymer beads.

high and fairly uniform loading of HFO particles (~9 to 12% of Fe by mass) within the polymeric beads.

Dispersing HFO nanoparticles inside anion exchangers is scientifically challenging because both ferric ions [Fe(III)] and the functional groups [quaternary amine, $(R_4N)^+$] are positively charged. Because of the high Donnan exclusion potential caused by the nondiffusible quaternary ammonium functional group inside the resin, it is difficult for the Fe(III) ions to diffuse inside the anion exchanger. However, an economically and technically viable way to prepare hybrid anion exchangers was devices at Lehigh University [64]. It is known that the Donnan exclusion potential increases with the increase in the valence of the co-ion. Thus, all other conditions remaining identical, equilibrium concentration of ferrous ions inside an anion exchanger will be greater than that of the ferric FE (III) ions. Subsequently, oxidation of the ferrous ions within the ion exchanger converts them to ferric ions that then form tiny precipitates of ferric hydroxide inside the anion exchanger when the pH is raised.

The salient steps for the method of dispersal of HFO inside anion exchange resins are as follows. First, the anion exchanger is contacted with a solution of potassium permanganate ($KMnO_4$) or calcium hypochlorite (CaOCl), transforming the anion exchanger to MnO_4^- or OCl^- form. Second, the anion exchange resin is contacted with 5% solution of ferrous chloride. The chloride ion replaces the MnO_4^- or OCl^- ion from the ion exchange site. The ferrous iron inside the exchanger is oxidized to ferric iron by reacting with the MnO_4^- or OCl^- ions released from ion exchange sites. Next, the passage of 5% solution of NaOH helps precipitate ferric hydroxide particles in the pores of the ion exchanger. Figure 7.15 depicts the loading of HFO particles on anion exchange resins.

In another synthesis process [65], a binary solution of $FeCl_3$–HCl was used as the starting solution. The complex $FeCl_4^-$ anion is formed because of the prevailing concentration and pH conditions in the solution. The anion can be preferentially taken up by the ion exchanger over Cl^- ions due to its significantly lower energy of hydration. The ion exchanger is subsequently contacted with a solution of NaCl and NaOH when hydrated ferric oxide nanoparticles is formed inside the ion exchanger. A third step involving washing of the hybrid ion exchanger with a solution of ethanol and water and mild thermal treatment was also included in the synthesis of the hybrid anion exchanger.

7.4.2 Characterization of Hybrid Nanosorbents

Figure 7.16a is a photomicrograph of a hybrid cation exchange resin; Figure 7.16b shows a hybrid anion exchanger, both loaded with HFO. The resin has turned brown or deep brown from the original white or pale yellow due to the deposition of brown HFO nanoparticles inside the resin. In both cases, the original spherical geometry of the resin was retained after the synthesis of the hybrid adsorbent. Figure 7.17a is a scanning electron micrograph (SEM) of freshly precipitated HFO; Figures 7.17b and 7.17c are SEMs of sliced beads of a parent gel type cation exchanger and HCIX beads, respectively. Because gel type ion exchangers do

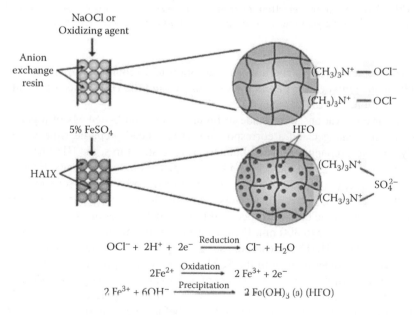

FIGURE 7.15 Two-step procedure for dispersing hydrated ferric oxide (HFO) particles inside anion exchange resins.

(a)   (b)

FIGURE 7.16 Photomicrographs of (a) hybrid cation exchange resin and (b) hybrid anion exchange resin beads.

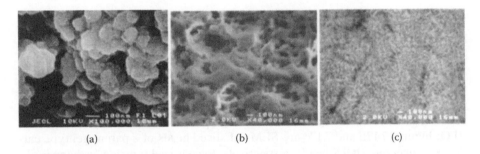

(a)   (b)   (c)

FIGURE 7.17 Scanning electron microscopy images of (a) freshly precipitated HFO (100,000×), (b) parent gel type cation exchanger (40,000×X), and (c) hybrid cation exchange resin (40,000×).

not have separate pore structures as macroporous exchangers do, no distinction can be made between the parent ion exchangers and the HFO phases even at high magnification.

Similar observations were made in Figures 7.18a and b—SEMs of a parent gel type anion exchanger and a corresponding HAIX bead prepared from it, respectively. For ion exchange resins with macroprorous structures, the HFO particulates are deposited throughout the beads in both gel phases and macropores. Figures 7.19 a and b are SEMs of a parent anion exchange resin and a hybrid sorbent prepared from it. Note that the nanoscale deposits of amorphous and crystalline HFO particles were accessible to the dissolved species through a network of macropores that originally had sizes of 20 to 300 nm. Figure 7.20 is a transmission electron micrograph (TEM) image of a HAIX nanosorbent suggesting that the apparent amorphous coatings of HFO particles as seen in the SEM images are composed of agglomerates of discrete nanoparticles of individual sizes of 5 to 10 nm. One study of HAIX particles reported a BET surface area of 120 m²/g with an average pore diameter of 17.4 nm [66]. However, as the resin shrinks from drying before such measurements are made, the data probably do not reflect the physical characteristics of the resin under actual operational conditions.

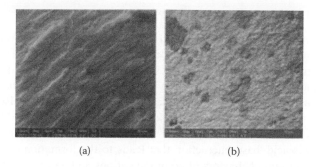

(a)   (b)

FIGURE 7.18 Scanning electron microscopy images of (a) parent gel type anion exchanger (15,000×) and (b) gel hybrid anion exchange resin (25,000×).

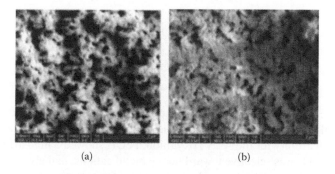

(a)   (b)

FIGURE 7.19 Scanning electron microscopy image of (a) parent macroporous anion exchanger (20,000×) and (b) macroporous hybrid anion exchange resin (20,000×).

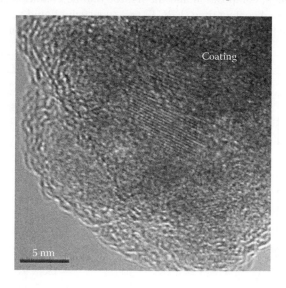

FIGURE 7.20 Tunneling electron micrograph image of macroporous hybrid anion exchanger.

7.4.3 Selective Sorption of Ligands: Phosphate and Arsenic

A hybrid polymeric sorbent should retain the selective adsorption property of the metal oxide nanoparticles dispersed within the host material. The HFO particles dispersed within hybrid nanosorbents can selectively remove dissolved ligands such as oxyacids and oxyanions of As(III), As(V), or phosphorus from a background of common anions like Cl^-, SO_4^{2-}, NO_3^-, etc. The HFO particles are also capable of selectively removing heavy metal cations such as zinc, copper, and lead from a background of cations like Na^+, K^+, Mg^{2+}, Ca^{2+}, etc. The selective sorption capability is due to the Lewis acid–base interaction that leads to the formation of inner sphere complexes; the sorption behaviors of common cations and anions are poor because they form only outer sphere complexes resulting from Coulombic interactions. Two protonation constants of the active functional groups present in HFO have been reported [67]:

$$\overline{FeOH_2^+} \rightleftharpoons H^+ + \overline{FeOH} \quad pK_{a1} = 5.3$$

$$\overline{FeOH} \rightleftharpoons H^+ + \overline{FeO^-} \quad pK_{a2} = 8.8$$

where the overbar represents the solid phase. The distribution of the three functional groups ($FeOH_2^+$, $FeOH$, and FeO^-) is depicted in Figure 7.21. Protonated $FeOH_2^+$ exhibited high affinity for the As(V) oxyanion; $FeOH$ showed a high affinity toward As(III), and FeO^- was selective toward heavy metal cations, namely, zinc.

Figure 7.22 shows complete effluent histories for selective removal of arsenate during a fixed bed column run using HCIX (prepared from macroporous resin C-145). The influent composition and the hydrodynamic conditions during the column run [empty bed contact time (EBCT) and superficial liquid velocity (SLV)] are also shown in the figure. Influent pH was slightly above neutral and no pH adjustment

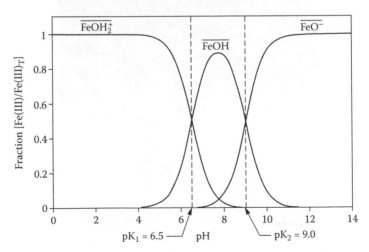

FIGURE 7.21   pH distributions of functional groups of hydrated Fe(III) oxides.

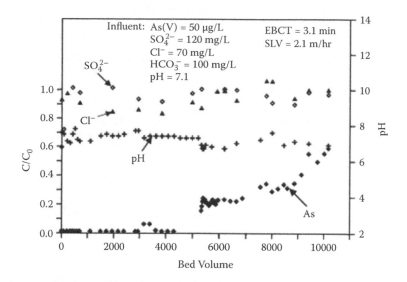

FIGURE 7.22 Complete normalized effluent history of sulfate, chloride, arsenic, and pH during fixed bed column run with HCIX. EBCT = empty bed contact time. SLV = superficial liquid velocity.

was made. The common anions, namely, sulfate and chloride, broke through almost immediately after the start of the column run, while 10 μg/L of arsenic breakthrough was observed only after about 4,000 bed volumes. The effluent and influent had almost the same pH values. Thus, besides the absence of dissolved arsenic, the influent composition and pH remained essentially unchanged after the passage of contaminated water through the HCIX column. This observation proves that HFO dispersed in a functional polymer retains its selectivity toward ligands even at trace concentrations compared to other commonly occurring anions.

At near-neutral pH, arsenate or As(V) exists in the aqueous phase as a mono- or divalent anion (Table 7.2). Previous studies and field trials used anion exchangers for removal of As(V)[68–71]. As the adsorption takes place due to nonselective Coulombic interaction only, the competition from other common anions like sulfate is quite fierce. Therefore, arsenic removal capacity for anion exchange resins is greatly reduced in the presence of competing anions, especially due to competition from divalent sulfate ions. Figure 7.23 shows As(V) effluent histories for two separate column runs using different sorbent materials: a commercially available anion exchanger (IRA-900, Rohm & Haas Co., Philadelphia, PA) and an HAIX containing hydrated ferric oxide nanoparticles dispersed within a macroporous anion exchanger.

The anion exchange resin broke through almost instantaneously, whereas HAIX continued to remove arsenic from the background of competing anions. A breakthrough amounting to only 10% of the influent arsenic concentration was observed after 10,000 bed volumes. This infers that the arsenic removal capacity of the parent anion exchanger is greatly enhanced with the dispersion of the HFO nanoparticles. It may be noted that for the IRA-900 column run, arsenic concentration in the treated water exceeded its influent concentration after breakthrough. This resulted from the

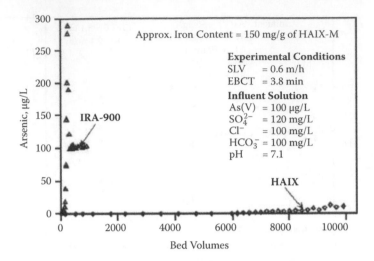

FIGURE 7.23 Comparison of As(V) effluent histories of a strong base anion exchanger (IRA-900) and HAIX under identical conditions.

chromatographic elution effect caused by higher sulfate selectivity for IRA-900 over arsenate. It further confirms that the high selectivity for arsenic stems from the HFO particles dispersed within the functional polymer.

Figure 7.24 provides As(III) effluent histories of two separate column runs using the same feed composition with an As(III) concentration of 100 μg/L. For one run, the IRA-900 anion exchanger was used in a fixed bed column; HAIX was the sorbent during the second run. At near-neutral pH, As(III) is nonionized (i.e., $HAsO_2$ or H_3AsO_3) and therefore IRA-900 was unable to remove As(III). A polymeric anion exchanger alone is thus not effective for As(III) removal. However, if As(III) has been

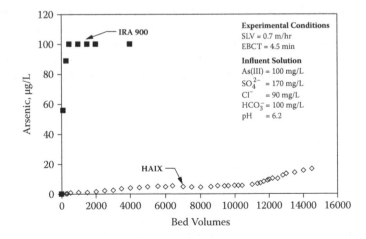

FIGURE 7.24 Comparison of As(III) effluent histories of strong base anion exchanger (IRA-900) and HAIX under identical conditions.

preoxidized to anionic As(V) with chlorine, manganese dioxide, ozone, or another innocuous oxidizing agent, the anion exchanger can show a small removal capacity in natural waters as indicated in Figure 7.23 [7,72,73]. Introducing a preoxidation step poses operational complexity and diminishes the overall viability of the fixed bed process. In comparison, As(III) was removed for a long period by the HAIX column. Total dissolved arsenic breakthrough at 10% of its influent concentration occurred after 12,000 bed volumes. Nitrogen was continuously sparged in the influent storage tank to eliminate any possible As(III) oxidation to As(V). Intermittent analyses of the influent based on the protocol prescribed by Ficklin [74] and Clifford [75] confirmed that As(V) was absent in the feed.

HFO particles possess high selectivities for another environmentally significant Lewis base: the phosphate anion[76–78]. Figure 7.25 shows the results of phosphate adsorption isotherm tests carried out with HAIX dispersed with HFO in two different concentrations of background sulfate ions. Other commonly occurring anions such as chloride and bicarbonate were also present. The phosphate adsorption capacity remained nearly the same even when the sulfate ion concentration was doubled. This proves that the binding mechanism of sulfate ions onto the HAIX was different from that of phosphate [79].

Figure 7.26 shows the results of two column runs with feed water containing trace concentrations of phosphate ions with a background of commonly occurring anions at nearly two orders of magnitude higher concentration. The column containing the IRA-900 resin was observed to break through almost immediately, while the column containing HAIX consistently removed phosphate even after 2,000 bed volumes. This observation is consistent with the earlier study with arsenate ions, suggesting that IRA-900 prefers divalent sulfate ions over monovalent ions such as phosphate and arsenate.

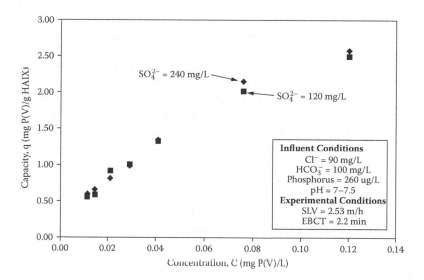

FIGURE 7.25 Comparison of phosphate isotherms for HAIX at two different background sulfate concentrations; all other conditions remained identical.

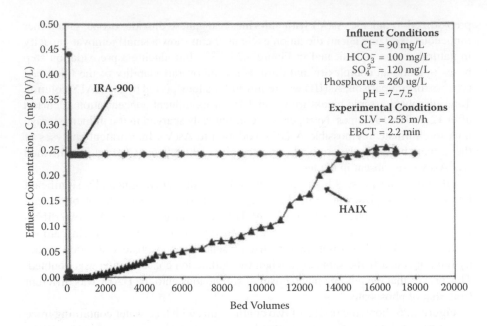

FIGURE 7.26 Comparison of effluent phosphate histories during column runs with parent IRA-900 and HAIX resin using the same influent and under identical hydrodynamic conditions. SLV = superficial liquid velocity. EBCT = empty bed contact time.

7.4.3 Hybrid Ion Exchangers for Ligand Sorption: Cation versus Anion

Two separate column runs were carried out with similar influent solutions under identical hydrodynamic conditions. In both cases, As(V) or arsenate was a trace species when compared to other competing electrolytes (sulfate, chloride, and bicarbonate anions). A hybrid cation exchanger (HCIX-G) gel and hybrid anion exchanger (HAIX-G) gel served as sorbents for the fixed bed runs. Table 7.4 summarizes their general characteristics. Figure 7.27 compares the As(V) effluent histories of the two column runs; the marked difference in the performances of the two sorbents is clear. Note that despite greater HFO content, HCIX-G was essentially unable to remove As(V), which broke through almost immediately after the start of the run. Conversely, HAIX-G with HFO nanoparticles dispersed in an anion exchanger showed excellent arsenic removal capacity. Only 10% of arsenic breakthrough occurred after ~10,000 bed volumes.

The results in Figure 7.27 are intriguing because the performance of HAIX was better than HCIX by at least two orders of magnitude. The results of the earlier column runs with anion exchange resins already proved that the anion exchange functional groups do not contribute to the sorption of arsenic to a significant extent. The high arsenic adsorption capacity of HAIX essentially results from the Donnan membrane effect exerted by the ion exchanger support. The next section explains the reason for the enhanced arsenic adsorption capacity of the anion exchanger in accordance with the Donnan membrane principle.

TABLE 7.4

General Properties of Hybrid Anion and Cation Exchangers Used in Study Cited in Figure 7.27

Designation	Type	Pore Structure	Functional Group	Matrix	Name and Source	Fe Loading (mg Fe/g)
HCIX-G	Cation	Gel	Sulfonate (SO_3^-)	Styrene–divinylbenzene	Purolite C-100, Philadelphia	70
HAIX-G	Anion	Gel	Quaternary ammonium (R_4N^+)	Styrene–divinylbenzene	Purolite A-400, Philadelphia	60

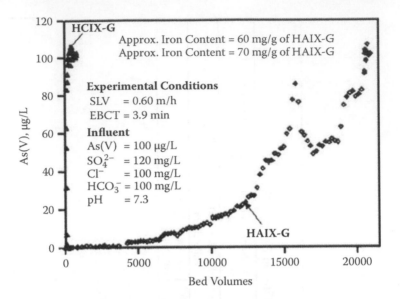

FIGURE 7.27   Comparison of As(V) effluent histories of HCIX and HAIX for two separate column runs under otherwise identical conditions.

7.4.4   Donnan Membrane Equilibrium and Co-Ion Exclusion Effect

The Donnan membrane equilibrium principle is essentially an extension of the second law of thermodynamics. Often called the Gibbs-Donnan equation, it deals specifically with completely ionized electrolytes in a heterogeneous system in which certain ions are unable to permeate from one phase to another through the interface[80,81]. The gel phase of an ion exchanger can be viewed as a polyelectrolyte in which functional groups (such as quaternary ammonium groups for anion exchangers and sulfonic acid groups for cation exchangers) are covalently attached to the matrix and hence are nondiffusible. Conversely, both the co-ions and counter-ions in the bulk liquid phase in contact with the ion exchanger are mobile and can move freely under a chemical or electrical potential gradient. Thus, the fixed nature of the functional group of an ion exchanger acts like a virtual semipermeable membrane that restricts the movement of one particular type of ion across the phase boundary. This gives rise to the development of the Donnan potential[82].

For a membrane that is completely permeable to both Na^+ and $H_2AsO_4^-$, the Donnan membrane principle provides the following equality at equilibrium assuming ideality:

$$\left[Na^+\right]_L \left[H_2ASO_4^-\right]_L = \left[Na^+\right]_R \left[H_2ASO_4^-\right]_R \qquad (7.1)$$

where subscripts L and R refer to the solutions at the left and right sides of the membrane, respectively, and the [ ] brackets represent molar concentration or activity under ideal conditions. If sodium arsenate is the only electrolyte present in the solution phase and the volume on both sides of the membrane is the same (e.g., 1.0 L), the constraint from electroneutrality requires:

$$\left[Na^+\right]_L = \left[H_2ASO_4^-\right]_L \tag{7.2}$$

$$\left[Na^+\right]_R = \left[H_2ASO_4^-\right]_R \tag{7.3}$$

Thus,

$$\frac{\left[H_2ASO_4^-\right]_L}{\left[H_2ASO_4^-\right]_R} = \frac{\left[Na^+\right]_R}{\left[Na^+\right]_L} = 1 \tag{7.4}$$

The above equality is understandably trivial, as shown in Figure 7.28 (Case I). The distribution of $H_2AsO_4^-$ on both sides of the membrane is greatly altered in the presence of an impermeable cation or anion. Consider Case II in which the Na_R salt is initially present on the left side of the membrane at 1.0 M concentration. The resulting anion (R⁻) cannot permeate the membrane. All other conditions are essentially the same as in case I (i.e., the membrane is completely permeable to both Na^+ and $H_2AsO_4^-$ and the initial concentration of NaH_2AsO_4 on the right side is 0.01 M). At equilibrium, the equality in Equation (7.1) will hold even in the presence of nonpermeating R⁻. Hence, Na^+ and $H_2AsO_4^-$ will redistribute to achieve the following equilibrium condition:

$$\frac{\left[H_2ASO_4^-\right]_L}{\left[H_2ASO_4^-\right]_R} = \frac{\left[Na^+\right]_R}{\left[Na^+\right]_L} = \frac{1}{99} \tag{7.5}$$

Note that the monovalent arsenate concentration on the left side of the membrane, $[H_2AsO_4^-]_L$, is nearly two orders of magnitude lower than the concentration of $[H_2AsO_4^-]_R$. Although the membrane is permeable to Na^+ and $H_2AsO_4^-$, the electrolytically dissociated Na_R at high concentration suppresses the permeability of $H_2AsO_4^-$ in one direction. This phenomenon is known as the Donnan co-ion exclusion effect and does not result from Coulombic or electrostatic interactions. The derivation of Equation (7.5) and Equation (7.6) below can be readily followed by consulting the recently published English translation of Donnan's original paper [80]. It is assumed that the volume on each side of the membrane is not altered by osmosis. Should HFO particles be added now to the solutions on both sides of the membrane at equilibrium, arsenic sorption capacity on the left side would be relatively low due to significantly lower aqueous phase arsenic concentration caused by the presence of nonpermeating R⁻.

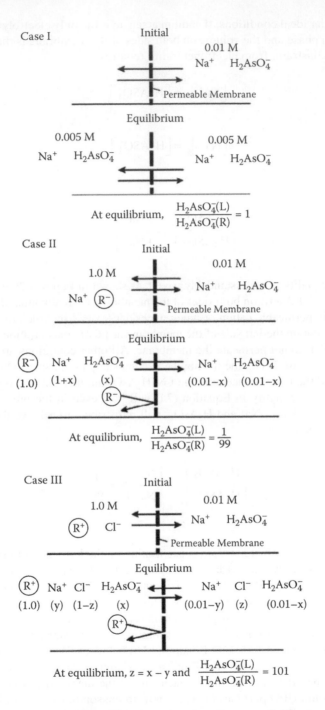

FIGURE 7.28 Three specific examples presenting Donnan distribution of arsenate ($H_2AsO_4^-$) when the membrane is permeable to (Case I) all the ions; (Case II) all the ions except R^-; and (Case III) all the ions except R^+.

Case III in Figure 7.28 is, in principle, similar to Case II except that the nonpermeating ion (R^+) is a cation. The relative distributions of Na^+, Cl^-, and $H_2AsO_4^-$ on both sides of the membrane, after necessary calculations, stand at equilibrium as follows:

$$\frac{\left[Na^+\right]_R}{\left[Na^+\right]_L} = \frac{\left[H_2AsO_4^-\right]_L}{\left[H_2AsO_4^-\right]_R} = \frac{\left[Cl^-\right]_L}{\left[Cl^-\right]_R} = \frac{101}{1} \tag{7.6}$$

Contrary to Case II, arsenic concentration on the left side $[H_2AsO_4^-]_L$, is nearly two orders of magnitude greater than that of $[H_2AsO_4^-]_R$. The addition of HFO particles in the LHS of the membrane will therefore offer very high sorption capacity due to enhanced arsenic concentration at equilibrium. The chloride concentration is also enhanced at the left side of the membrane, but HFO particles have negligible sorption affinities for chloride anions. The fixed nature of charged functional groups residing on the ion exchanger makes the phase boundary between the exchanger and bulk solution act like a virtual semipermeable membrane described above.

To elucidate the inability of the HCIX hybrid cation exchanger to remove arsenic, let us consider a typical cation exchanger bead of 0.5 mm or 500 μm diameter (d_p). Considering a bead density (ρ_b) of 1100 kg/m³, the mass (m) of a single bead is 7.2×10^{-5} g. The estimated capacity of C-100 cation-exchange resin is $q = 4.0$ eq/kg. The number of fixed negative charges (Ne) in a single bead is

$$N_e = \frac{qm}{1000} X6.02X10^3 = 1.73X10^{17} \text{ charges}$$

where 6.02×10^{23} is Avogadro's number.

Thus, inside a tiny 500 μm cation exchange bead are 1.73×10^{17} covalently attached sulfonic acid groups with negative charges that cannot permeate from the exchanger phase to the aqueous phase. Comparing this situation with Case II in Figure 7.28, it may be readily inferred that the monovalent and divalent arsenate species will be completely prevented from permeating into the polymer phase due to the Donnan exclusion effect. HFO nanoparticles dispersed inside the cation exchanger are thus inaccessible to arsenate. That is why HCIX did not reveal arsenic removal capacity for the column run presented in Figure 7.27. It is worth mentioning that activated carbon, zeolite, alginate, etc. also contain significant concentrations of negatively charged functional groups, namely carboxylate and aluminosilicate. These substrates may be easily dispersed with HFO nanoparticles, but arsenic removal capacity will not be fully attained due to the Donnan exclusion effect. In a previous study, alginate loaded with HFO showed poor arsenic removal capacity during fixed bed column runs[54].

Conversely, polymeric anion exchangers are excellent substrates because they allow enhanced permeation of anions within the polymer phase due to high concentrations of fixed positive charges. Figure 7.29 illustrates the difference between cation and anion exchangers as substrate materials. Note that both

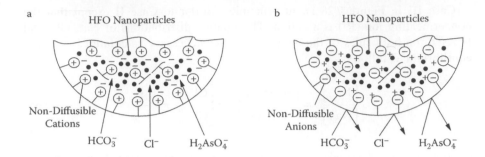

FIGURE 7.29 (a) enhanced permeation of anions into hybrid sorbent in the presence of nondiffusible cations (anion exchanger) and (b) exclusion of anions from hybrid sorbent in the presence of nondiffusible anions (cation exchanger).

cation and anion exchanger beads are electrically neutral. Electrostatic repulsion or attraction is not the underlying reason for the difference in permeation of arsenate into the polymer phase. The high concentrations of nondiffusing fixed charges (R^+ or R^-) in the polymer phase act as highly permeable or impermeable interfaces for arsenate, thus influencing its sorption onto the HFO particles embedded in the polymer phase. Thus, a polymeric substrate acts as a robust and hydraulically suitable support material and also influences the adsorption capacity of the hybrid ion exchange resin. Judicious design of a hybrid polymeric sorbent with an appropriate functional group provides a synergy leading to the enhancement of sorption capacities of metal oxide nanoparticles not otherwise achievable. The Donnan principle can be successfully used to design and develop novel materials and engineered processes[83].

7.4.5 Simultaneous Removal of Perchlorate and Arsenate

Both perchlorate (ClO_4^-) and arsenate ($H_2AsO_4^-$ or $HAsO_4^{2-}$) exist as anions in the aqueous phase; however, their chemical characteristics are very different. While perchlorate is a hydrophobic anion with poor ligand characteristics, arsenate is a fairly strong ligand. Several groundwater sources in the western United States are contaminated with trace concentrations of both arsenic and perchlorate. Figure 7.30a shows the effluent histories of both perchlorate and arsenic for an HAIX column run using simulated groundwater.

For nearly 10,000 bed volumes, both arsenic and perchlorate were completely removed. The parent strong base anion exchange resin was unable to remove arsenic as demonstrated earlier. For comparison, simultaneous perchlorate and arsenic removal was also examined independently with a commercially available iron oxide based sorbent medium [granular ferric hydroxide (GFH), U.S. Filter Co.]. Figure 7.30b presents effluent concentration profiles of perchlorate and arsenic through a GFH column. Note that the GFH was unable to remove perchlorate but arsenic removal was satisfactory with it. HAIX has two distinctly different binding sites within the polymer phase: (1) covalently attached quaternary ammonium

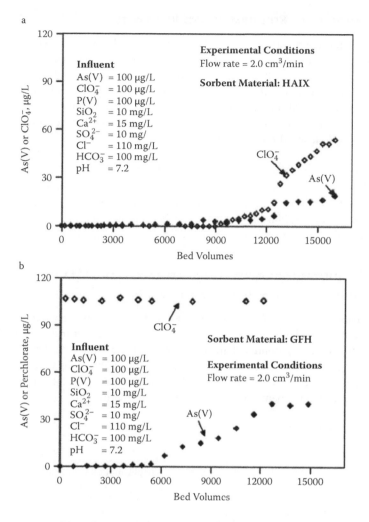

FIGURE 7.30 (a) Effluent histories of arsenic and perchlorate during fixed bed column run with HAIX. (b) Effluent histories of arsenic and perchlorate during a fixed bed column run using granulated ferric hydroxide (GFH) sorbent.

functional groups have a high affinity toward hydrophobic anions such as perchlorate; (2) surface hydroxyl groups of HFO have a high affinity toward ligands such as arsenites and arsenates. These two classes of sorption sites are also independent of each other. During the run shown in Figure 7.30a, quaternary ammonium groups were responsible for selective sorption of perchlorate ions while HFO surface groups bound the arsenate via the formation of inner sphere complexes. Thus, a hybrid ion exchanger, if designed appropriately, can achieve simultaneous removal of multiple contaminants with drastically different chemical characteristics. Hybrid ion exchangers with dual zones combined with magnetic activities were synthesized earlier [84].

7.4.6 Efficiency of Regeneration and Reusability

Because of their chemical stability and durable physical structure, hybrid polymeric sorbents are amenable to regeneration. Regeneration and subsequent containment of the treatment residuals make the removal process cost effective and environmentally sustainable. HAIX materials can be reused over many cycles and its regeneration brings down the cost of the treated water. It is also possible to concentrate the arsenic or phosphate removed in a small volume of spent regenerant that can subsequently be transformed into a small mass of solids[85,86].

Regeneration of both the exhausted resins used for phosphate and arsenic removal was accomplished using a solution containing 2 wt% of NaCl and 2 wt% of NaOH. Figure 7.31 shows the eluent concentration profile for the HAIX used for arsenic removal. Nearly 95% of the adsorbed arsenic exited the hybrid ion exchanger phase within 15 bed volumes. Phosphate showed similar characteristics. Figure 7.32 shows the eluent concentration profiles of phosphate over the bed volumes of the regenerant used.

7.4.7 Commercialization and Field Applications of HAIX Nanosorbents: India and United States

HAIX nanosorbents have been used for treatment of arsenic contaminated water around the world including the remote villages of West Bengal in India. Millions of people on the Indian subcontinent are at risk of developing arsenic-related health hazards from ingestion of high concentrations of arsenic in drinking water drawn from underground sources. The arsenic removal units are essentially adsorption columns housing hybrid nanosorbents. The unique feature in the design of the column is that

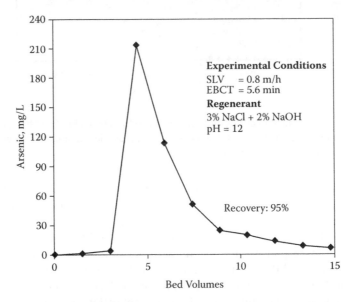

FIGURE 7.31   Dissolved arsenic concentration profile during desorption of HAIX using 2% NaOH and 3% NaCl as the regenerant.

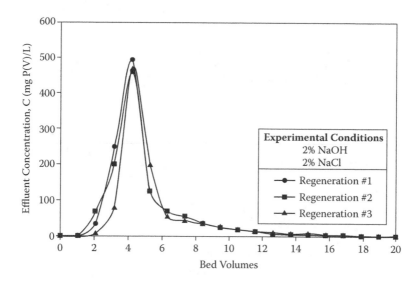

FIGURE 7.32 Phosphate elution profiles during regeneration of HAIX resin with high phosphate recovery (>95%) in 12 bed volumes.

the adsorption column is deliberately kept open to the atmosphere; dissolved iron in the raw water is oxidized to form HFO particulates that aid in removing some of the arsenic in the water. The remaining arsenic is removed by the HAIX nanosorbents[87].

Each community-scale adsorption column provides arsenic-safe drinking water to an average of 1,000 people for about a year. After its capacity is exhausted, the medium is regenerated at a central regeneration facility by a group of trained villagers. To date, about 200 such community-based units have been installed in the state. While most of the units use activated alumina as the adsorbent, HAIX has been used in about 20 different places. They all run very well; the length of a run often exceeds 10,000 bed volumes. Figure 7.33 shows a representative breakthrough profile of arsenic for a removal unit at Nabarun Sangha in Ashoknagar in the N 24 Parganas District. Phosphate and silica broke through at the initial stage of the column run. Under identical conditions, the arsenic removal units using the hybrid anion exchanger exhibited superior performance compared to those using activated alumina as the adsorbent medium. Details about the performance of the arsenic removal units, the sustainability of the process via regeneration and reuse of the media, and ecologically sound management of treatment residuals can be found in the literature[85–87].

HAIX nanosorbents are commercially available in the United States. They are manufactured and sold by SolmeteX, Inc. under the trade name Layne[RT]. This product is a modified version of the first generation of commercial HAIX (ArsenXnp). To date, more than one million pounds of the hybrid nanosorbent are in use in the United States and around the world to abate arsenic poisoning from naturally contaminated groundwater. Figure 7.34a is a photograph of a plant in Sahuarita, Arizona that uses Layne. Figure 7.34b represents the effluent histories of a pilot scale run at the same location where the hybrid adsorbent was observed to perform better than other iron oxide-based media.

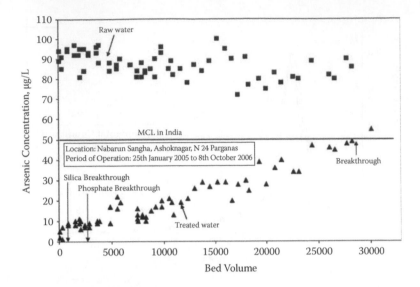

FIGURE 7.33 Breakthrough history of arsenic removal unit at Ashoknagar in West Bengal, India using hybrid anion exchange resin.

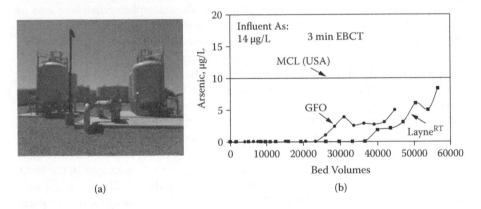

FIGURE 7.34 (a) Plant in Sahuarita, Arizona using Layne[RT] for arsenic removal. (b) Arsenic breakthrough profiles from pilot plant run prior to installation. GFO = granulated ferric oxide. MCL = maximum contaminant limit. (Courtesy of SolmeteX, Inc.)

7.5 TUNABLE HYBRID POLYMERIC NANOSORBENTS

Hydrated Fe(III) oxides (HFOs) and other metal oxides such as zirconium(IV), titanium(IV), and aluminum(III) oxides exhibit amphoteric sorption behaviors; i.e., they can selectively bind Lewis acids or transition metal cations (e.g., Cu^{2+}) and Lewis bases or anionic ligands (e.g., arsenates and phosphates) through the formation of inner sphere complexes. Thus, Cu^{2+}, which is a Lewis acid and environmentally regulated heavy metal, is sorbed in preference to other competing but innocuous alkaline and alkaline–earth metal cations: Na^+, Ca^{2+}, and Mg^{2+}.

Similarly, the sorption of arsenate, which is an anionic ligand, is preferred over commonly encountered anions such as sulfate, chloride, and bicarbonate. The point of zero charge (PZC) of crystalline or amorphous iron oxide colloids in an inert electrolyte (sodium nitrate, sodium perchlorate, or equivalent) resides within a pH range of 6.5 to 8.5. At near-neutral pH, iron oxide nanoparticles sorb both Cu^{2+} and $H_2AsO_4^-$ simultaneously and selectively in the presence of common competing ions, such as Na^+, Ca^{2+}, Cl^-, SO_4^{2-}, and HCO_3^- that can form only outer-sphere complexes through Coulombic interactions. The metal oxide nanoparticles have very high specific surface areas that make them excellent adsorbents.

It was conceptualized that if the HFO or other metal oxide nanoparticles were dispersed in a cation or anion exchanger, the anions or cations would be rejected by the ion exchanger in accordance with the Donnan co-ion exclusion effect. Consequently, amphoteric HFO nanoparticles will selectively bind transition metal cations (e.g., Cu^{2+}) or anionic ligands ($HAsO_4^{2-}$). The ion-exchanger host material for HFO nanoparticles can be made to completely reject the transition metal cations while allowing enhanced selective sorption of anionic ligands and vice versa. Thus, in principle, a hybrid sorbent amphoteric metal oxide nanoparticles can be tailored to behave strictly as metal-selective sorbents or as containing ligand-selective exchangers. It is likely that such tunability for complete rejection or selective uptake of target ions may also be easily incorporated into ion exchange membranes.

7.5.1 Evidence of Tunability

We performed three separate fixed-bed column runs, using (1) commercially available GFH from U.S. Filter Co. without any ion exchanger support material; (2) HFO dispersed in a cation exchanger (referenced as HCIX-Fe); and (3) HFO dispersed in an anion exchanger (referenced as HAIX-Fe). The feed compositions were identical in all three cases, and trace concentrations of both anionic As(V) or $H_2AsO_4^-$ and Cu^{2+} were present as target solutes along with other electrolytes. The hydrodynamic conditions, namely, superficial liquid velocity (SLV) and empty bed contact time (EBCT) were also identical. Figure 7.35 shows the column-run results for the three runs: (1) GFH removed both As(V) anions and Cu^{2+} significantly; (2) HCIX-Fe removed only Cu^{2+} very selectively for over 2,000 bed volumes but rejected As(V) anions completely; and (3) HAIX-Fe showed extraordinary As(V) sorption with a minimum breakthrough for nearly 5.000 bed volumes while Cu^{2+} broke through immediately.

The parent cation exchanger with sulfonic acid functional groups can sorb cations only through electrostatic interaction and thus offers no specific selectivity for Cu^{2+} in the presence of greater concentrations of other competing cations such as calcium and sodium. Similarly, anion exchangers with quaternary ammonium functional groups exhibit no specific selectivity toward anionic arsenate in the presence of competing sulfate anions. Figure 7.36a demonstrates that the sorbed copper may be efficiently desorbed from HCIX-Fe at a slightly acidic pH. HAIX-Fe with arsenate was amenable to efficient regeneration with an alkaline solution (Figure 7.36b). Thus, Cu^{2+} and $HAsO_4^{2-}$ may be separated and recovered quantitatively from the same solution

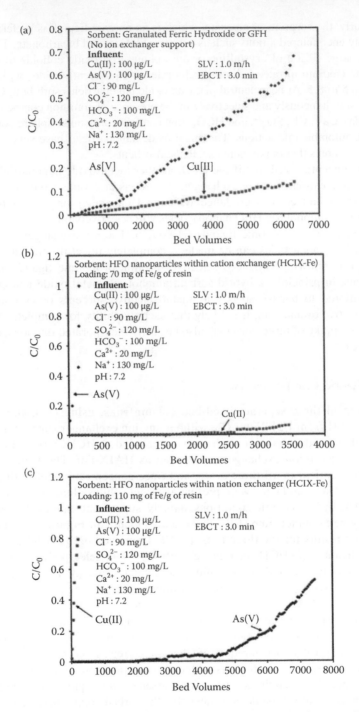

FIGURE 7.35 Concentration profiles of Cu(II) and As(V) during column runs for (a) granulated ferric hydroxide (GFH), (b) HCIX-Fe, and (c) HAIX-Fe. SLV = superficial liquid velocity. EBCT = empty bed contact time.

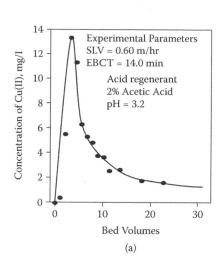

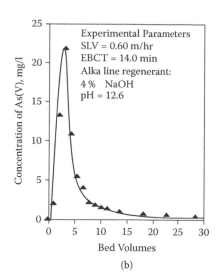

(a)   (b)

FIGURE 7.36 Concentration profiles of (a) Cu(II) and (b) As(V) during regeneration of HCIX-Fe and HAIX-Fe. Cu(II) recovery was 90%; AsV recovery was 96%. SLV = superficial liquid velocity. EBCT = empty bed contact time.

using HFO nanosorbents through appropriate choice of ion exchanger materials as support. There was no measurable loss of iron from HCIX-Fe or HAIX-Fe following regeneration.

To further confirm that the sorption behaviors of other amphoteric metal oxide nanoparticles may be tuned similarly, we studied zirconium oxide (ZrO_2) by dispersing zirconium nanoparticles within a cation exchanger, using a procedure described in the literature [88]. The resulting hybrid cation exchanger (HCIX-Zr) was used in a batch sorption study in which both copper and arsenate were present in trace concentrations. Figure 7.37 shows that while copper concentration decreased to almost zero within an hour, As(V) concentration remained essentially unchanged.

7.6 HYBRID ANION EXCHANGE FIBERS

Like arsenic, perchlorate is also viewed as a trace contaminant of major environmental concern. It produces adverse health effects even at extremely low concentrations. Both arsenic and perchlorate exist in water as oxyanions, but their chemistry and genesis in groundwater are different. Unlike arsenic, perchlorate is a poorly hydrated inert anion [91]. It is more mobile in subsurface water than arsenic. In several arsenic-contaminated groundwater aquifers in the Western United States and elsewhere, other trace contaminants including perchlorate have been reported [92]. HAIX resins used in fixed bed sorption processes have proven efficient for removing trace contaminants like arsenic and also are amenable to efficient regeneration and reuse over multiple cycles of operation.

Unlike arsenate, perchlorate has no sorption affinity toward HFO surface binding sites. However, anion exchanger host materials with polystyrene divinylbenzene

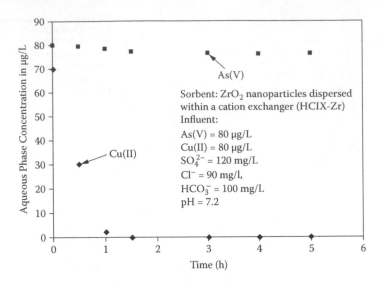

FIGURE 7.37 Aqueous phase arsenic and copper concentration profiles during a batch sorption test using HCIX-Zr.

matrices containing quaternary ammonium functional groups exhibit high sorption affinities for perchlorate [93,94]. HAIX has two characteristically different binding sites: HFO surfaces with high affinity toward anionic ligands and quaternary ammonium functional groups with preferences for hydrophobic anions. That is why HAIX resin efficiently removes both arsenate and perchlorate anions simultaneously from synthetic contaminated water, as shown in Figure 7.30, while GFH was unable to remove perchlorate. This observation has been confirmed by others [82].

The selective sorption of arsenate ligand onto the HFO surfaces is strongly dependent on pH. Hence, ligands like arsenic can be easily and rapidly desorbed from binding sites by raising the pH with an innocuous desorbent like 2% NaOH. On the other hand, the sorption and desorption of perchlorate are kinetically controlled by intraparticle diffusion and are independent of pH. However, for weak base anion exchange resins, sorption–desorption behavior can be kinetically controlled by "swinging" the pH, as with other anionic solutes. In previous efforts, the perchlorate could not be efficiently regenerated from the exhausted HAIX resins by using an innocuous brine solution over the entire range of pH. The possible cause of this inefficient regeneration may have been the significantly long diffusion path lengths for spherical ion exchanger beads of an average size of 600 μm [82, 92]. Reduction in size of the ion exchange resins may prove effective, but the major disadvantage is that smaller resin size would drastically increase the fixed bed pressure drop. An efficient regeneration technique for perchlorate utilizing $FeCl_4^-$ under very high acidic conditions was demonstrated recently [96]. However, it is not suitable for HAIX because extremely acidic pH promotes dissolution of HFO nanoparticles from HAIX beads.

Cylindrical ion exchange fibers with diameters of 20 to 50 μm have short diffusion path lengths and are amenable to use in fixed bed columns [89,97,98]. Because of very high void fractions, no increase in the head loss across the bed occurs. Fibers with polystyrene–divinylbenzene matrices and quaternary ammonium functionality, if impregnated with HFO nanoparticles, should exhibit high selectivity toward both arsenic and perchlorate, as do their resin counterparts. They should also show efficient regeneration capability from perchlorate with innocuous regenerants such as NaOH and NaCl solutions. A superior regenerability was expected because of the shorter diffusion path lengths offered by the ion exchange fibers. Fiban A-1 fibers were obtained from the Institute of Physical Organic Chemistry of the National Academy of Science in Belarus and used for laboratory study. The properties and structure of the fibers are described in the literature [98]. Dispersal of HFO nanoparticle on the fiber was accomplished via a two-step process similar to the one used to prepare HAIX resin beads. HFO content was approximately 110 mg/g as Fe.

Figure 7.38a shows an enlarged view of the HAIX-F hybrid anion exchange fiber. Figure 7.38b is an SEM image of the surface of the parent Fiban A-1 fiber at 2000× magnification. Figure 7.38c shows the surface of the hybrid fiber at 2000× magnification. The change in the surface roughness due to HFO loading can be readily observed. Figure 7.39a shows a cross-sectional SEM image of a sliced HAIX-F. Figure 7.39b shows the energy dispersive x-ray mapping of iron across the cross section. A greater concentration of HFO nanoparticles can be observed at the periphery of the fibers.

Fixed bed column runs with 1 g of hybrid ion exchange fibers were performed with synthetic feed water containing both arsenate and perchlorate ions along with other commonly occurring anions present in natural water. The exhausted HAIX-F column was regenerated in two consecutive steps using 2% NaOH followed by 10% NaCl. Prior to resuming the column run for the next cycle, the bed was protonated using carbon dioxide-sparged water. No excess pressure drop or head loss was observed during the lengthy column runs.

Figure 7.40 shows the effluent history for arsenic and perchlorate for three consecutive column runs. The bed containing the fibers was regenerated between the column runs following the protocol mentioned above. The results indicate only a minor change in the perchlorate and arsenic effluent histories during consecutive runs.

Figure 7.41 shows the overall uptake capacities of the hybrid fibers for arsenate and perchlorate for each run. The overall sorption uptakes remained practically unchanged even after two regenerations. This confirms that the HFO nanoparticles are irreversibly

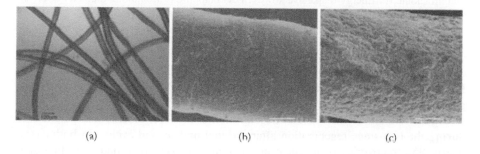

(a)   (b)   (c)

FIGURE 7.38   (a) Photograph of HAIX fiber at 10× magnification. (b) SEM image of surface of parent fiber at 2,000× magnification. (c) SEM image of surface of hybrid fiber at 2000× magnification.

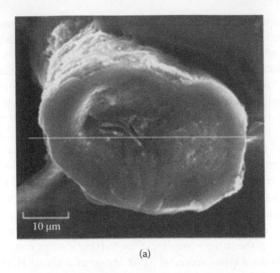

(a)

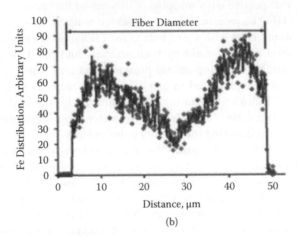

(b)

FIGURE 7.39   (a) SEM image of cross-section of hybrid anion exchange fiber at 2,000×
magnification. (b) Energy-dispersive x-ray (EDX) mapping of iron along diameter of cross-
section of hybrid fiber.

contained within the gel phase of the ion exchange fiber and the regeneration process
did not adversely affect the concentration or activity of the hybrid ion exchange fiber.
For thousands of bed volumes, breakthroughs were undetectable for both arsenate and
perchlorate, even for the third run.

Figure 7.42 shows the elution concentration profiles for arsenic and perchlorate
during the two-stage regeneration after the first and second cycles. In both cases,
nearly 95% recoveries of arsenate and perchlorate were recorded. The first step
using 2% NaOH solution was effective in regenerating the adsorbed arsenic from the
exhausted hybrid fiber. The rise in pH caused the surface functional groups of HFO
to change their polarity. As a result of the polarity reversal, the arsenic was rejected

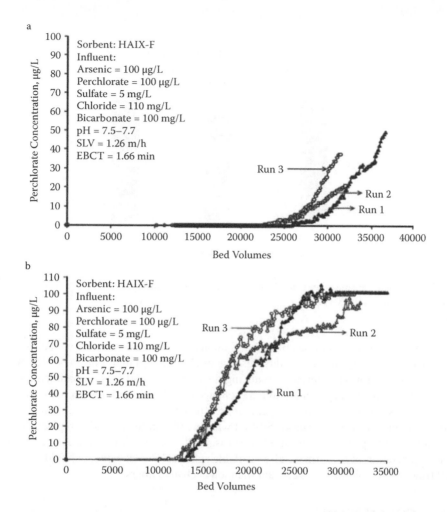

a

Sorbent: HAIX-F
Influent:
Arsenic = 100 µg/L
Perchlorate = 100 µg/L
Sulfate = 5 mg/L
Chloride = 110 mg/L
Bicarbonate = 100 mg/L
pH = 7.5–7.7
SLV = 1.26 m/h
EBCT = 1.66 min

Run 3

Run 2

Run 1

b

Sorbent: HAIX-F
Influent:
Arsenic = 100 µg/L
Perchlorate = 100 µg/L
Sulfate = 5 mg/L
Chloride = 110 mg/L
Bicarbonate = 100 mg/L
pH = 7.5–7.7
SLV = 1.26 m/h
EBCT = 1.66 min

Run 3

Run 2

Run 1

FIGURE 7.40 (a) Effluent history of arsenate during three consecutive column runs with HAIX-F. (b) Effluent history of perchlorate during three consecutive runs with HAIX-F.

by the adsorbent. The mechanism of desorption of arsenic is similar to its desorption from HAIX beads and is detailed in the literature [82,87,89].

It is interesting to note that regeneration allowed almost complete separation of arsenic-rich effluent from that of perchlorate. Conceptually, high concentrations of common anions like Cl^-, SO_4^{2-}, and OH^- can desorb perchlorate ions from an exhausted resin. The strong base anion exchangers are least selective toward hydroxyl ions. Hence, the first step of regeneration involving 2% NaOH solution could not desorb any significant perchlorate concentration from the anion exchange sites, although it was effective in desorbing arsenate ions from the HFO nanoparticles. Cl^- ions at high concentrations (10%) were chosen as effective regenerants over SO_4^{2-} ions because, at such a high ionic concentrations, chloride ions have greater selectivity than sulfate ions due to a phenomenon known as selectivity reversal [99,100].

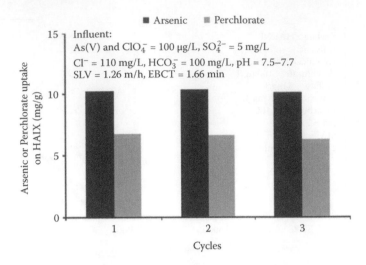

FIGURE 7.41 Arsenate and perchlorate uptake during multiple cycles of sorption and desorption using HAIX-F.

However, for the common strong base anion exchange resins, very low intraparticle diffusivity of perchlorate in the presence of chloride is a major obstacle affecting efficient regeneration. Therefore, reducing the intraparticle diffusion path length by replacing relatively large spherical ion exchange resin beads with ion exchange fibers is a clever strategy to improve regeneration efficiency. It is calculated that the time required for desorption deceases 50-fold when a 500 μm resin bead is replaced by a resin fiber of 50 μm diameter. In other words, if all other conditions remain identical, the regeneration of a commercially available spherical ion exchanger should require 50 times more volume of regenerant than ion exchange fibers [90].

7.7 CONCLUSIONS

Polymeric inorganic hybrid ion exchangers represent a new class of porous material ion exchangers presenting two heterogeneously combined distinct phases: (1) a functionalized polymeric phase that primarily acts as a host and (2) an inorganic metal or metal oxide phase that is finely dispersed within the host. The properties of the hybrid material often depend on the natures of functional groups on the polymeric phase and on the sizes and natures of the nanoparticles to be dispersed. The diverse natures of the functional groups and the different factors controlling the sizes of the dispersed nanoparticles often make effective dispersal and control of the properties of the hybrid material challenging tasks.

The hybrid material retains the properties of both of its phases and offers a synergy that is not otherwise attainable by the phases individually. In the more familiar physical world, a hybrid material can be compared with an alloy that consists of different metals; the metals retain their individual identities on a microscopic scale but their union produces or enhances significantly different properties not present in the

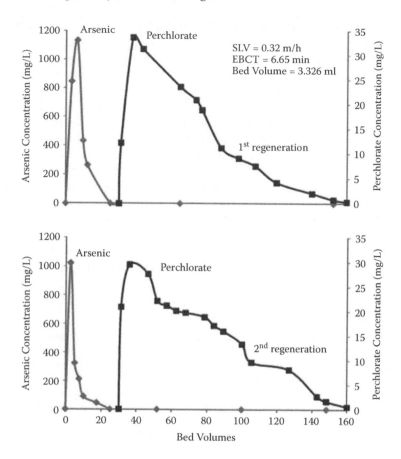

FIGURE 7.42 Elution concentration profiles of arsenate and perchlorate during two-stage regeneration after first and second column runs.

individual metals. Dispersal of magnetite nanoparticles imparts magnetic properties to polymeric sorbents that make them effective for many applications that would not be otherwise possible. A few examples of such specialized applications include environmental forensics to monitor the sources of water-contaminating pollutants and sequestration of target contaminants from backgrounds of complex matrices such as sludges, slurries, viscous liquids, and biomass systems.

The Donnan membrane effect arising from the inability of a charged functional group to diffuse from one phase to another is effectively exploited to develop a new class of hybrid nanosorbents. With an anion exchanger as a support, a hybrid sorbent significantly enhances the ligand sorption capacities of the nanoscale metal oxide particles dispersed inside it. A change of the functionality of the polymeric support renders the hybrid material selective for the sorption of trace heavy metals from contaminated water. Thus, the sorption behaviors of the hybrid nanosorbents can be tuned by changing the natures of the polymer functional groups so that complete rejection or selective uptake of target species is possible.

Hybrid nanosorbents also offer excellent regenerability that is not possible for bare metal oxide nanoparticles. Irreversible dispersion of the nanoparticles of metal oxides in conjunction with shorter diffusion path lengths make hybrid ion exchange fibers suitable candidates for simultaneous removal of two distinctively different trace anions, such as arsenate and perchlorate, and provide ease of regeneration, allowing multiple cycles of operation. The multiple cycle use of hybrid fibers can be extended to the treatment of water or wastewater containing other hydrophobic anions such as aromatic and linear long chain sulfonates, pentachlorophenols, and other ligands such as phosphate, vanadate, oxalate, and chromate.

To date, only a few combinations have been tested as hybrid ion exchangers; many other opportunities remain. Nevertheless, hybrid ion exchangers have already found practical applications in several fields including environmental remediation. Although major advancement in the environmental field concerned selective sorption of target contaminants, ample opportunities exist to develop new types of hybrid ion exchangers, for example, for environmental catalysis, selective degradation of organic and inorganic trace contaminants, photocatalytic reactions, disinfection, and others.

REFERENCES

1. Wang, C.G. and Zhang, W. Synthesizing nanoscale iron particles for rapid and complete dechlorination of TCE and PCBs, *Environ. Sci. Technol.* **31** (1997), 2154–2156.
2. Matheson, L.J. and Tratnyek, P.G. Reductive dehalogenation of chlorinated methanes by iron. *Environ. Sci. Technol.* **28** (1994), 2045–2053.
3. Miehr, R., Tratnyek, P. G., Bandstra, J. Z. et al. The diversity of contaminant reduction reactions by zero-valent iron: Role of the reductate. *Environ. Sci. Technol.* 38 (2004), 139–147.
4. Cheng, F., Muftikian R., Fernando, Q. et al. Reduction of nitrate to ammonia by zero-valent iron. *Chemosphere* **35** (1997), 2689–2695.
5. Manning, B.A., Fendorf, S.E., and Goldberg, S. Surface structures and stability of arsenic(III) on goethite: Spectroscopic evidence for inner-sphere complexes. *Environ. Sci. Technol.* **32** (1998), 2383–2388.
6. Cotton, F.A. and Wilkinson, J. *Advanced Inorganic Chemistry*, Wiley Interscience, New York (1998).
7. Bajpai, S. and Chaudhury, M. Removal of arsenic from ground water by manganese dioxide-coated sand. *Environ. Eng. Div. J. ASCE* **125** (1999), 782–784.
8. Xiong, Z., Zhao, D. and Pan. G. Rapid and complete destruction of perchlorate in water and ion-exchange brine using stabilized zero-valent iron nanoparticles. *Water Res.* 41 (2007), 3497–3505.
9. Chen, S., Cheng, C, Li, C. et al. Reduction of chromate from electroplating wastewater from pH 1 to 2 using fluidized zero valent iron process. *J. Haz. Mater.* 142 (2007), 362–367.
10. Lee, Y., Rho, J., and Jung, B. Preparation of magnetic ion exchange resins by the suspension polymerization of styrene with magnetite. *J. Appl. Polym. Sci.* 89 (2003), 2058-2067.
11. Eldridge, R.J., Norret, M., Dahlke, T.W. et al. Complexing resins and method for preparation thereof. U.S. Patent 7,514,500, April 7, 2009.
12. Shenhar, R., Norsten, T.B., and Rotello, V.A. Polymer-mediated nanoparticle assembly: structural control and applications. *Adv. Mater.* 17 (2005), 657–669.
13. Schubert, U., Gao, Y., and Kogler, F.R. Tuning properties of nanostructured inorganic–organic hybrid polymers obtained from metal oxide clusters as building blocks. *Prog. Solid State Chem.* 35 (2007), 161–170.

14. Balazs, A.C., Emrick, T., and Russell, T.P. Nanoparticle polymer composites: Where two small worlds meet. *Science* 314 (2006), 1107–1110.
15. Xu, Z.Z., Wang, C.C., Yang, W.L. et al. Encapsulation of nanosized magnetic iron oxide by polyacrylamide via inverse miniemulsion polymerization. *J. Magnet. Magnet. Mater.* 277 (2004), 136–143.
16. Li, P. and SenGupta, A.K. Genesis of selectivity and reversibility for sorption of synthetic aromatic anions onto polymeric sorbents. *Environ. Sci. Technol.* 32 (1998), 3756–3766.
17. Zhao, D. and SenGupta, A.K. Ultimate removal of phosphate using a new class of anion exchangers. *Water Res.* 32 (1998), 1613–1625.
18. SenGupta, A.K., Zhu, Y., and Hauze, D. Metal ion binding onto chelating exchangers with multiple nitrogen donor atoms. *Environ. Sci. Technol.* 25 (1991), 481–88.
19. Browski, A.D., Hubicki, Z., Podkoscielny, P. et al. Selective removal of heavy metal ions from waters and industrial wastewaters by ion-exchange. *Chemosphere* 56 (2004), 91–106.
20. Ghurye, G.L., Clifford, D.A., and Tripp, A.R. Combined arsenic and nitrate removal by ion exchange, *J. AWWA* 91 (1999), 85–96.
21. Kim. J. and Benjamin, M.M. Modeling a novel ion exchange process for arsenic and nitrate removal. *Water Res.* 38 (2004), 2053–2062.
22. Bolto, B., Dixon, D., Elridge, R. et al. Removal of natural organic matter by ion exchange. *Water Res.* 36 (2002), 5057–5065.
23. Furusawa, K., Nagashima, K., and Anzai, C. Synthetic process to control the total size and component distribution of multilayer magnetic composite particles. *Colloid Polym. Sci.* 272 (1994), 1104–1110.
24. Tanyolac, D. and Ozdural, A.R. Preparation of low-cost magnetic nitrocellulose micro-beads. *React. Funct. Polym.* 45 (2000), 235–242.
25. Tanyolac, D. and Ozdural, A.R. A new low cost porous magnetic material: Magnetic polyvinylbutyral microbeads. *React. Funct. Polym.* 43 (2000), 279–286.
26. Ren, J., Hong, H., Ren, T. et al. Preparation and characterization of magnetic PLA–PEG composite nanoparticles for drug targeting. *React. Funct. Polym.* 66 (2006), 944–951.
27. Yanase, N., Noguchi, H., Asakura, H. et al. Preparation of magnetic latex particles by emulsion polymerization of styrene in the presence of a ferrofluid. *J. Appl. Polym. Sci.* 50 (1993), 765–776.
28. Kondo, A., Kamura, H., and Higashitani, K. Development and application of thermo-sensitive magnetic immunomicrospheres for antibody purification. *Appl. Microbiol. Biotechnol.* 41 (1994), 99–105.
29. Horak, D. Magnetic polyglycidylmethacrylate microspheres by dispersion polymerization *J. Polym. Sci. A Polym. Chem.* 39 (2001), 3707–3715.
30. Yang, C., Liu, H., Guan, Y. et al. Preparation of magnetic poly(methylmethacrylate–divinylbenzene–glycidylmethacrylate) microspheres by spraying suspension polymerization and their use for protein adsorption. *J. Magn. Magn. Mater.* 293 (2005), 187–192.
31. Lee, W. and Chung, T. Preparation of styrene-based, magnetic polymer microspheres by a swelling and penetration process. *React. Funct. Polym.* 68 (2008), 1441–1447.
32. Drever, J.I. *Geochemistry of Natural Water.* Prentice-Hall, Englewood Cliffs, NJ, 1988.
33. Leun, D. and SenGupta, A.K. Preparation and characterization of magnetically active polymeric particles (MAPPs) for complex environmental separations. *Environ. Sci. Technol.* 34 (2000), 3276–3282.
34. Cumbal, L., Greenleaf, J.E., Leun, D. et al. Polymer supported inorganic nanoparticles: Characterization and environmental applications. *React. Funct. Polym.* 54 (2003), 167–180.
35. Helfferich, F.G. *Ion Exchange.* McGraw Hill, New York, 1962.
36. Blum, P. Magnetic susceptibility. In *PP Handbook.* 1997, Chap. 4. Downloaded from www-odp.tamu.edu/publications/tnotes/tn26/CHAP4.PDF on 4/21/2010.

37. Diniz, C.V., Ciminelli, V.S.T. and Doyle, F.M. The use of the chelating resin Dowex M-4195 in the adsorption of selected heavy metal ions from manganese solutions. *Hydrometallurgy* 78 (2005), 147–155.

38. SenGupta, A.K. and Zhu, Y. Metals sorption by chelating polymers: A unique role of ionic strength. *AIChE J.* 38 (1992), 153–157.

39. Brown, C. J. and Dejak, M. J. Process for removal of copper from solutions of chelating agent and copper. U.S. Patent 4,666,683.

40. Zhu, Y., Millan, E., and SenGupta, A. K. Toward separation of toxic metal (II) cations by chelating polymers: Some noteworthy observations. *React. Polym.* 13 (1990), 241–253.

41. Stum, W. and Morgan, J.J. *Aquatic Chemistry: Chemical Equilibria and Rates in Natural Waters.* Wiley Interscience, New York, 1996.

42. Ghosh, M. M. and Yuan, J.R. Adsorption of inorganic arsenic and organicoarsenicals on hydrous oxides. *Environ. Prog.* 3 (1987), 150–157.

43. Dutta, P.K., Ray, A.K., Sharma, V.K. et al. Adsorption of arsenate and arsenite on titanium dioxide suspensions. *J. Colloid Interface Sci.* 278 (2004), 270–275.

44. Pierce, M.L. and Moore, C.B. Adsorption of As(III) and As(V) on amorphous iron hydroxide. *Water Res.* 6 (1982), 1247.

45. Jang, J.H. Surface chemistry of hydrous ferric oxide and hematite as based on their reactions with Fe(II) and U(VI). Ph.D. Dissertation, Pennsylvania State University, 2004.

46. Dixit, S. and Hering, J.G. Comparison of arsenic(V) and arsenic(III) sorption onto iron oxide minerals: Implications for arsenic mobility. *Environ. Sci. Technol.* 37 (2003), 4182–4189.

47. Driehaus, W., Jekel, M., and Hildebrandt, U. Granular ferric hydroxide: A new adsorbent for the removal of arsenic from natural water. *J Water SRT Aqua* 47 (1998), 30–35.

48. McNeil, L.S. and Edwards, M. Soluble arsenic removal at water treatment plants. *J. AWWA* 87 (1995), 105–114.

49. Dzombak, D.A. and Morel, F.M. *Surface Complexation Modeling: Hydrous Ferric Oxide.* Wiley Interscience, New York, 1990.

50. Kartinen, E.O. and Martin, C.J. Overview of arsenic removal processes. *Desalination* 103 (1995), 79–88.

51. Gao, Y., SenGupta, A.K. and Simpson, D. A new hybrid inorganic sorbent for heavy metals removal. *Water Res.* 29 (1995), 2195–2205.

52. Manning, B.A., Fendorf, S.E., and Goldberg, S. Surface structures and stability of As(III) on goethite: Spectroscopic evidence for inner-sphere complexes. *Environ. Sci. Technol.* 32 (1998), 2383–2388.

53. Min, J.M. and Hering, J. Arsenate sorption by Fe(III)-doped alginate gels. *Water Res.* 32 (1998), 1544–1552.

54. Zouboulis, A.I. and Katsoyiannis, I.A. Arsenic removal using iron oxide loaded alginate beads. *Ind. Eng. Chem. Res.* 41 (2002), 6149–6155.

55. Guo, X. and Chen, F. Removal of arsenic by bead cellulose loaded with iron oxyhydroxide from groundwater. *Environ. Sci. Technol.* 39 (2005), 6808–6818.

56. Xu, Y., Nakajima, T., and Ohki, A. Adsorption and removal of arsenic(V) from drinking water by aluminum-loaded Shirasu zeolite. *J. Haz. Mat.* 92 (2002), 275–287.

57. Chen, W., Parette, R., Zou, J. et al. Arsenic removal by iron modified activated carbon. *Water Res.* 41 (2007), 1851–1858.

58. Vaughan, R.L., Jr. and Reed, B.E. Modeling As(V) removal by an iron oxide-impregnated activated carbon using the surface complexation approach. *Water Res.* 39 (2005), 1005–1014.

59. Munoz, J.A., Gonzalo, A., and Valiente, M. Arsenic adsorption by Fe(III)-loaded open-celled cellulose sponge. *Environ. Sci. Technol.* 36 (2002), 3405–3411.

60. Onyango, M.S., Kojima, Y., Matsuda, H. et al. Adsorption kinetics of arsenic removal from groundwater by iron-modified zeolites. *J. Chem. Eng. Jpn.* **36,** (2003), 1516–1522.

61. DeMarco, M.J., SenGupta, A.K. and Greenleaf, J.E. Arsenic removal using a polymeric/inorganic hybrid sorbent. *Water Res.* **37** (2003), 164–176.
62. Katsoyiannis, I.A. and Zouboulis, A.I. Removal of arsenic from contaminated water sources by sorption onto iron-oxide-coated polymeric materials. *Water Res.* **36** (2002), 5141–5155.
63. SenGupta, A.K. and Cumbal, L.H. Hybrid anion exchanger for selective removal of contaminating ligands from fluids and method of manufacture thereof. U.S. Patent 7,291,578, 2007.
64. Gupta, A., Chauhan, V.S., and Sankararamakrishnan, N. Preparation and evaluation of iron–chitosan composites for removal of As(III) and As(V) from arsenic contaminated real life groundwater. *Water Res.* **43** (2009), 3862–3870.
65. Zhang, Q.J., Pan, B.C., Chen, X.Q. et al. Preparation of polymer-supported hydrated ferric oxide based on Donnan membrane effect and its application for arsenic removal. *Sci. China B* **51** (2008), 379–385.
66. Sylvester, P., Westerhoff, P., Möller, T. et al. A hybrid sorbent utilizing nanoparticles of hydrous iron oxide for arsenic removal from drinking water. *Environ. Eng. Sci.* **24** (2007), 104–112.
67. Chou, S., Huang, C., and Huang, Y. Heterogeneous and homogeneous catalytic oxidation by supported γ-FeOOH in a fluidized-bed reactor. *Environ. Sci. Technol.* 35 (2001) 1247–1251.
68. Vagliasandi, F.G. and Benjamin M.M. Arsenic removal in fresh and NOM-preloaded ion exchange packed bed adsorption reactors. *Water Sci. Technol.* 38 (1998), 337–343.
69. Korngold, E., Belayev, N., and Aronov, L. Removal of arsenic from drinking water by anion exchangers. *Desalination* 141 (2001), 8184.
70. Ghurye, G. L., Clifford, D.A., and Tripp, A.R. Combined arsenic and nitrate removal by ion exchange. *J. AWWA* 91 (1999), 85–96.
71. Chwirka, J.D., Thomson, B.M., and Stomp, J.M. Removing arsenic from groundwater. *J. AWWA* 92 (2000), 79–88.
72. Driehaus, W., Seith, R., and Jekel, M. Oxidation of arsenic(III) with manganese oxides in water treatment. *Water Res.* 29 (1995), 297–305.
73. Frank, P and Clifford, D. As(III) oxidation and removal from drinking water. EPA Project Summary Report 600/S2-86/021, Water Engineering Research Laboratory, Cincinnati, OH, 1986.
74. Ficklin, W. Separation of As(III) and As(V) in groundwaters by ion exchange. *Talanta* 30 (1983), 371.
75. Clifford, D., Ceber, L., and Chow, S. As(III)/As(V) separation in chloride form by ion exchange resins. AWWA WQTC, Norfolk, VA, 1983.
76. Genz, A., Kornmuller, A., and Jekel, M. Advanced phosphorus removal from membrane filtrates by adsorption on activated aluminum oxide and granulated ferric hydroxide. *Water Res.* 38 (2004), 3523–3530.
77. Zeng, L., Li, X., and Liu, J. Adsorption removal of phosphate from aqueous solutions using iron oxide tailings. *Water Res.* 38 (2004), 1318–1326.
78. Seida, Y. and Nakano, Y. Removal of phosphate by layered double hydroxide containing iron. *Water Res.* 36 (2002), 1306–1312.
79. Blaney, L.M., Cinar, S., and SenGupta, A.K. Hybrid anion exchanger for trace phosphate removal from water and wastewater. *Water Res.* 41 (2007), 1603–1613.
80. Donnan, F.G. Theory of membrane equilibria and membrane potentials in the presence of non-dialysing electrolytes. *J. Membr. Sci.* 100 (1995), 45–55
81. Donnan, F.G. *Zh. Electrochem. Angew. Phys. Chem.* 17572, 1911, 581.
82. Cumbal, L. and SenGupta, A.K. Arsenic removal using polymer-supported hydrated iron(III) oxide nanoparticles: Role of Donnan membrane effect. *Environ. Sci. Technol.* 39 (2005), 6508–6515.

83. Sarkar, S., Prakash, P., and SenGupta, A.K. Donnan membrane principle: opportunities for sustainable engineered processes and materials. *Environ. Sci. Technol.* 44 (2010), 1161–1166.
84. Cumbal, L. and SenGupta, A.K. Preparation and characterization of magnetically active dual-zone sorbent. *Ind. Eng. Chem. Res.* 44 (2005), 600–605.
85. Sarkar, S., Gupta, A., Biswas, R. et al. Well head arsenic removal units in remote villages of Indian Subcontinent. *Water Res.* 39 (2005), 2196–2206.
86. Sarkar, S., Blaney, L.M., Gupta, A. et al. Arsenic removal from groundwater and its safe containment in a rural environment. *Environ. Sci. Technol.* 42 (2008), 4268–4273.
87. Sarkar, S., Blaney, L.M., Gupta, A. et al. Use of ArsenXnp, a hybrid anion exchanger for arsenic removal in remote villages in the Indian Subcontinent. *React. Funct. Polym.* 67 (2007), 1599–1611.
88. Puttamaraju, P. and SenGupta, A.K. Evidence of tunable on–off sorption behaviors of metal oxide nanoparticles. *Ind. Eng. Chem. Res.* 45 (2006), 7737–7742.
89. Greenleaf, J.E., Lin, J.C., and SenGupta, A.K. Two novel applications of ion exchange fibers: Arsenic removal and chemical-free softening of hard water. *Environ. Prog.* 25 (2006), 300–311.
90. Lin, J.C. and SenGupta, A.K. Hybrid anion exchange fibers with dual binding sites: Simultaneous and reversible sorption of perchlorate and arsenate. *Environ. Eng. Sci.* 26 (2009), 1673–1683.
91. Marcus, Y. *Ion Solvation.* Wiley Interscience, London, 1985.
92. Hristovski, K., Westerhoff, P., Moller, T. et al. Simultaneous removal of perchlorate and arsenate by ion-exchange media modified with nanostructured iron (hydr)oxide, *J. Haz. Mater.* 152 (2008), 397.
93. Tripp, A.R. and Clifford, D.A. Selectivity considerations in modeling the treatment of perchlorate using ion-exchange processes. In *Ion Exchange and Solvent Extraction*, SenGupta, A.K. et al., Eds. Marcel Dekker, New York, 2004.
94. Tripp, A.R. and Clifford, D.A. Ion exchange for the remediation of perchlorate contaminated drinking water. *J. AWWA* 98 (2006), 105.
95. Xiong, Z., Zhao, D., and Harper, W.F. Sorption and desorption of perchlorate with various classes of ion exchangers: A comparative study. *Ind. Eng. Chem. Res.* 46 (2007), 9213.
96. Gu, B., Brown, G.M., Maya, L. et al. Regeneration of perchlorate loaded anion exchange resins by novel tetrachloroferrate displacement technique. *Environ. Sci. Technol.* 35 (2001), 3363.
97. Soldatov, V., Pawlowski, L., Shunkevich, A. et al. *New Material and Technologies for Environmental Engineering: I. Synthesis and Structure of Ion Exchange Fibers.* Polska Akademia Nauk, Lublin, Poland, 2004.
98. Greenleaf, J.E. and SenGupta, A.K. Environmentally benign hardness removal using ion exchange fibers and snow melt. *Environ. Sci. Technol.* 40 (2006), 370–376.
99. Sarkar, S. and SenGupta, A.K. A new hybrid ion exchange-nanofiltration (HIX-NF) separation process for energy-efficient desalination. *J. Membr. Sci.* 324 (2008), 76–84.
100. Liberti, L., Petruzzelli, D., Helfferich, F.G. et al. Chloride–sulfate ion exchange kinetics at high solution concentration. *React. Polym.* 5 (1987), 37.

Index